Crop Production and Management

Crop Production and Management

Edited by **Shirley Doy**

SYRAWOOD
PUBLISHING HOUSE
New York

Published by Syrawood Publishing House,
750 Third Avenue, 9th Floor,
New York, NY 10017, USA
www.syrawoodpublishinghouse.com

Crop Production and Management
Edited by Shirley Doy

International Standard Book Number: 978-1-68286-348-0 (Hardback)

Printed in China.

Contents

Preface

Crop production is affected by varied factors such as soil, water, weather, etc. This book discusses such aspects of production, their effects and various tools and techniques to use these factors for better production. It comprises researches which bring forth diverse practices for an improved quality and increased quantity of yield. The management methods elucidated in this book will benefit students, agriculturists and anyone else associated with this field.

Significant researches are present in this book. Intensive efforts have been employed by authors to make this book an outstanding discourse. This book contains the enlightening chapters which have been written on the basis of significant researches done by the experts.

Finally, I would also like to thank all the members involved in this book for being a team and meeting all the deadlines for the submission of their respective works. I would also like to thank my friends and family for being supportive in my efforts.

Editor

1

Determinants of sustainable development dimension among leaders of rural community: A case study of Malaysia

Bahaman Abu Samah, Jeffrey Lawrence D' Silva and Hayrol Azril Mohamed Shaffril*

Laboratory for Rural Advancement and Agriculture Extension, Institute for Social Science Studies, Universiti of Putra, Malaysia.

Rural development researchers have identified a number of factors that affect sustainable development among the rural community. Nevertheless, mix results were reported in literature. Hence, this study was conducted to investigate on the constructs that affect sustainable development among members of Village Development and Security Committee (VDSC) under the Vision Village Movement (VVM). A total of 235 respondents from 18 villages were involved encompassing three states in Malaysia. The final model derived from this study showed that four pertinent constructs that fit into the sustainable development model are spirituality, leadership, beautification, and education. It is believed that the findings of this study would add new perspective in understanding the complexities associated with sustainable development among the rural community.

Key words: Village development and security committee (VDSC), vision village movement (VVM), sustainable development, rural community.

INTRODUCTION

In line with globalization and modernization, the Malaysian government has implemented numerous efforts to develop rural community so that it would enable this community to successfully enhance their quality of life similar to the urban community. Development strategy in Malaysia has started since 1957 whereby this strategy emphasized on planned development. Rural development is one of the key areas placed in this strategy whereby the government had initiated Village Vision Movement (VVM) program launched in 1996 by the ex-Prime Minister of Malaysia to expedite rural development in this country. The main focus of VVM is to cultivate consciousness and positive attitude among rural community so as to be in tandem with government's effort to create villages that are more beautiful, developed, and competitive. Koharudin (2005), in his study concluded that VVM has the ability to create positive attitude and paradigm shift among the villagers towards development. However, the success of VVM is very much dependable on the committee members of Village Development and Security Committee (VDSC) since they are the change agents that would be able to cultivate competitive and succcessful rural community (Shamsul, 1998), through rural sustainable development.

Sustainable development is an important element that has a wide array of views under its umbrella. In the 1980s, increasing concern on the effects of economic development on health, natural resources and the environment led the United Nations to publish the Brundtland Report. This report defined sustainable development as "development which meets the needs of the present without compromising the ability of future generations to meet their own needs." In June 1992, the Rio Earth Summit declared that "the right to development must be fulfilled so as to equitably meet developmental and environmental needs of present and future generations." Sustainable development is not just about the environment, but also encompasses the economy and society at

*Corresponding author. E-mail: hayrol82@gmail.com.

large (The Brundtland Report, 1987).

Literature analysis

There is extensive research done that portray constructs which contribute to rural sustainable development. Studies implemented by Shiblaq and Fielden (2008), and Hopkins and McKeown (2002), concluded that Information and Communication Technology (ICT) and education will improve rural community economically, socially and academically while according to Ngcobo and Herselman (2007), ICT and education will empower rural poor people and community to lead their own development. On top of it, beautification and health are also important factors that contribute toward sustainable development whereby Hazell and Lutz (1998), stressed that rural community should be exposed to the importance of beautifying their environment and focusing on the cleanliness of their village since their conciousness towards environment is degrading. Moreover, a study done by Day (1998) identified that the attainment of sustainable development in rural areas depends upon effective leaders that would create social and institutional networks that would boost economic activities in villages besides enabling the villagers to enhance on agricultural skills. Adil et al. (2005), on the other hand, emphasized on the aspect of spirituality that should not be ignored in the context of sustainable development so that the interests of every member in the rural community is being taken care of and more efforts toward creating a culture of collaboration and teamwork can exist in villages.

In spite of the above stated facts, the question that arises is whether the VDSC members perceive all the aspects mentioned above as important dimension of sustainable development? Thus, the main aim of this study is to determine the significant higher order constructs that significantly explain the factors of sustainable development among the VDSC members in their respective villages. It is hoped that this study would contribute to the existing literature by examining different variables that might have been neglected previously. Consequently, this would add to the body of knowledge pertaining to rural sustainable development.

MATERIALS AND METHOD

This study used a descriptive framework. The items for the instrument of this study were developed by the researchers based on previous literature. Pre-test was carried out among 13 VDSC members located in the VVM village in the area of Kajang, Selangor and the value of cronbach alpha was 0.6 - 0.9 indicating that the instrument is valid and reliable. Respondents were interviewed using a questionnaire for the purpose of getting the data needed. The sample (list of VVM village) was gained from the Department of Community Development (DCD) based on the selected states. The sample framework was then strengthened by only selecting the members of VDSC that were listed in the VVM villages. Those who were selected were from three districts from each state and from every district selected, six VVM villages will represent each district. A total of 235 respondents who were members of Village Development and Security Committee (VDSC) from three states were randomly selected. The three states were Selangor, Negeri Sembilan and Pahang.

The research instrument used in this study consisted of three parts. The first part seeks information on the respondents' socio-demographic profiles: gender, zone, grade of village, job category, level of education, position in VDSC, age, income per month, period of staying in VVM village and period of holding a position in VDSC. The second part of the instrument measured the sustainable development of respondents' in terms of social, environment and monetary elements. The third part of the instrument measured the constructs that influence sustainable development. These constructs consisted of seven dimensions: 1) Beautification, 2) Economy, 3) Leadership, 4) Information and Communication Technology, 5) Education, 6) Health and 7) Spirituality. Overall, these seven dimensions consisted of 70 items or statements.

The data collection process involved a group of trained enumerators and was conducted for a period of two months from April 2008 until May 2008. Data collected was analyzed using SPSS and AMOS software. SPSS was used to reveal the frequency, percentage, and Cronbach Alpha, while AMOS software was used for analysis of CFA. The CFA analysis will enable to utilize a suitable method to measure the construct validity and to determine the "goodness of fit" of a model determined.

RESULTS AND DISCUSSION

Table 1 and Table 2 depict the socio-demographic profile of respondents. As in Table 1, data revealed that men are still the main dominant force of village administration accounting 86.0% of the respondents. Majority of respondents stayed in central zone (38.3%), followed by southern (31.9%) and only 29.8% respondents represented east coast zone. Nearly half of the respondents lived in VVM grade C village (46.4%). Results gained also showed that respondents who work on their own are involved in village administration matters compared to other job groups (43.8%), There is a possibility that this group of respondents were less tied to job committments unlike those who work in public and private sector. In terms of level of education, more than two fifth (44.7%) of the respondents had SPM (Malaysia Certificate Education) / SPMV (Malaysia Vacational Educational certificate) certificate. Table 1 also signals that only a minority of respondents had pre-university and tertiary level of education (13.6%). Besides, more than three quarter of respondents (78.3%) are holding the position of committee members of VDSC.

The data obtained as in Table 2 showed that "senior villagers" still dominate village administration based on the respondents age mean score recorded which was 52.61 years. More than one third of respondents (37.9%) were among those whose age were 56 years and above.

More than four fifth of the respondents gained income less than RM2,000 per month. Majority of the respondents had stayed in the VVM village for a period of 26 - 45 years (41.2%). The mean score for staying period in VVM village was 37.1 years while the mean score for holding a position in CVD was 7.33 years. Slightly more

Table 1. Respondents profile (n = 235).

Respondents profile	Frequency	Percentage
Gender		
Male	202	86.0
Female	33	14.0
Zone		
Central	90	38.3
Southern	75	31.9
East coast	70	29.8
Village grade		
Grade A	57	24.2
Grade B	69	29.4
Grade C	109	46.4
Job category		
Public sector	57	24.3
Private sector	33	14.0
Work on their own	103	43.8
Unemployed	32	17.9
Level of education		
Primary school	55	23.4
PMR/SRP/LCE*	43	18.3
SPM/SPMV/MCE/Certificates	105	44.7
Pre-University/ University	32	13.6
Position in VDSC		
Chairman/ Vice chairman	19	8.1
Secretary	17	7.2
Treasury	15	6.4
Members	184	78.3

*Malaysia Lower Education Certificate.

Table 2. Respondents profile.

Respondents profile	Frequency	Percentage	Mean	S.D
Age (n = 235)				
≤45 years	67	28.5	52.61	10.38
46 - 55 years	79	33.6		
≥56 years	89	37.9		
Income per month (n = 213) (Value in Ringgit Malaysia)				
≤RM1,000	85	39.9	1663.00	1239.85
RM1,001-RM2,000	86	40.4		
≥RM2,001	42	19.7		
Period of staying in VVM village (n = 221)				
≤25 years	57	25.7	37.10	15.10
26 - 45 years	91	41.2		
≥46 years	73	33.1		
Period of holding A position in VDSC (n = 209)				
≤5 years	106	50.7	7.33	6.09
6 - 10 years	62	29.7		
≥11 years	41	19.6		

than half of the respondents held the VDSC position for ≤ 5 years.

Instrument development

AMOS software was used to conduct the Confirmatory Factor Analysis (CFA) in determining instrument that measure constructs of sustainable development. In CFA analysis, Chi Square, GFA Index Value, CFI and RMSEA were applied. An instrument is considered reliable if the Chi Square value is low and its significance value is high (larger than 0.05), GFI index and CFI value must be larger than 0.9 and RMSEA value must be lower than 0.08. Analysis conducted involved two phases. For the first phase, it involved testing on model of measurement for each dimension and for the second phase to test on the constructs of sustainable development model.

Beautification

The five items listed for this construct were:
1) Emphasize on the beauty of the village was made.
2) Initiative to enhance the beauty of the village was taken.
3) Involvement in the majority of programs conducted for the beautification and cleanliness of the village.
4) VDSC organization encourages all villagers to enhance beauty in VVM villages.
5) Unhappy response to those who do not care about the cleanliness of the village.

From these five items only four were selected based on the result of CFA ($\chi^2 = 0.318$, $p = 0.853$, GFI = 0.999, NFI = 0.999 and RMSEA = 0.000), which are items number 1, 2, 3 and 4.

Economy

A total of ten items consisted the economy dimension which are:

1) There are undeveloped land in this village.
2) Generally, the villagers economy status has been uplifted out from the poverty level.
3) Every household in this village has their own transport like bicycle, motorcycle and car.
4) Establishing strategic coorperation with number of government and private agencies.
5) Agriculture, industry and services are the main activities of the villagers.
6) There are a number of entreprenuership courses conducted in this village for the purpose of enhancing this local economy status.
7) Agriculture, industry and services provide job opportunities for the villagers.
8) Entreprenuership / commercial activities increase the number of rural entreprenuers.
9) There are new/ unique sources and products that can be commercialized.
10) There is zero unemployment rate in this village

However, from the CFA analysis conducted, only four items were included in the dimension of economy ($\chi^2 = 1.971$, $p = .373$, GFI = 0.996, NFI = 1.00 and RMSEA = 0.000). The four items identified were items number 2, 5, 7 and 8.

Leadership

There are 16 original items used to measure leadership which are:

1) VDSC has initiative to overcome issues regarding social problems, family problems, drug addict problems and community unity problem.
2) Members opinion are accepted by all VDSC members in any discussion or meeting.
3) CVD leaders take care of members charity.
4) Team spirit in this village CVD is high
5) Generally, there is a satisfaction in the VDSC leadership in this village.
6) The way of this village CVD administration can be accepted by any party.
7) There are cooperation between CVD and government/ private / NGO agencies
8) CVD is alert on changes in government policy.
9) Satisfaction with the relationship among VDSC members.
10) All CVD members have equal chances of attending training and courses.
11) There are scheduled program/activity conducted by CVD for the purpose of changing the mindset and attitude of the villagers.
12) There are effective information disseminations to the villagers.
13) This village leadership provides complete and useable village profile and action plan.
14) There are management of emergency issues such as unexpected programs that must be conducted instantly.
15) There are complete and useable village activity schedule.
16) There are unhealthy relations in CVD organizations.

After conducted CFA, only four items supported the leadership model ($\chi^2 = 1.223$, $p = 0.542$, GFI = 0.996, NFI = 1.00 and RMSEA = 0.000). The four items selected were items number 1, 2, 6 and item number 9.

Education

A total of ten items were prepared to calculate sustainable level from the education aspect which are:

1) There exist the concept of one house/one degree in this village.
2) There are increasing numbers of this villagers son/ daughter who gain the opportunity to continue their study at university level.
3) Villagers are open to any program conducted to increase the education level of this village.
4) There exist information on villagers son or daughter who are successful.
5) Villagers education level is at a good level.
6) There exist life-long education programs such as reciting Yaasin, fardu ain class and others.
7) There are no kids in this vilage who do not go to school.
8) There are villagers who have illiteracy problem.
9) There exist annual educational visit organized by VDSC.
10) There exist contribution of succesful villagers to the village, to conduct community , motivational and other beneficial programs.

CFA analysis conducted revealed that five Items have the ability to construct the education model (χ^2 = 9.374, p = 0.095, GFI = 0.985, NFI = 0.974 and RMSEA = 0.061). The five items selected were items number 1, 4, 6, 9 and 10.

Information and comunication technology (ICT)

The original number of items included for ICT was 12. The items included were:

1) There exist courses/ trainings regarding ICT usage for the villagers provided by VDSC.
2) VDSC encourages its members to enhance their knowledge and skills concerning ICT
3) This village has its own website.
4) Villagers are easy to get information regarding development happening outside Malaysia through ICT.
5) Villagers were able to get lots of information through ICT devices provided
6) Majority of the youth in this village know how to use computer.
7) There exist computer training regarding computer management and computer software.
8) Possesion of at least a computer per home at the VVM village.
9) VDSC members and villagers follow computer classes and computer literacy programs.
10) CVD and the community were able to join computer learning class that was conducted.
11) ICT gave negative impact to the villagers.
12) Internet services were used widely by the villagers

From the CFA analysis done, it was identified that only a total of four items have been included in the ICT dimension (χ^2 = 0.318, p = 0.853, GFI = 0.999, NFI = 0.999 and RMSEA = 0.000) which were items number 1, 3, 7 and 9.

Health

Under this construct ten items were identified for the purpose of measurement which were:

1) The Ministry of Health frequently conducts free health check and monitor the cleanliness of the village.
2) Emphasizes on health care aspect to avoid any infectious disease in this village was made.
3) Satisfactory level of cleanliness and health of this village.
4) The village drainage system has been serviced systematically.
5) Septical toilet is used in this village.
6) There are rural clinics in this village for free treatment for the villagers.
7) Total of the villagers that have chronic disease is low
8) This village waste management has been managed systematically
9) This village has won "the most clean village" award at the state level.
10) There are cleanliness campaign and tasks to beautify the road, alley, ditch and public places.

CFA analysis towards the ten original items measuring health aspect has resulted in a total of only four items being included in the health model (χ^2= 1.445, *p* = .486, GFI=.997, NFI =.996 and RMSEA=.000) and the items were items number 1, 4, 5 and item number 10.

Spirituality

Seven items were prepared to measure this construct:

1) Emphasizes that the spirituality aspect should be cultivated among the community especially among the youths in the village.
2) The believe that the good and bad that happened to one is determined by God.
3) There are a number of religious activities being conducted by VDSC and mosque commitee members for the villagers.
4) There are percentage of involvement from gender generations in the program conducted.
5) There are document on scheduled planning/implementation of spirituality/ religious programs.
6) Emphasize on the importance and good deeds of other people in various matters were made.
7) Organizing safety campaign and good deeds such as working together and overcoming crimes in village.

From the seven original items prepared for the dimension of spirituality, only four items were found to be significant. CFA analysis had produced the following outcome: χ^2= 2.828, *p* = .243, GFI=.994, NFI =.992 and RMSEA = .042

Table 3. CFA analysis result for sustainable development dimension.

Instrument/ Items	**β**	χ^2	p	**GFI**	**NFI**	**RMSEA**
Beautification		0.318	0.853	0.999	0.999	0.000
I emphasize on the beauty of village.	0.55					
I take initiative to enhance the beauty of village.	0.79					
I get involve in majority of programs conducted for the beauty and cleanliness of village.	0.88					
VDSC organization encourages all villagers to enhance beauty in VVM village.	0.80					
Economy		1.971	0.373	0.996	1.000	0.000
Generally, the villagers economy status has been uplifted out from the poverty level.	0.28					
Agriculture, industry and services are the main activities of the villagers.	0.73					
Agriculture, industry and services provide job opportunities for the villagers.	0.89					
Entreprenuerhip/commercial activities increase the number of rural entreprenuers	0.67					
Leadership		1.223	0.542	0.996	1.000	0.000
VDSC has initiative to overcome isues regarding social problem, family problems, drug addict problems and community unity problem.	0.79					
Generally, I'm satisfied with the VDSC leadership in this village.	0.83					
I'm satisfied with the relationship among VDSC members.	0.66					
Members opinion are accepted by all VDSC members in any discussion or meeting.	0.67					
ICT		2.309	0.315	0.999	0.995	0.026
There exist courses/trainings regarding ICT usage for the villagers provided by VDSC.	0.72					
This village has its own website.	0.68					
There exist computer training regarding computer management and computer software.	0.94					
VDSC members and villagers follow computer class and computer literacy programs.	0.84					
Education		9.374	0.095	0.985	0.974	0.061
There exist the concept of one house/ one degree in this village.	0.70					
There exist information on villagers son or daughter who are successful.	0.75					
There exist life-long education programs such as reciting Yaasin, *fardu ain* class and others	0.60					
There exist annual educational visit organized by VDSC.	0.75					
There exist contribution of succesful villagers to the village to conduct community, motivation and other beneficial programs	0.62					
Health and Cleanliness		1.445	0.486	0.997	0.996	0.000
The village drainage system has been serviced systematically	0.65					
The Ministry of Health frequently conduct free health check and monitor the cleanliness of the village.	0.77					

Table 3. contd.

Septical toilet is used in this village	0.66					
There are cleanliness campaign and tasks to beautify the road, alley, ditch and public places	0.86					
Spirituality		2.828	0.243	0.994	0.992	0.042
I'm emphasizing that the spirituality aspect to be cultivated among the community especially among the youths in the village.	0.84					
There are a number of religious activities being conducted by VDSC and mosque commitee members for the villagers.	0.72					
I emphasize on the importance and good deeds of other people in various matters.	0.78					
Orgnaizing safety campaign and good deeds such as working together and overcoming crimes in village.	0.59					

Table 4. The instrument reliability.

Instrument	Number of Items		Cronbach alpha value
	Previous	Current	
Beautification	5	4	0.837
Economy	10	4	0.710
Leadership	16	4	0.836
ICT	12	4	0.868
Education	10	5	0.810
Health and Cleanliness	10	4	0.813
Spirituality and Civic	7	4	0.813
Overall		29	0.922
Sustainable Development Model		17	0.902

The significant items were items number 1, 3, 6 and item number 7

Instrument reliability

The instrument will be considered as reliable if Cronbach Alpha value – 0.70 or more. Seven dimensions which were analyzed using CFA analysis, then was once again analyzed in order to reveal its reliability level. Data gathered is presented in Table 4. Based on the obtained data, all the seven dimensions were found to be reliable, the Cronbach Alpha value exceeding the minimum of 0.70. The economy dimension proved to be the lowest dimension (0.710) while the ICT dimension was found to be the highest dimension (0.868). Overall, the seven dimensions which consisted of 29 items produced Cronbach Alpha value of 0.922.

The Malaysia sustainable development model for rural community

This study was designed to develop a firm model to measure sustainable development from the seven dimensions that have been identified. CFA analysis was applied for this purpose and the result obtained from Table 3 is presented in Figure 1.

From the seven original dimensions, only four dimensions were found to be fit into the sustainable development model. In order for a data to fit the model the requirements that must be met are GFI INDEX and CFI = > 0.9 and RMSEA = < 0.8. After CFA was run, only four dimensions were found to fit the models which were beautification, leadership, education and spirituality. This result is in tandem with what have been done by Hazell and Lutz (1998), Ngcobo and Herselman (2007), Adil et al. (2005) and Day (1998) where all of them agreed that all of these items are the key for rural development success. Not only will these four dimensions develop the area but also those who lived in that area. From the previous studies it was also proven that these four items have the ability and power to enhance the quality of life among the rural people especially on the aspects of economy, social and academic. The standardized estimated value for each was 0.72, 0.72, 0.60 and 0.84. The statistic estimation of fit model indicate that this model

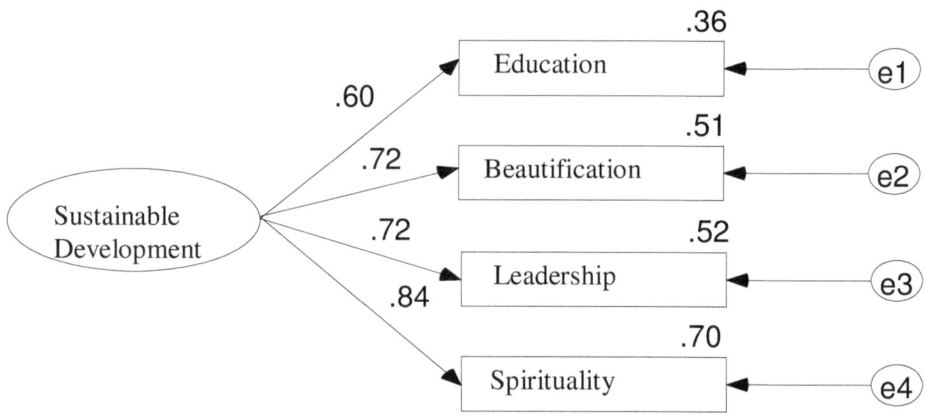

Figure 1. Malaysia rural sustainable development model.
Chi-Square = 1.801, DF = 2, p = .406, CFI = 1.000, GFI =.996, AGFI = .981, NFI = .994, TLI = 1.002, RMSEA = .000

had failed to be rejected, which means that a model that consists of this four dimensions fit the data ($\chi^2 = 1.801$, p = 0.406, GFI = 0.996, NFI = 0.994 and RMSEA = 0.0001.

Conclusion

The findings of this study imply that rural community leaders perceive the dimension of spirituality as an important element towards sustainable development. This is an important factor that needs to be taken into serious consideration by policy makers. It is vivid that rural people do not wish to sacrifice their lives by the tidal waves of unfettered materialism that would lead them to live in a world that can best be described as a dystopia. It becomes a surety that rural people would lucratively accommodate development if it is packaged with the component of spirituality as they believe that acknowledgement of spiritual worth would lead to creating responsible sustainable development that would be beneficial and bring true happiness to a greater number of people.

Secondly, the study implies that rural community leaders perceive leadership to be imperative towards sustainable development. The study identified that to ensure sustainable development among rural community effective leadership helps villages to develop through times of peril to overcome numerous social and community problems and to accommodate the wishes of the people in the community. Thus, it is of utmost importance that sustainable development is accompanied by visionary leaders that would help to conquer obstacles and implement decisions that are timely, complete and correct. Without good leadership, sustainable development might be hindered as the activities in villages would be moving too slowly, stagnate and might lose their way.

Further implication that can be drawn from this study is that beautification too is an important factor towards sustainable development. The study depicts that the realization on the need to take care of the environment has been imprinted in the minds and souls of rural community. It is clear that rural leaders feel that sustainable development should be accompanied with the factor of beautification as they believe that this will increase the quality of life and economic development in rural areas. Moreover, developing a safe and clean environment apparently seem to create an awesome feeling for rural people as this would help them to preserve their heritage and make their village to look stronger than ever.

It is also undeniable that education plays an important role towards sustainable development. The traditional way of having rural community with good manners but poor education is considered as antiquated. The study suggests that rural people believed on the importance of education as an important tool to lift them out of the state of chronic poverty, and their focus is on two main areas - having their children educated till the tertiary level and to equip themselves with relevant knowledge and skills. The rural leaders believe that via education rural people would be instilled with a long-term vision and they will be more responsible in utilizing the resources available in villages.

In conclusion, it is recommended that policies on sustainable development among the rural community would need a different approach from the policy makers since this community does not wish to live a disintegrated life at an insidious pace. Future research is recommended to encompass a wider scope of constructs pertaining to sustainable development that incorporate socio-political sustainability and cultural diversity.

REFERENCES

Adil SA, Ghaffor A, Nadeem AM (2005). Investigations into the human resource development status and its impact on farm output in rural areas of district Jhang, Pakistan. Agric. Soc. Sci. 1:10-13.

Day G (1998). Working with the Grain? Towards sustainable rural and community Development. Rural Stud. 14: 89-105.

Hazell P, Lutz E (1998). Integrating environmental and sustainability concerns into rural development policies. In Lutz, E., Binswanger, H., Hazell, P., and Mc Calla, A (eds.). Agriculture and the environment. The World Bank Publication. Washington D.C, United States of America pp: 9-21.

Hopkins C, McKeown R (2002). Education for sustainable development: An international perspectives. In Tilbury, D., Stevenson, R.B., Fien, J., and Schrueder, D (eds.). Education and Sustainability: Responding to Global Challenges. IUCN Publisher. Gland, Switzerland pp: 150-156.

Koharudin MB (2005). Rural development in Malaysia: Village Vission Movement (VVM) as a mechanisme for development of rural community. J. Technol. 42: 31- 48.

Ngcobo P, Herselman ME (2007). Evaluating ICT provision in selected communities in South Africa. Issues Informing Sci. Info. Tech. 4: 714-724.

Shamsul AB (1998). Development and change in rural Malaysia: The role of Village Development Committee. South East Asian Stud. 26: 1-9.

Shiblaq F, Fielden K (2008). *ICT* provision in rural community: A New Zealand case study. Paper presented at IADIS International Conference ICT, Society and Human Beings, Amsterdam, The Netherland.

The Brundtland Report (1987). http://www.ace.mmu.ac.uk-/eae/Sustainability/Older/Brundtland_Report.html [17 July 2009].

2

An innovative public-private partnership for irrigation extension in Fergana valley of Central Asia

Kahramon Jumaboev[1]*, J. Mohan Reddy[1], Sh. Muhammedjanov[2], Oyture Anarbekov[1] and Davron Eshmuratov[1]

[1]International Water Management Institute Tashkent, Uzbekistan.
[2]Scientific Information Center of the Interstate Committee for Water Coordination Tashkent, Uzbekistan.

A 'farmer-centric' innovative institutional mechanism, a public-private partnership, was created and strengthened, in the Fergana valley of Central Asia, for facilitating communication between farmers and researchers, and to disseminate knowledge on improved agronomic and irrigation management practices to improve water productivity at field level. As a result, yields of cotton from the twenty five demonstration sites in the three countries of Fergana valley - Kyrgyzstan, Tajikistan and Uzbekistan – were, on the average, 28% higher than the average yield of cotton in the valley, suggesting that the proposed institutional mechanism was very effective in dissemination of information to farmers. Yields from neighboring farmers of demonstration fields were 14% higher than the average yields. In addition, demonstration site farmers used, on the average, 20% less water than the non-project farmers. Two independent external reviewers stated that this innovative public-private mechanism was very effective in disseminating information on improving water productivity at plot level to farmers, and suggested that the focus in the future should be on devising effective policy and economic instruments for financial sustainability of the innovation cycle after the donor support is withdrawn.

Key words: Agricultural extension, irrigation extension, innovation cycle, public-private partnership, water productivity.

INTRODUCTION

Irrigation extension advisory services can play an important role in assisting farmers to adopt new techniques and technologies and to increase productivity, minimizing environmental risks and contributing to the sustainability of the agrarian sector (Smith and Munoz, 2002). Agricultural extension services have evolved over time, following patterns across the globe. The training and visit system stated 40 to 50 years ago, and the system experienced success in a number of countries, but later it was considered as a supply driven system and not a demand driven one. Agricultural extension evolved as pluralistic models and modes (Birner and Anderson, 2007; Neuchâtel Group, 2000). New thinking included delivery of extension services in the context of decentralization, aspects such as outsourcing, cost recovery and involvement of private sectors and NGOs.

Birner et al. (2006) argues that there is no single best method for providing need-specific, purpose-specific and target-specific extension advice. The right approach depends on the policy and infrastructural environment, capacity of potential service providers, characteristics of local communities, including their willingness to co-operate with agricultural extension agents. Raabe (2008) argues that to fit a particular situation, agricultural extension needs to be flexible and able to accommodate local needs.

Agriculture is at the forefront of development objectives of the three Central Asian Republics of Kyrgyzstan, Tajikistan and Uzbekistan. Since independence, these

*Corresponding author. E-mail: k.jumaboev@cgiar.org.

countries have undergone transition from centrally planned economies to market-oriented economy. The needs of the three Central Asian countries for improved agricultural extension are not yet similar. Over the years, there have been several sporadic attempts to revitalize agricultural extension, but with no sustainable effect. Hence, there is an urgent and clear need for establishing and strengthening agricultural extension services in Central Asia (Kazbekov and Qureshi, 2011). Also, after the Land Reform Acts of early 2000s, more than 30 to 40% of the farmers are new to irrigated agriculture, and are hungry for information on irrigated agriculture.

After independence from the former Soviet Union (in 1991), the State/Collective farms disintegrated, with nobody to claim the ownership of irrigation and drainage infrastructure. Land was distributed to local people, irrespective of their prior background in agriculture. In Kyrgyzstan, Kazakhstan and Tajikistan, farmers own their land, whereas in Turkmenistan and Uzbekistan farmers lease their land from the government for 50 years. Disintegration of large farms has increased the number of farmers the majority of whom have inadequate knowledge/skills of irrigated agriculture. There was insufficient on-farm irrigation infrastructure to distribute water to individual farmers. During the Soviet era, every State/Collective farm had professional agronomists and irrigation specialists for providing advisory services for irrigated agriculture. However, with the collapse of the system, some of this expertise has been lost. This has created a big 'vacuum' for knowledge dissemination to thousands of individual farmers in these countries specifically on irrigation water management from on-farm level to WUA level. It was difficult to government institutions such as Basin Water Management Organizations (Oblvodhoz) and Irrigation System Management Organizations (Rayvodhoz) to deliver water to thousands of farmers in Kyrgyzstan and Tajikistan, and hundreds of water users in Uzbekistan. Lands of farmers varied from 5 to 50 hectares in Uzbekistan and 0.1 to 2 ha in Kyrgyzstan and Tajikistan. In Kyrgyzstan and Tajikistan, the number of WUA members varies from 3000 to 5000 users and in Uzbekistan from 100 to 120 users. All the above mentioned issues have created issues on equitable water distribution among farmers and led to conflicts on water use between tail and head-end farmers.

As a consequence, crop yields have decreased, and a threat of food security is looms large in the region. To avert the potential threat of food security in the region, the Swiss Agency for Development and Cooperation (SDC) initiated a Water Productivity Improvement at Plot Level (WPI-PL) project. The objectives of the project were to design and evaluate an institutional mechanism for generation, synthesis and dissemination of knowledge related to improvement of water productivity; identify a set of agronomic and irrigation practices (interventions) to improve water productivity; and demonstrate increase in water productivity as a result of the proposed interventions. The institutions consist of public and private organizations. Public organizations such as research institutes receive financial support from the state, whereas private organizations such as information centers and disseminators are self-supporting. The proposed innovation cycle integrated the public and the private organizations in order to deliver knowledge to farmers to improve water productivity. Main aim of the innovation cycle was to channel knowledge on irrigation and agronomic practices based on needs of farmers.

This project was implemented over a three-year period (2009, 2010 and 2011) in the Fergana valley of Kyrgyzstan, Tajikistan and Uzbekistan. This paper describes the institutional mechanism developed and applied through the innovation cycle, and its effect on improving crop water productivity at plot level.

Description of the innovation cycle

The proposed innovation cycle was 'farmer-centric', that is, the knowledge generated, translated, and disseminated was based upon the identified needs of the farmers in the project area. These needs were identified through direct contact with the farmers and through simple survey forms. The proposed technologies or interventions included both agronomic as well as irrigation water management practices at the field level.
The cycle begins by identifying farmer's problems, by project disseminators, at field level, and conveying this information to the research centers through information centers (Figure 1). The A to H steps involved in the Innovation Cycle are explained as follows:

Step A: Information on the knowledge the farmers are seeking arrives at the research center. The research center searches for solutions in their database. If a solution already exists, the solution (a research report or a published paper, etc) is passed on to Step D. If not, it is sent to step B for adaptive research.
Step B: Adaptive research trials are conducted to provide a solution to farmers' problem.
Step C: Adaptive research results are analyzed, synthesized, and sent to information center.
Step D: Information received from either Step A or Step C is simplified into farmers' language, and readied for dissemination. This information is disseminated through brochures, monthly bulletins, newspaper articles, etc.
Step E: Information center prepares training materials to train disseminators.
Step F: Information center conducts training to disseminators.
Step G: Disseminators conduct training to farmers, and distribute dissemination materials in target areas. At the same time, they also gather information on the new information needs of the farmers.

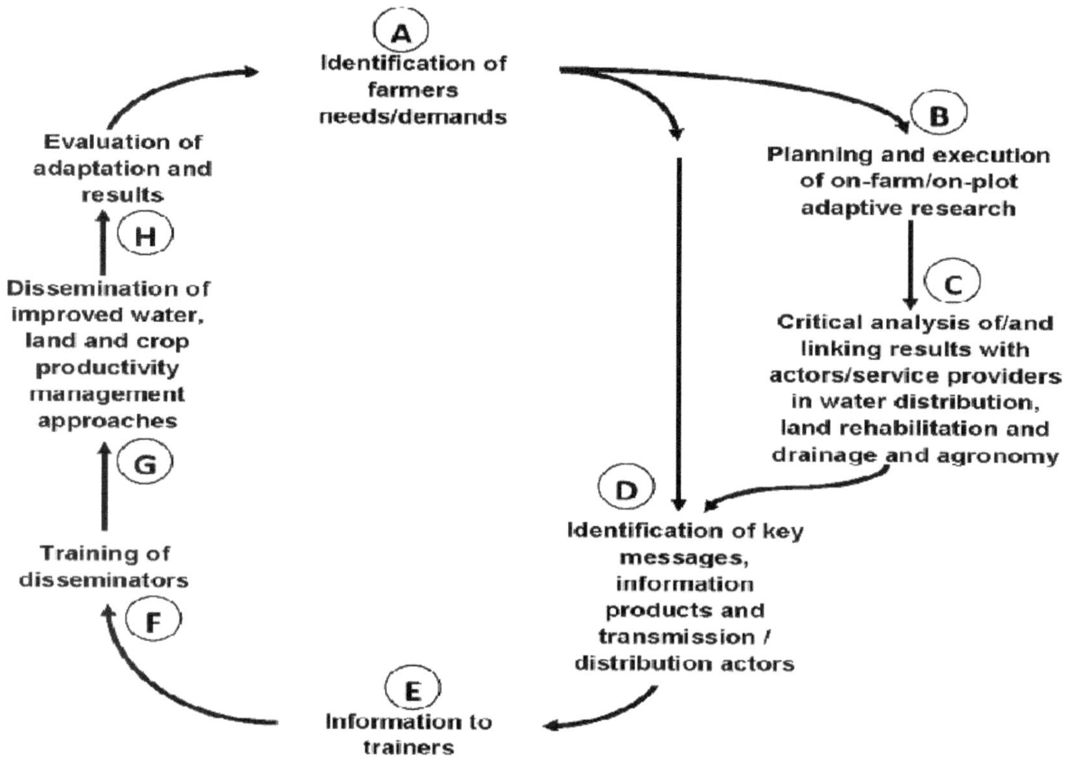

Figure 1. Innovation cycle.

Step H: Evaluate the effect of recommended interventions on crop yields and water savings, and document it for further analysis by researchers.

The above cyclical procedure continues, year-after-year. The same cycle is followed in all the three countries of the project. The institutional structure of the innovation cycle was the same for all the three countries, and is presented in Figure 2. The institutional members for each country are as follows:

1. Uzbekistan: It consisted of the Central Asian Scientific Research Institute for Irrigation (SANIIRI) as the research institute, the Naryn-Kara Darya Basin Irrigation System Authority (BISA) as the information center, and the Syr Darya-Sokh, the Naryn-Syr Darya and the Naryn-Kara Darya BISAs as disseminators. The project area in Uzbekistan covered three provinces of Andijan, Fergana and Namangan. Though there were three provinces to cover, only one BISA was selected to be the information center because farmers in all the three provinces speak the same language. In addition, the arrangement saved time and money in producing training and dissemination materials.
2. Tajikistan: The institutional structure here consisted of the Tajik Water Design Research Institute-Sogd Branch as the research institute, SOF as the information center, and ZarZamin and Irrigation Agrarian Consulting as disseminators of knowledge to farmers.
3. Kyrgyz Republic: The institutional arrangement consisted of the Kyrgyz Research Institute for Irrigation as a research institute, the Training Advisory and Innovation Center (ZOKI) as an information center, and the Osh Rural Advisory Service (Osh-RAS) and Osh WUA Support Unit as disseminators.

STUDY AREA AND METHODOLOGY

In order to assess the functioning of the innovation cycle and effectiveness of delivery of information, and finally its effect on improving water productivity at field level, 18 demonstration sites were selected in the Fergana valley of Central Asia. Fergana valley is located in the Southeastern part of Central Asia region and the Eastern part of Aral Sea basin, and its territory is shared by three countries - Kyrgyzstan, Tajikistan and Uzbekistan. Map of Fergana valley is given in Figure 3.

In Fergana valley, the annual precipitation varies from 109 to 502 mm, whereas the evaporation ranges from 1133 to 1294 mm throughout Fergana valley. Fergana valley is home for 11,342,000 people over an area of 124,200 km^2. In the framework of the project, recommendations and technologies proposed to farmers were tested in the demonstration sites. The agronomic and irrigation interventions proposed to the demonstration farmers, and the high level of implementation of these practices is shown in Table 1. For most of these interventions, brochures were prepared in the national languages of Kyrgyzstan, Tajikistan and Uzbekistan so that the farmers would follow it easily. In addition, trainings were conducted for farmers based on their needs. An impact assessment interview was conducted in the framework of the project in order to assess the performance of the partners involved in the innovation cycle (Figure 4). The interview mostly concentrated on how well the

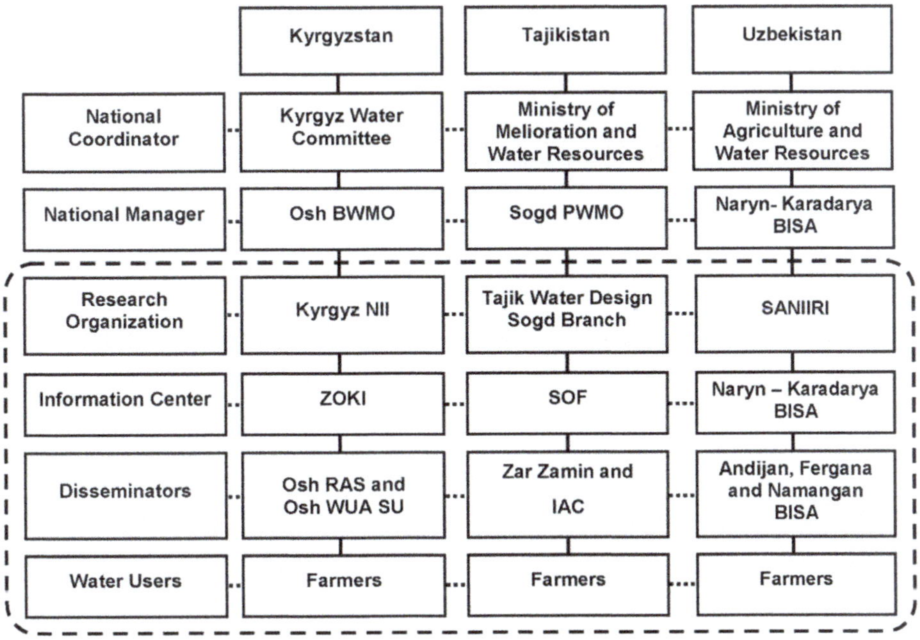

Figure 2. Institutional structure of innovation cycle.

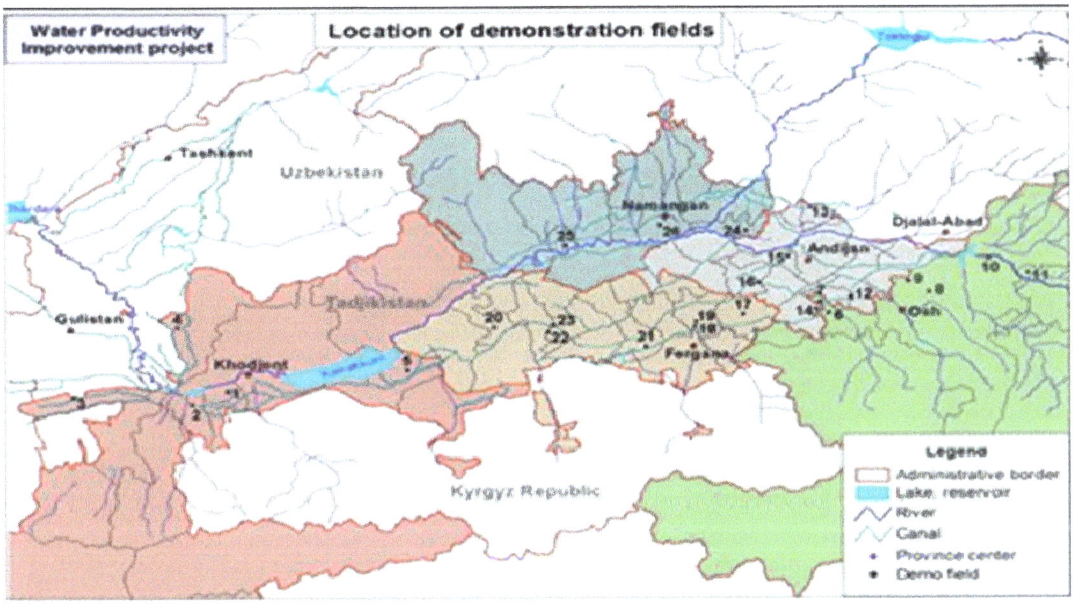

Figure 3. Location of demonstration sites.

institutions in innovation cycle performed in the three countries of Kyrgyzstan, Tajikistan and Uzbekistan. In addition, the interaction between researches institutes, information is shown in Figure 4. Components of impact assessment centers, and farmers were also assessed. Adoption of proposed technologies by farmers was also covered in the interview. Furthermore, the satisfaction of farmers with the knowledge and training provided by the institutions, and the overall contribution of the project in increasing farmers' incomes was assessed. Meetings were organized with research institutes, information centers, disseminators, and farmers both inside and outside of the project area. Semi-structured interviews were used to gather information. Interviews of this type are suited to working with small number of people and useful for studying specific situation (Institute National de Santé Publique, 2009). Before starting interviews with target groups, a topic guide was prepared tentatively for the discussion and short introduction was provided for respondents, and the objectives of interview were explained. Interview time was fixed for a maximum of one and half hours. The interview consisted of three main activities: interview among trained farmers, non-trained farmers, which are outside of project area and

Table 1. Recommendations and technologies

Site №	Leveling of irrigated land	Agro-ameliorative certification of farms	Selecting technological irrigation scheme	Application of mineral fertilizers (soil fertility improvement) for cotton and other crops in conditions of Fergana Valley (to irrigate through pits filled with organic fertilizers)	Fertilizer irrigation through application of liquid mineral fertilizers with irrigation water (fertigation)	Crop pest control	Adoption of volumetric water delivery method	Irrigation scheduling	Mechanism of efficient irrigation water use in small farms (Sokolok)	Short furrows	Alternate furrow irrigation	Cutback furrow irrigation	Installation of plastic films at furrow heads	Use of plastic bottle heads and siphons	Re-use of runoff water from fields	Water rotation	Inter-row cultivation	Leaching of saline soils with different types of salinity
2009																		
1			1	1	1	1			1	1	1		1		1	1	1	
2			1	1	1	1					1		1			1	1	
3				1	1	1				1	1		1		1	1	1	
4			1	1	1	1					1		1			1	1	
5			1	1	1	1				1	1		1					
6	1			1		1				1	1		1		1		1	
7	1		1	1	1	1				1	1		1		1	1	1	
8			1	1		1									1			1
9	1	1		1	1	1		1		1	1		1		1	1	1	
10	1		1	1		1		1		1	1		1		1		1	
11				1		1				1	1		1		1	1	1	
12				1		1				1	1		1		1		1	
13				1	1	1				1	1		1		1			
14	1		1	1		1	1				1	1	1		1	1	1	
15	1		1	1		1	1				1	1	1		1	1	1	
16	1		1	1		1	1				1	1	1		1	1	1	
17	1		1	1		1	1				1	1	1		1	1	1	
18	1		1	1		1	1				1	1	1		1	1	1	
2010																		
1			1	1	1	1			1	1	1		1		1	1	1	
2			1	1	1	1				1	1		1			1		
3			1	1	1	1		1		1	1		1	1	1	1	1	
4			1	1	1	1					1		1			1	1	
5			1	1	1	1		1		1	1		1	1				
6	1			1		1				1	1		1		1		1	
7	1		1	1	1	1				1	1	1	1		1	1	1	
8	1		1	1	1	1				1			1		1			1
9	1	1		1	1	1		1		1	1		1		1	1	1	
10	1		1	1		1		1		1	1		1		1		1	
11	1	1		1	1	1					1		1		1		1	

Table 1. Contd.

12	1	1		1	1	1				1	1		1		1			
14	1		1	1		1	1				1	1	1		1	1	1	
15	1		1	1		1	1		1	1	1	1	1		1	1	1	
16	1		1	1		1	1		1	1	1	1	1		1	1	1	
17	1		1	1		1	1		1	1	1	1	1		1	1	1	
18	1		1	1		1	1		1	1	1	1	1		1	1	1	
2011																		
1			1	1	1	1			1	1	1		1		1	1	1	
2			1	1	1	1		1		1	1		1			1		
3			1	1	1	1		1		1	1		1	1	1	1	1	
5			1	1	1	1		1		1	1		1	1		1	1	
7	1		1	1	1	1				1	1	1	1		1	1	1	
8	1		1	1	1	1				1			1		1			1
9	1	1		1	1	1		1		1	1		1		1		1	
10	1		1	1		1		1		1	1		1		1		1	
11	1	1		1	1	1					1		1		1		1	
14	1		1	1		1	1	1	1	1	1	1	1		1	1	1	
15	1		1	1		1	1	1	1	1	1	1	1		1	1	1	
17	1		1	1		1	1	1	1	1	1	1	1		1	1	1	
18	1		1	1		1	1		1	1	1	1	1		1	1	1	

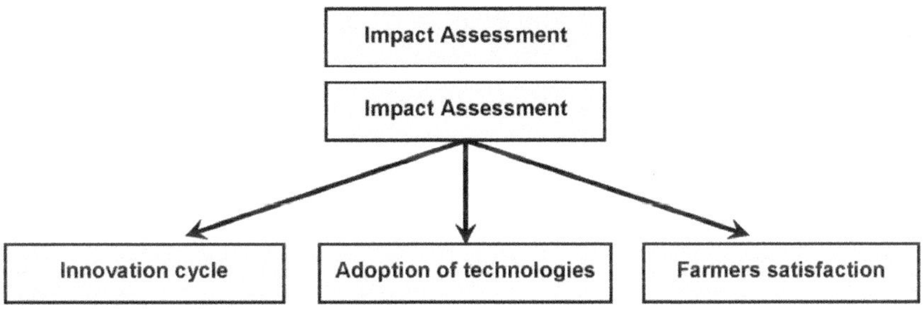

Figure 4. Components of impact assessment.

interviews with disseminators and discussion with representatives from the information centers and research institutes were also conducted. To conduct the interview between farmers and disseminators, a list of questions were prepared. Interview questionnaires comprised of the following main questions, and the main questions were usually followed by sub-questions:

1. Has the project established appropriate partnerships (innovation actors, research institutes, information centers and disseminators) that fit the agricultural innovation cycle?
2. Do all partners clearly understand their roles in the innovation cycle?
3. What is the number of technologies identified, disseminated and adopted by farmers in the project zone?
4. To what extent is the project contributing to reduce tensions, and facilitating current competing water use among farmers?
5. What is the satisfaction of farmers with the given knowledge and trainings?
6. To what extent improved technologies are disseminated among farmers and communities outside of the project area?
7. Are the farmers reporting economic benefits? Is there a noticeable saving in irrigation water use?

The topic guide and interviews had to assess the performance of institutions involved in the innovation cycle, the adoption of proposed technologies by farmers and the overall satisfaction of farmers. For the interviews, 6 trained farmers and 6 non-trained farmers from each country were selected randomly. In total, 36 farmers (18 trained and 18 non-trained) were involved. The monitoring covered 3 districts in Tajikistan, 6 districts in Uzbekistan and 2 districts in Kyrgyzstan. A total of 11 districts from the three countries were covered. Trained farmers for the interviews were chosen randomly from the lists provided by the disseminators. The non-trained farmers were chosen in the neighborhood of the trained farmers. The interviews were conducted during the spring (April, 2011) of the 3rd year of the project. Representatives of the disseminators accompanied the interviewers to identify the farmers. Interviews with disseminators, information centers and research institutes were conducted separately in the same form.

RESULTS AND DISCUSSION

During the interviews, all the 18 farmers that were trained

Table 2. How farmers know about water saving technologies.

Please indicate how do you know about efficient water use techniques?	Kyrgyzstan		Uzbekistan		Tajikistan	
	Trained	Non-trained	Trained	Non-trained	Trained	Non- trained
I saw installed water weirs	0	1	0	0	0	0
personal experience	2	5	0	1	0	1
From project trainers	4	1	6	1	6	1
From trained farmers	0	2	0	1	0	0
Other service providers	1	1	0	0	0	1

confirmed that their knowledge increased and they used the proposed recommendations and technologies. But while checking their detailed knowledge, it became clear that additional trainings were necessary. Trained farmers reported that appropriate partners were chosen for the innovation cycle in the three countries even though the partners were located far from each other they communicate and addressed the issues of farmers. Though the research institute in Kyrgyzstan (Kyrgyz SRII) is located in Bishkek, very far from the project area, it has good linkages with other institutes, which are in water and agricultural development sector. This research center provided recommendations and technologies that are related not only to water but also to agronomic issues. The information center in Kyrgyzstan is also located in Bishkek, allowing for good interaction between research institute and information center. The information center (ZOKI) has a branch office in the project area and has close contact with Osh Rural Advisory Services and Osh WUA Support Unit. In Kyrgyzstan the institutions involved in the innovation cycle clearly know their roles and responsibilities. Trained farmers reported that they know how much water to supply and pay for volume of water they use in their fields, whereas farmers that are outside of project area still follow hectare-based tariff. Partners in Kyrgyzstan reported that they disseminate knowledge and experience the project even though it is not in their mandate. Farmers that are outside of project area reported that they seek information from trained farmers.

In Uzbekistan, the research institute is located in Tashkent, quite far from the project area, but has a branch office in Fergana. This institute has highly qualified specialists in water and agronomy, and has good interaction with other project partners. The information center is located in the Basin Water Management Organization office in Andijan. This center worked closely with the Andijan Agricultural Institute to transform scientific materials to farmer-friendly language. The disseminator's role was played by three pilot WUAs in the three Provinces- Fergana, Andijan and Namangan- of the project. These disseminators worked through these pilot WUAs which had an irrigation engineer and an agronomist on their staff. Their roles were to disseminate the proposed recommendations and technologies to demonstration sites and to farmers within and outside of the WUA. All trained farmers in Uzbekistan confirmed that they are direct beneficiaries of the institutes involved in the innovation cycle. Farmers reported that after training they used less water and knew how to measure discharge with water flume meters. All trained farmers applied improved methods of irrigation such as short-furrow irrigation and measured the discharge of water received.

In Tajikistan, the research center, the information center, and the disseminators were located in Khujand city of Sogd province. Hence, there was plenty of interaction between the partners of the innovation cycle. Partners know their roles and responsibilities in innovation cycle. With respect to the research centers the farmers reported that:

1. It has very good archive of materials and good scientists, which provided recommendations for farmers based on local conditions.
2. Materials of research institutes were written in scientific language and it took too much time for information centers to simplify to farmer-friendly language;
3. Occasionally, advanced technologies were recommended, which did not consider farmers capacity and affordability.

Finally, the following observations were made regarding the disseminators:

1. Farmers had easy access to information and advice when the disseminators were located within the WUAs, as in Uzbekistan.
2. Provided irrigation advice to individual farmers which required a lot of time.

In Tajikistan and Kyrgyzstan, about 70% of all technologies proposed were linked to irrigation and 30% to agro-techniques. In Uzbekistan, 36% of the recommendations were related to irrigation and 64% to agricultural production issues. Tables 2 and 3 show the results of the survey on how farmers learnt about different types of water saving technologies. It is clear that the dissemination materials were reaching the farmers within and outside the project area. Trained farmers reported yield increases: sixty six percent reported yield increases of 6 to 20%, twenty seven percent reported increases of 20 to 30%, and seven percent reported

Table 3. Technologies known to farmers.

Technologies	Kyrgyzstan		Uzbekistan		Tajikistan	
	Trained	Non- trained	Trained	Non- trained	Trained	Non-trained
Efficient furrow irrigation	8	7	7	5	10	3
Using of water weirs	5	3	3	0	6	0
Water harvesting	0	0	1	0	0	2
Drip and sprinkler irrigation	2	0	1	1	0	0
Other technologies	0	0	2	0	1	0

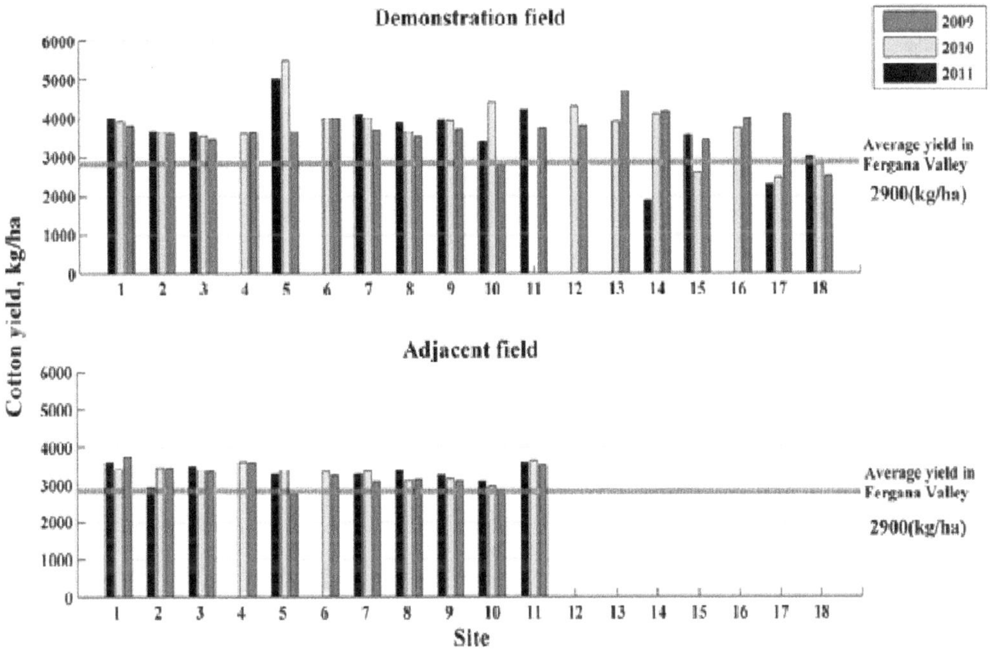

Figure 5. Cotton yields from demo fields and adjacent fields.

increases of more than 30%.

In addition to the interviews data on actual yields of cotton from the 18 demonstration sites (13 in Uzbekistan, and 5 in Tajikistan) were collected. During the 2010 and 2011 irrigation seasons, all the demonstration farmers did not grow cotton on their demonstration fields. Hence, the number of cotton fields was less than 18. Yields of cotton from the demonstration sites ranged from 2000 to 5500 kg/ha. This difference in yields was due to a combination of factors such as the quality and quantity of seed used, the availability and quality of inputs received or applied by farmers, crop variety, the irrigated cotton production knowledge-base of the farmers, etc. Two things are obvious from Figure 5. First, the average yield of cotton from the demonstration sites in Tajikistan was lower (less than 3000 kg/ha) than the average yield of cotton from demonstration sites in Uzbekistan (about 3500 kg/ha). This difference in yields may be partly explained by the availability of credit for purchasing agricultural inputs plus application of land use practices which are also supported and monitored by the provincial officials in Uzbekistan. In the case of Tajikistan, no such monitoring is done. Secondly, as mentioned elsewhere, 2009 and 2010 (more rainfall and more water was available for irrigation during the vegetation period) were considered as wet years whereas 2011 was a dry year (less rainfall and less amount of irrigation water was available during the vegetation period). Yet, on the average, there was no significant difference in the yield of cotton between the wet and dry years (Reddy et al., 2012).

Cotton yields from the adjacent fields are also shown In Figure 5. No data was available from the adjacent fields in Tajikistan. As expected, the average yields from the adjacent fields were lower (around 3300 kg/ha) than the average yields (3700 kg/ha) obtained from the demonstration sites in Uzbekistan. The demonstration field farmers implemented a variety of ‘innovations’ or improved agronomic and irrigation practices, as shown in Table 1, whereas the adjacent farmers were using one or more of the innovative practices implemented by the demonstration farmers. The average cotton yields in Fergana valley were about 2900 kg/ha, indicating that the average crop yields from the demonstration fields were 28% higher than the average crop yields in the area. It is evident that there is a substantial opportunity to increase crop yields in Fergana valley through a combination of agronomic and irrigation water management interventions (Table 1). Also, the yields of adjacent fields have already increased by 14% above the average for Fergana Valley suggesting that, with time, more farmers would adapt these interventions to raise the average yield of cotton in Fergana valley. The following observations about the innovation cycle were made by external reviewer (Christoplos, 2011).

1. The technologies being promoted have proven extremely relevant for the participating farmers and for more efficient water productivity in the Fergana valley more generally.
2. WPI-PL project has been very successful in developing and applying the innovation cycle among participating institutional partners and farmers.
3. The project has made good progress in a limited period of time in encouraging scaling up, but the current structure may not be appropriate for mobilizing broader technological adoption and adaptation.
4. Some success has been achieved in moving toward sustainability, but uncertainty regarding the WPI-PL project mandate in including WUAs in the innovation cycle and how to relate to government and other actors to explore avenues for collaboration, alternative financing and expanded diffusion are obstacles to moving further toward sustainability. The review mission suggested some avenues to explore to make the innovation cycle financially sustainable.

SUMMARY AND CONCLUSION

An innovative institutional mechanism, called a public-private partnership, was developed and strengthened for dissemination of information on improving water productivity at field level from research centers to farmers. This approach was 'farmer-centric' and promoted the interaction between researchers and farmers. This innovative institutional mechanism was called innovation cycle, and was implemented in the three countries of Fergana valley of Central Asia. The institutional structure was identical in each country, and consisted of research centers, information centers, and disseminators. In general, the research centers were public institutions, and the information centers and disseminators were private entities (NGOs) with several years of experience in advising farmers. The role of the research centers was to generate answers to farmers questions, whereas the role of information centers was to synthesize information from research centers into farmer-friendly language, and produce dissemination materials(brochures, bulletins, etc), and conduct training to information disseminators (training-of-trainers). Finally, the role of the disseminators was to provide training to farmers and disseminate materials produced by the information centers. Disseminators also worked with demonstration site (25 sites) farmers on implementing the improved agronomic and irrigation practices over a period of three years. On the average, crop (cotton) yields from the demonstration sites were 28% higher than the average yield of cotton in Fergana valley. In addition, the average application efficiency of the demonstration farms was at least 20% higher than the average application efficiency of fields in Fergana valley. The proposed innovation cycle was very effective in disseminating knowledge on improved agronomic and irrigation practices to farmers, and this innovation cycle must be further strengthened and streamlined for adoption by the respective governments.

Altogether, there were 15 different organizations in three countries, on the average 5 organizations in each country, worked as a team with excellent interaction, understanding and cooperation between the organizations. During the course of the WPI-PL project, all the organizations involved received financial support from the donor. However, since the donor support will cease sooner or later, strategies for making the innovation cycle self-sustaining financially should be investigated.

ACKNOWLEDGEMENT

The funding for this research was provided by the Swiss Agency for Development and Cooperation (SDC) Tashkent office, Uzbekistan, as part of the Water Productivity Improvement at Plot Level (WPI-PL) project. Their financial support is highly appreciated.

REFERENCES

Birner R, Davis K, Pender J, Nkonya E, Anandajayasekeram P, Ekboir J, Mbabu A, Spielman D, Benin S (2006). From best practice to best fit: A framework for analyzing agricultural advisory services worldwide. Development Strategy and Governance Division, Discussion p. 39, IFPRI, Washington, DC.

Birner R, Anderson J (2007). How to make agricultural extension demand driven? The case of India's agricultural extension policy. IFPRI Discussion Paper 00729, Development Strategy and Governance Division, IFPRI, Washington, DC.

Christoplos I (2011). External Review of Water Productivity Improvement of Plot Level Project, Project internal report. Swiss Agency for Development and Cooperation, Tashkent, Uzbekistan.

Institute National de Santé Publique (2009). Guide to Organizing Semi-Structured Interviews with Key Informants. p. 11. www.inspq.qc.ca

Kazbekov J, Qureshi A (2011). Agricultural Extension in Central Asia: Existing Strategies and Future Needs. IWMI Working Paper 145, Colombo, Sri Lanka.

Neuchatel Group (2000). Guide for monitoring, evaluation and joint analysis of Pluralistic extension support. http://www.neuchatelinitiative.net/images/guide_for_monitoring.pdf

Raabe K (2008). Reforming the agricultural extension system in India – what do we know about what works where and why? IFPRI Discussion Paper 00775, Development Strategy and Governance Division, IFPRI, Washington, DC.

Reddy JM, Muhammedjanov S, Jumaboev K, Eshmuratov D (2012). Analysis of cotton water productivity in Fergana Valley of Central Asia. Agric. Sci. 3(6):822-834, doi:10.4236/as.2012.36100.

Smith M, Munoz G (2002). Irrigation Advisory Services for effective water use: a review of experiences. Workshop on Irrigation Advisory Services and Participatory Extension in Irrigation Management, FAO-ICID. Proceedings on-line: http://www.enorasis.eu/uploads/files/Business%20modelling/irrigation_adv_serv.pdf 24th July, Montreal, Canada.

Seed handling system and its implications on seed quality in South western Nigeria

J. A. Adetumbi[1]*, J. O. Saka[1] and B. F. Fato[2]

[1]Institute of Agricultural Research and Training, Obafemi Awolowo University, P. M. B. 5029, Moor Plantation, Ibadan, Nigeria.
[2]Federal College of Agriculture, P. M. B. 5029, Moor Plantation, Ibadan, Nigeria.

There are several operations between seed production and distribution. This study analysed the prevailing situation surrounding seed business and considered its implication on seed quality in Southwestern Nigeria. Surveys were conducted in 2009 on seed marketers and public Institutions handling seeds within the zone using structured questionnaires. Data were collected on socio-economics, seed procurement, handling and distribution activities of marketers, as well as seed production, storage and distribution activities of seed producing institutions. Well packaged clean seeds of maize and cowpea, kept among the seed lot of seed dealers and public institutions for six months were evaluated. Level of deterioration during storage was determined. Data analysed by means of descriptive statistics of simple proportion and percentages showed that plant breeding and seed production activities are concentrated within the Research Institutes while Agricultural Development Programmes (ADPs) are limited to seed multiplication and marketing. Private individuals handling seed are educated and experienced in the business but majority cannot afford the provision of conducive seed storage environment; hence seed deteriorate faster in their hand and ADPS than when with the NARIs. Periodic training, notably on seed storage methods to empower the seed handlers in coping with the challenges of efficient seed delivery is recommended.

Key words: Seed deterioration, seed handling, seed inspection, seed quality.

INTRODUCTION

Seed is the core of the "Technology package" needed to achieve the much desired increased agricultural production. The use of good quality seed is important to increase food production and agricultural productivity. Therefore, timely availability of good seeds of improved varieties to farmers should be a priority in farming system because quality of seed determines the crop response to other inputs. Seed supply system is broadly divided into formal and informal sector. Formal seed system is characterized by organized public and private sector participating in seed industry and has been able to supply only 4% of the seed sown by farmers in most of the African countries while the remaining 96% originate from informal sector which is largely unorganized traditional seed supply system (Anon, 2001). Marketing of field crop seed may be simple or complex with one or more contributive functions (Kelly, 1988). Seed of agricultural crop whether produced by farmers or seed companies in special areas follows many paths from the producer to the consumers (farmers). Seed is required to be moved from where it is produced to where it is processed and stored before finally moving from warehouse to farms. Adetumbi and Olakojo (2010) have reported that seed deterioration is inevitable but rate can be reduced when utmost attention is paid to storage temperature and moisture. Durtschi (1999) also reported that some seed store better in air, while others store better in nitrogen and others do better in a vacuum, carbon dioxide or organ. These indicate that seed moisture content and

*Corresponding author. E-mail: jaadetumbi@yahoo.com.

storage temperature are the two (2) major factors affecting seed during storage. Ability of seed to maintain a high germination percentage and vigour both at harvest until its return into the soil is the desired primary aim of seed handlers. Desai et al. (1997) stated that seed that failed to germinate or give a healthy and vigorous crop stand upon planting are worthless. Since the quality and value of seed mostly cannot be assessed only by sight, smell or touch, farmers therefore have to put a great deal of trust in the supplier and the dealer, however, Daniel and Adetumbi (2004) has reported that the dynamics of seed system for production of seed is not well understood in South Western Nigeria at small holders level that dominates farming activities. Seeds are vulnerable to damage by temperature, moisture, chemical and biotic factors during transport and storage and since a great deal of several operations is involved between seed production and seed sales, it therefore becomes very important to study and analyse the prevailing situation surrounding seed handling in terms of personnel handling seed marketing activities and available materials for seed storage and consider its implication on seed quality. Therefore the objectives of this study are to (1) study the Socio-economic characteristics of seed sellers; (2) identify different seed storage facilities and equipment available to the seed sellers; (3) determine the rate of seed deterioration at the different marketing outlets, and (4) identify other activities involved in seed marketing.

MATERIALS AND METHODS

Data for this study was generated from two set of field surveys conducted on seed marketers and public institutions dealing with seed distribution selected from Southwest Agro-ecological zone. The respondents for the survey on marketers were selected by multi-stage sampling technique. Southwest Agro-ecological zones comprises of eight States namely Delta, Edo, Ekiti, Lagos, Ogun, Ondo, Osun and Oyo. The first stage of the sampling involved the selection of 4 States (Ekiti, Ogun, Osun and Oyo) randomly from the list of States that makes up the zone. Secondly, each of the selected States were stratified into the existing ADP zones with two zones selected randomly from the list of existing ADP zones in each of the States. Finally, twenty seed marketers located in different towns (Figure 1) were selected from the list of seed marketers obtained from each of the State ADPs for a total of 80 respondents but 76 questionnaires were found suitable for analysis. The second set of survey was conducted on six (6) public Institutions involved in seed distribution comprising of two research institutes, three ADPs and the National Seed Agency that has the National mandate for seed releases and distribution in Nigeria. Data were collected on socio-economic and seed procurement, storage materials and equipment available to seed handlers, handling and distribution activities of the marketers and the institutions using structured questionnaire and were subjected to descriptive statistics. 0.5 kg each of clean good seed samples of maize and cowpea were packaged in closed polyethylene bag and kept among the seeds of the seed dealers and seed store of the public Institutions for 6 months to evaluate level of seed deterioration during storage in each place. The seed samples was tested for seed viability before storage and after storage using International Seed Testing Association (ISTA) procedure of planting 100 seeds in soil substrate in four replicates. Germinated seeds were counted after seven days and percentage viability was evaluated by using the formula:

$$V(\%) = \frac{N_{germ}}{N_{plted}}.100 \quad (1)$$

Where $V(\%)$ = percentage viability; N_{germ} = Number of seeds that germinated after 7 days of planting in the soil substrate, and N_{plted} = Number of seeds planted in the soil substrate (ISTA1985).

Stored seed was also evaluated for percentage seed damaged. Percentage seed damaged and percentage seed viability loss are given by Equations 2 and 3, respectively. The equations were developed by personal consultations on mathematical procedure. The percentage seed damage is given as:

$$S_{dmg}(\%) = \frac{N_{dmg}}{N_{strd}} \times 100 \quad (2)$$

Where: $S_{dmg}(\%)$ = percentage of seed damaged after retrieval; N_{dmg} = Number of seed damaged after retrieval, and N_{strd} = Initial number of seed stored.

$$V_{loss}(\%) = \frac{V_{init} - V_{final}}{V_{init}} \times 100 \quad (3)$$

Where: $V_{loss}(\%)$ = viability loss percentage; V_{init} = Initial viability percentage before storage; V_{final} = Final viability percentage after retrieval from storage

RESULTS AND DISCUSSION

Demographic characteristics of seed marketers

The distribution of the respondents by their socio-economic characteristics (Table 1) indicated that seed marketing is male dominated with about 91% of the marketers been male while female constituted about 9%. Majority of the respondents (82.90%) aged between 41 and 60 years while 17.10% were between 21 to 40 years of age with an average age of 49.8 years. All the seed marketers are educated with majority (72.40%) having read up to tertiary level of formal education. However, majority of the seed marketers engaged in seed marketing as a secondary source of income with average years of experience in the trade estimated as 13.55 years. The distribution of the respondents implies that the population of seed marketers in the zone is dominated by men who were not only within the productive age but possessed adequate level of education and years of experience that could enhance their managerial capabilities in coping with the most technical aspect of seed handling.

In addition, the study also revealed that about 82% of the marketers have attended training in seed handling

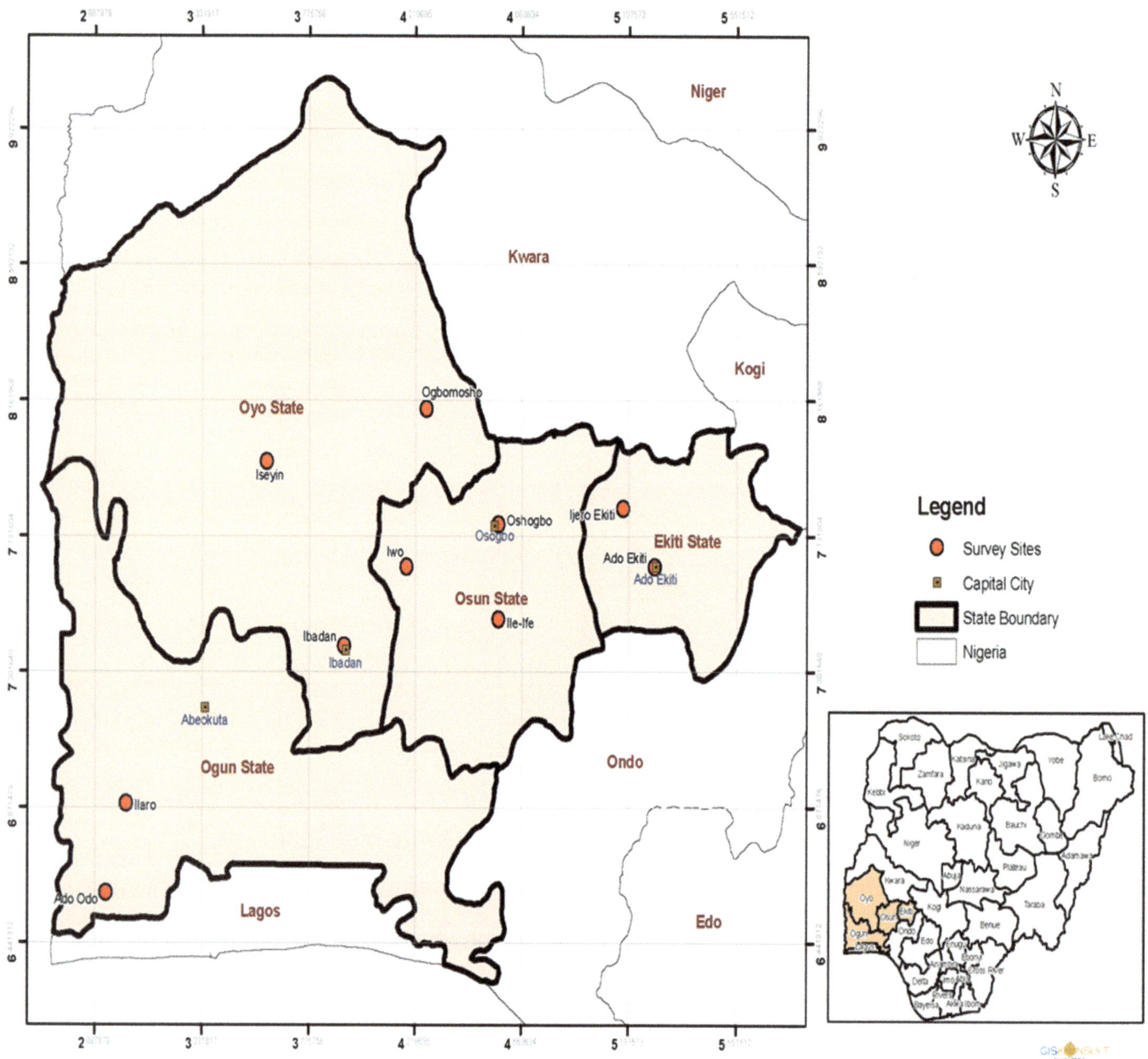

Figure 1. Map showing the study sites and states.

organized by both governmental and non-governmental agencies (Table 2). A greater percentage of these training were organized by governmental agencies like Agricultural Development Programmes (ADPs) (27.60%), and non-governmental agencies like IFDC (18.40%), private seed firms (17.11%). However, the training was attended mostly not later than 6 years ago by about 54% of the seed marketers. These implies that most seed handlers needs to be trained more regularly on techniques of seed handling operations that will reduce rate of seed deterioration.

Seed handling activities

One of the major factors affecting seed quality at the down-stream is the sources and quality of the storage method used along the distribution channels. Sourcing seed from certified sources has for decades being a prominent recommendation to farmers towards ensuring that healthy seeds are planted for improved productivity. Considering the importance of seed marketers in the distribution channels, getting supply from certified sources becomes an important ingredient in ensuring that

Table 1. Socio-economic characteristics of seed dealers in South western Nigeria.

Characteristics	Frequency	(%)
Sex		
Male	69	90.80
Female	07	9.20
Age group		
< 20 years	0	0
21 - 40 years	13	17.10
41 - 60 years	63	82.90
Mean	49.8 (11.33)	
Educational level		
Primary education	14	18.40
Secondary education	07	9.20
OND/NCE/HND/University	55	72.40
Major occupation		
Seed marketing	35	46.10
Others	41	53.90
Seed business experience		
< 5 years	14	18.40
6 - 10 years	20	26.30
> 10 years	42	55.30
Mean	13.55 (6.38)	

Figures in parentheses are standard errors.

Table 2. Training in seed technology for seed dealers.

Training	Frequency	(%)
Attendance		
Attended	62	81.60
Not attended	14	18.40
Training organizers		
IFDC	14	18.40
State government/ADPs	21	27.60
University	07	9.20
NASC	O7	9.20
Private seed firm	13	17.11
Year of training		
5 years and less	21	27.63
Between 6 - 10 years	14	18.42
Above 10 years ago	27	35.53

seed are sourced from certified sources and properly handle to guarantee the supply of healthy seeds to farmers. The distribution of the crop species whose seed were marketed by the respondents is shown in Table 3.

Table 3. Crop seed handled by seed dealers and sources of such seed.

Crop seed handled	Frequency	Percentage
Maize	76	100.00
Cowpea	63	82.89
Soybean	21	27.63
Vegetables	63	82.89

Sources of seed	Maize		Cowpea		Soybean		Vegetables	
	Frequency	(%)	Frequency	(%)	Frequency	(%)	Frequency	(%)
Research Institutes	56	73.68	35	46.05	21	27.63	14	18.42
ADPs	35	46.05	14	18.42	07	9.21	0	0
Private Firm	35	46.05	14	18.42	07	9.21	28	36.24
Out growers	14	18.42	07	9.21	0	0	07	9.21
Open Market	9	27.63	14	18.42	0	0	21	27.63
Personal farms	24	53.95	21	27.63	0	0	28	36.24

Table 4. Type of Storage method and Materials available to seed handlers.

	Seed dealers	NARIs	Agency	ADPs
Storage method				
Natural(ambient)	69 (90.79)	0 (0)	100 (1)	100 (3)
Air conditioned	07 (9.21)	100 (2)	0 (0)	0 (0)
Materials and equipment available				
Sampling probes	35 (46.05)	50 (1)	100 (1)	100 (3)
Trolley	21(27.63)	100 (2)	100 (1)	100 (3)
Pallets	48 (63.16)	100 (2)	100 (1)	100 (3)
Bag closer	69 (90.79)	100 (2)	100 (1)	100 (3)
Sprayer	76(100.00)	100 (2)	100 (1)	100 (3)
Thermometer	32 (42.11)	100 (2)	0 (0)	66 (2)
Hygrometer	08 (10.53)	50 (1)	0 (0)	0 (0)

Figures in parenthesis are percentages of respondents.

The distribution indicated that all the respondents (100%) handled maize while 82.9% each were involved in marketing of cowpea and vegetable seeds. Only 27.3% of the respondents were selling soybean seeds owing to greater sensitivity and fragility of soybean seeds to storage environment. However, prominent sources of seed offered for sale differs by crop species. While seeds of food crops were mostly sourced from research institutes (maize 73.68%, cowpea 46.05%, soybean 27.63%), seed of vegetables were mainly sourced from private seed firms and personal farms of the seed marketers (36.24%) thereby depicting the low level of attention given by government seed agencies to this category of crop. In addition, maize seeds were also sourced from personal farms of the seed marketers (53.95%), ADPs (46.05%), private seed firms (46.05%) and open market (27.63%). Similar results were observed for cowpea.

The study showed that seed are prominently stored under ambient condition by dealers (90.90%), Seed Agency and ADPs (Table 4). Only a few (9.10%) of the dealers and all the Research Institutes (100%) stored their seed stock under air-conditioned environment. This can however be attributed to the demand of air-conditioned environment in terms of cost of energy and maintenance of the facilities especially considering the additional constraint constituted by the current epileptic power supply. The study also showed that of all the equipments required for effective management of seed stocks, spraying pump (100%) and bag closer (90.79%) were prominent among the dealers while 46.05, 42.11 and 10.53% possesses sampling probe, thermometer and hygrometer respectively. Similarly, about 28 and 63% of the dealers possess trolleys and pallets, respectively. It is obvious from this study that the dealers were in least possession of equipments that are relevant

Table 5. Supportive services rendered by seed agents.

Supportive services	Frequency	Percentage
Information Dissemination	69	90.80
Seed Dressing	62	81.60
Credit Facility	07	9.20
Pest Management Service	41	53.90

Table 6. Seed quality assessment.

	Maize		Cowpea	
	Mean % seed damaged	Mean % viability loss	Mean % seed damaged	Mean % viability loss
Seed dealers	46.89	27	22.14	37
NARIs	2.15	2	12.25	21
Seed agency	5.00	10	16.50	23
ADPs	10.42	13	17.58	25
Mean	16.12	13	17.12	26.5
SD	32.85	17.70	22.14	3.55

in monitoring the condition within the immediate environment of the stored seed and this is capable of making quality management or assurance on stored seeds difficult. However, all the governmental agencies saddled with handling of larger quantity of seed were found to posses these equipments with the exception of hygrometer which were not available with the ADPs and corporate seed agencies.

Supportive services rendered by seed marketers

The study (Table 5) showed that in addition to seed marketing, majority of the seed marketers (90.80%) were also involved in information dissemination, seed dressing (81.60%) and pest management practices to clients. However, only a few of the seed agents (9.20%) offer credit facilities to clients. Such services when efficiently delivered are capable of further strengthening the delivery system through enhanced patronage of certified seed channels rather than farmers resorting to buying seeds from the local (grain) market.

Seed quality test

In order to appraise the effectiveness of the handling methods by each categories of seed agent/agency, a viability test was carried out on sample of seed stored with each category of seed handlers to determine the viability loss. The result of the test (Table 6) showed that seed damage for maize was highest for seed stored with the seed marketers (46.89%) followed by the ADPs (10.42%), Seed Agency (5.00%) and NARIs (2.15%). Similarly, average cowpea seed damage was highest for seed dealers 22,14%, followed by ADPs (17.58%), Seed Agency (16.50%) and NARIs (12.25%). Consequently, the seed damage was found to have resulted in maize seed viability loss amounting to 27.00, 13.00, 10.00 and 2.00% for seed stored with seed marketers, ADPs, Seed Agency and NARIs respectively. In the same vein, cowpea seed viability loss was highest for seed marketers (37.00%) followed by ADPs (25.00%), Seed Agency (23.00%) and NARIs (21.00%) respectively. This result suggest that seed handling methods adopted by NARIs were more effective in minimizing damage and maintaining viability than the methods adopted by Seed agencies, ADPs, and marketers. Tekrony (2002) has earlier stated that the rate of seed deterioration depends majorly on the storage environment and physiological quality of the seed.

Seed inspection

One of the quality control mechanism instituted by the National Seed System is Regular Seed Inspection whereby inspectors are required to pay routine inspection to handlers at the different stratum of the distribution channels. However, an assessment of the visit of the Seed Inspectors to the handlers showed that inspection was not popular among the dealers who represent the final and largest outlet to farmers. Table 7 shows that seed inspection was not prominent among seed dealers with about (11%) claiming to have been visited by the inspectors in addition to the fact that the visit was made

Table 7. Seed inspections activities.

	Seed dealers	NARIs	Agency	ADPs
Seed inspectors visit				
Yes	08(10.53)	50 (1)	100 (1)	100 (3)
No	68(89.47)	50(1)	0 (0)	0 (0)
Frequency of visit				
Monthly	0 (0)	0 (0)	0 (0)	0 (0)
Quarterly	0(0)	50 (1)	100 (1)	0 (0)
Yearly	8(10.53)	0 (0)	0 (0)	100 (3)

Figures in parenthesis are the percentage of respondent.

Table 8. Training need of the commercial seed agents.

Training needs	Frequency	Percentage
None	28	36.80
All aspects	06	7.9
Seed production	07	9.2
Seed storage	21	27.6
Variety Identification	07	9.2
Seed production and storage	07	9.2

once in a year. However, quarterly visit were made to the NARIs and other seed agencies while it was once in a year for the ADPs. The level of visit was found to be lower and inadequate both in coverage and frequency to be able to carry out expected duties as stated in the National seed rules and regulations of Nigeria. The rules permits routine and uninterrupted visits and control in terms of labeling, tagging and drawing of samples for test at all time to ensure minimum standard are met throughout the year (Federal Ministry of Agriculture and Rural Development, 1994).

Training needs

Although the study has shown that majority of the farmers have attended training programmes on seed handling, Table 8 shows that majority of the respondents (27.6%) indicated that they needed additional training in seed storage methods while 9.2% each required training on seed production and variety identification respectively. Similarly about 8% required training in virtually all aspect of seed handling while 9.2% required training in seed production and storage. However, 36.80% of the respondents did not require training in any aspect of seed handling. The results have evidently shown the importance that the seed marketers attach to storage methods and as such, majority display it to acquire more skills in this respect.

Conclusion

This study has shown that the populations of seed agents in the study area are educated and could be described as having adequate orientation in seed handling by virtue of their education and experience. Although, majority of the marketers had attended trainings in seed handling, the fact that only few attended such trainings in recent years could undermine their opportunity of dealing effectively with newly emerging challenges in seed handling. It is obvious from this study that majority of those involved in seed business store their seed under ambient conditions that is considered less conducive for maintaining the viability of seed, hence seed deteriorate faster in the stores of the commercial seed agents and the ADPS than with the NARIs. The study has also shown that seed inspection has grossly been inadequate to facilitate effective monitoring of the distribution channels and this is capable of undermining the effectiveness of the delivery system in guaranteeing quality seed delivery to farmers as end users.

It is hereby recommended that the system be further strengthened to accommodate effective monitoring through registration of all seed handlers with regulatory agency at the grass root. This will allow formal interaction and linkages of private and public seed handlers, which has been identified by Mutlu (2000) to be the most effective and most sustainable way to secure the supply of adequate quality seed of all crops to small and large scale farmers in addition to facilitating virile forum for updating the skills of the agents in the identified and other areas of training need.

REFERENCES

Anonymous (2001). Spore Number 94, August 2001, p. 3.

Adetumbi JA, Olakojo SA (2010). Storage potentials and tolerance of high protein maize (hpm) and quality protein maize (qpm) to seed storage pests in controlled environment. J. Plant Protection Res., 50: 36-40

Daniel IO, Adetumbi JA (2004). Seed supply system for vegetable production at small holder farm in South Western Nigeria. *Euphytica* 140: 189-196.

Desai BB, Kotecha PM, Salunkhe DK (1997). Seeds handbook. Biology, Production, Processing and Storage. Marcel Decker Inc. NewYork U.S.A. pp. 223-234.

Durtschi A (1999). Introduction to seed viability http:\\ www.waltonfeed.com

FMARD (1994). National Seed Rules and Regulations of Nigeria.

ISTA (1985). International rules for seed seed testing. Seed Sci and Technology, (13): 348-350.

Kelly AF (1988). Seed planning and policy for Agricultural production. Belhaven Press. London. p. 289.

Mutlu P (2000). Seeds are life, seed sector projects in German development cooperation publication of Deutsche Geselischaft for Technicheo Zusrammenarbeit (GTZ), Germany. p. 13.

Tekrony DM (2002). What is seed vigour? University of Kentucky, ISTA Compact Disc Inc.

4

Ratification of the Ramsar convention and sustainable wetlands management: Situation analysis of the Kilombero Valley wetlands in Tanzania

Felister Mombo[1,2], Stijn Speelman[2], Guido Van Huylenbroeck[2], Joseph Hella[1], Munishi Pantaleo[1]* and Stein Moe[3]

[1]Sokoine University of Agriculture, P.O Box 3011, Morogoro, Tanzania.
[2]Ghent University, Coupure Links 653, 9000 Ghent, Belgium.
[3]Norwegian University of Life Science, P.O Box 5003, Aas, Norway.

In recent years, the Kilombero Valley wetlands in Tanzania was designated and added to the Ramsar convention's list in an attempt to improve its social, economic and environmental values. This study, carried out in selected sites within the Kilombero Valley wetlands, uses a participatory approach to analyze the existing situation and to reflect upon the quest for sustainable management as defined by Ramsar convention. The empirical findings reveal that the Kilombero Valley wetlands are an important source of livelihood for 87% of the dwellers in the area. Moreover, the wetlands also contribute significantly to welfare outside the area as 70% of the country's hydropower depends on water regulation functions of this wetland and they are the source of a diverse number of streams, adding to their biological and ecological value. The study furthermore revealed that the current institutional arrangement for the site threaten the sustainability of the wetlands, despite its addition to the Ramsar convention's list. A more detailed study on appropriate incentive mechanisms for the sustainable management of the wetlands, which would help to internalize the negative effects created by the users, is recommended.

Key words: Wetlands, sustainable management, incentive mechanisms, institutions.

INTRODUCTION

Wetlands are among the world's most biologically productive ecosystems, usually rich in species diversity. Nevertheless they are often undervalued and consequently mismanaged, leading to their degradation (Turner et al., 2000; Baron et al., 2002; Schulte-Hostedde, 2007). In Tanzania about 10% of the land surface is covered by various wetland systems ranging from deltaic mangrove formation to lakes and river flood plains. Due to their high fertility and moist soil, wetlands in Tanzania usually have a high potential for agriculture and contribute significantly to rural livelihoods in terms of cash and food security. In a country such as Tanzania where about 80% of the people are rural poor and depend on agriculture for their livelihoods, wetlands are therefore very important (MNRT, 2004a, 2007). However, due to differences in environmental and socio-economic conditions in Tanzania's rural communities there are significant variations in the patterns of use between the various wetland systems. These variations exist due to differences in the existing interactions between humans and the ecosystems (Hanna et al., 1996; Ostrom and Schagler, 1996; Ostrom, 2005; Hagedorn, 2008). The resulting patterns are complex and have an impact on the natural performance of the wetlands. This impact varies from place to place depending on the extent to which local communities are in contact with or dependent on the wetland resources.

Most wetlands in Tanzania are in fairly good conditions, although some like the coastal mangroves deltaic formations, the Kilombero Valley, the Pangani River and the Usangu plains are highly degraded. A thorough understanding of the causes of this degradation and the possible measures to revert the situation is important to

*Corresponding author. E-mail: pmunishi2001@yahoo.com.

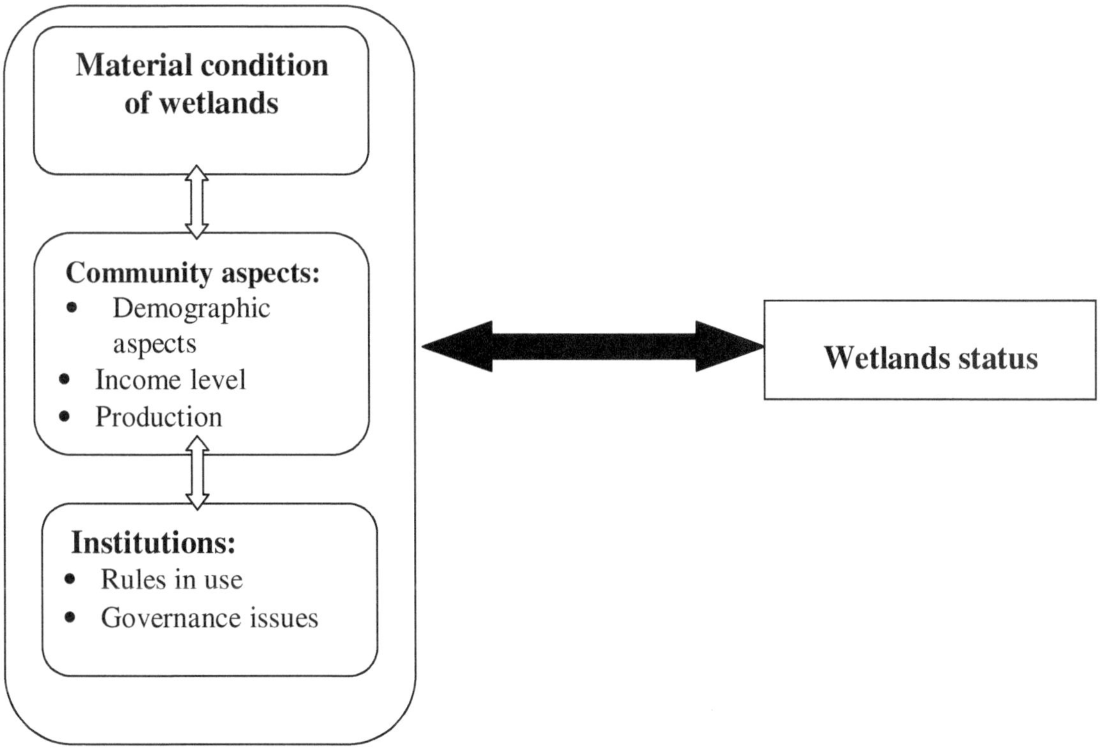

Figure 1. Conceptual frame work for situation analysis of Kilombero Valley wetlands. Source: Authors' construct with much contribution from IAD and IoS frame works.

avoid further degradation (Munishi and Kilungu, 2004). In order for the country to continue having these vital resources for economic development there should be a wise use of wetland resources through sustainable management (ESP, 2003). The Government of Tanzania (GoT) has shown its commitment towards wise use and recognition of wetlands as significant natural resources with important ecosystem services and biological values by ratifying the Ramsar convention on wetlands in August, 2000 (ESP, 2003; MNRT, 2004a). Together with efforts towards wise use as the Ramsar convention (1971) suggests there are still limits and challenges in attaining such objective. Some of the limitations and challenges include compromise between the long term and short term benefits based on the fact that short term solutions might provide a burst in income over a short time period but are seldom sustainable (Fischer, 2002; Baron et al., 2002). On the other hand, long-term sustainable planning and utilization is often more difficult, although it has been proven to give the largest and continuous benefits over time. The later is widely used according to Ramsar convention, but to the rural poor where their focus is on day to day survival, it is a great challenge.

Although some studies have already been conducted with regard to problems of wetland ecosystems in Tanzania and elsewhere, few have concentrated on explaining why the existing interactions between humans and wetland ecosystems have continued to be the cause of degradation of these ecosystems even after the country's ratification of the Ramsar convention. Moreover, Ostrom (2007) argues that most existing studies regarding degradation of common pool resources, from which wetlands are a special case, have used simple linear models which are not able to thoroughly explain the existing complex pattern of variations in their degradation. It is in this light that this study comes with some insights on why and how such interactions impact on the sustainability of wetlands. Kilombero Valley wetland, one of the most prone wetland to degradation in Tanzania, is used here as a case study. This study will be useful to inform the policy process and other development agents concerning issues to be considered when designing strategies for sustainable wetlands management.

THEORETICAL FRAMEWORK AND METHODS

A conceptual framework for this study (Figure 1) has been developed based on an extensive literature review and the authors' firsthand knowledge of the context in Tanzania. The resulting framework is closely related to the Institutional Analysis and Development (IAD) framework by Ostrom (2005) and the Institution of Sustainability (IoS) framework by Hagedorn (2008). The IAD model introduces a specific set of variables regarding persons, institutions and the physical world, which constitute the important drivers of the unsustainable wetland resources management causing the degradation in the Kilombero Valley wetlands.

Moreover, in the IoS framework it is claimed that when people interact with their physical environment through rules and existing governance structures, these interactions have impacts in both directions.

This means that while humans design the rules governing these interactions, sometimes the nature itself determines what kind of rules suits these interactions, thereby ultimately determining if the status of the wetlands is sustainable or unsustainable. On the other hand the status of the wetlands has an impact on the status of the communities' welfare which then affects the material condition of the wetlands.

The study area

The situation analysis was carried out between September 2009 and January 2010 in Kilombero District of Morogoro region where the Kilombero Valley wetlands which contain the largest freshwater wetland at low altitude are situated. The region is in the eastern part of Tanzania. The wetland area is located between the Udzungwa Mountains and the Mahenge escarpment. The valley is divided by the Kilombero River and falls within two districts (Kilombero and Ulanga). The wetland area covers 7,967 Km^2, with a catchment's area of about 40,000 Km^2. Many rivers, permanent and seasonal, feed the floodplains (Figure 2). The area is characterized by a sub-humid tropical climate with relative humidity ranging from 70 to 80% with an annual rainfall of about 1200 to 1400 mm. It has two rainy seasons: a long rainy season in March to May and a shorter one around October to December. Temperatures normally vary from 20 to 30°C (MNRT, 2007).

Sample selection

The target population comprised all primary users of the wetlands in the Kilombero Ramsar site. These were further categorised into local people, a group of companies and organizations which enjoy the products and services of the wetlands. For the locals a sample size of 5% was picked from the total household population in nine randomly selected villages (Sonjo, Mofu, Idete, Mahutanga, Kivukoni, Milola Mwaya, Sanje and Kikwawila) in Ifakara and Ulanga Districts in Morogoro region. The sample consisted of 160 households and for this case the sampling frame was a village and sampling unit was a household, defined as all people living under one roof and are sharing the same pot for cooking their dishes. The companies and organizations were randomly picked to represent all the companies and organizations, who are primary users or managers of the wetlands.

Data collection

Secondary and primary data were collected for this study. The secondary data sources included web resources, scientific papers and documents from NGOs and Government departments. The primary data was collected to fill in the gaps to whatever was not found from secondary data or when those data appeared to be out of date. The primary data were obtained through a Participatory Rural Appraisal (PRA) as a standard social survey data capture tool combined with semi structured questionnaires and checklists.

During the surveys, various wetland beneficiaries were contacted, including local communities in the selected areas, relevant NGOs and Community Based Organizations (CBOs); companies and enterprises. Qualitative data were collected through the appropriate PRA technique to understand the situation of the wetlands in terms of location, utilization, perception, trends and conservation aspects. Focus group discussions with key informants were also used in this aspect. Quantitative data were obtained by questionnaires and literature surveys, yielding information on wetland related products and services. This provided understanding on the way these products and services benefit the users and how the daily use of the services and products impacts on the wetlands' sustainability.

Data analysis

The analysis is mainly based on the conceptual framework, with as main assumption that wetlands would be functioning at 100% capacity if there would not be any anthropogenic activities. The exploration of human activities and their impacts can thus be used to assess the biophysical condition of the wetlands. While the community is the central point of these interactions, the attributes of the community have also been explored to quantify their impacts on the status of the wetlands through the rules established and the governance structures.

RESULTS AND DISCUSSION

Biophysical analysis of wetlands

Wetlands offer a variety of products and services to local households living in and around the valley. When respondents in the area were asked about the services and products having impact on their welfare and the rank of the products and services according to their usefulness and importance in their daily life, water for domestic use was ranked first, wetlands as source of food and cash (agricultural crops and fishing) was ranked second and the third was wetlands as sources of timber and non timber forest products. The ranking of these top three was the same in six of the nine surveyed villages indicating that these are the most important wetlands related products and services for the communities in Kilombero Valley. These three most important products and services are therefore referred to in order to explore how the access to and use of the resources have impacted the biophysical state of the wetlands. The average ranking of other products and services in the eight villages is listed in Table 1.

Wetlands as source of crops production

Crop production in the Kilombero Valley flood plain is practiced both at small and large scale. For several years as documented by MNRT (2004b) the plain has attracted a number of large scale farming investors. Their relative importance and distribution is presented in Figure 3a and b. Of the total area under cultivation only 23% is cultivated by small scale farming. An estimated 87% of all the people living in the valley depend on this area for their livelihood. The farming is usually done individually and sometimes in association under private farms. Majority of these farms are small, with an average size of 1 ha. About 10% of the farmers have landholdings exceeding 4 ha. The land is primarily acquired through inheritance, purchase or through allocation by the village

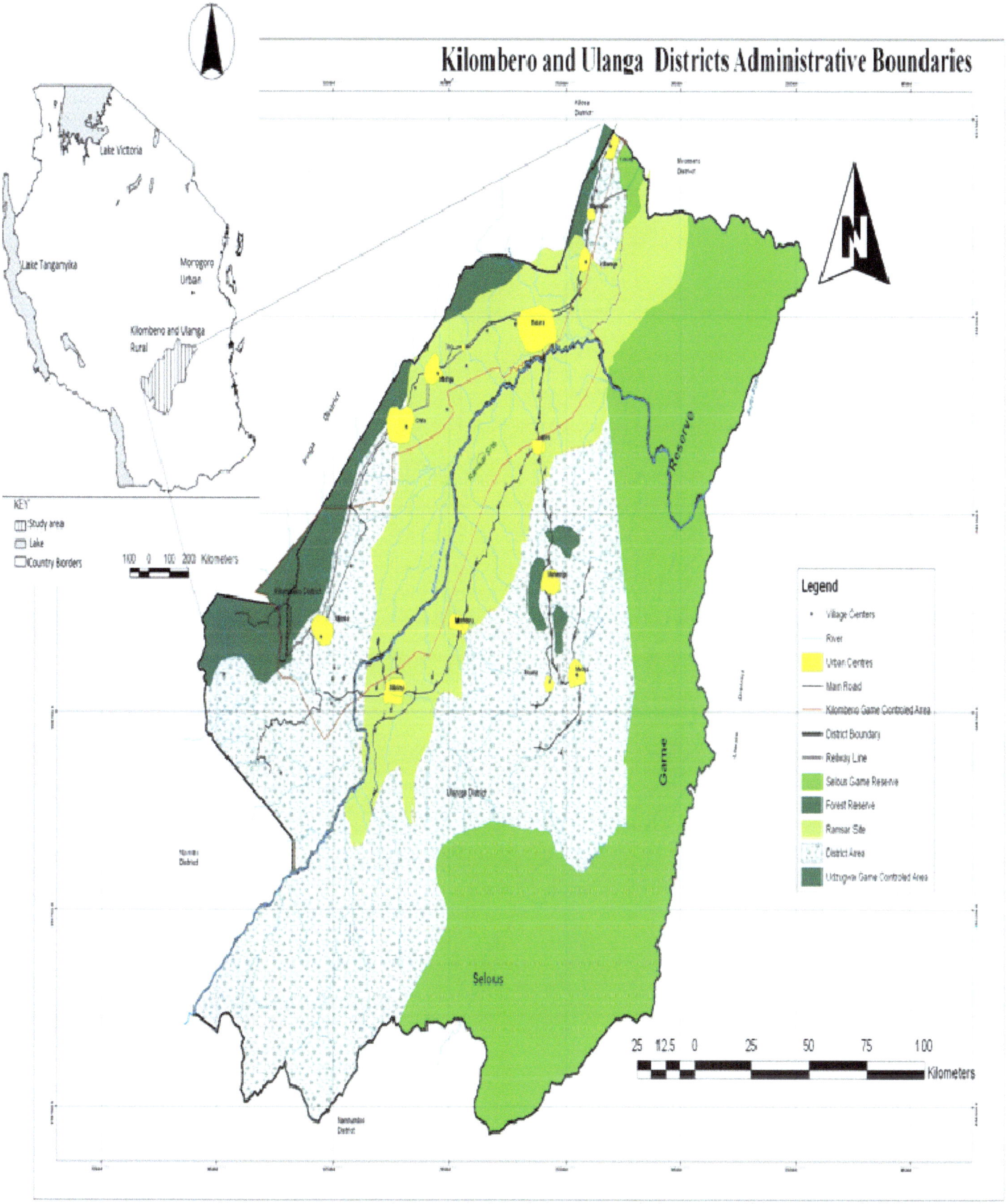

Figure 2. The map showing the location of Kilombero Valley Wetlands in Tanzania and the administration boundaries (KVRSP, 2009).

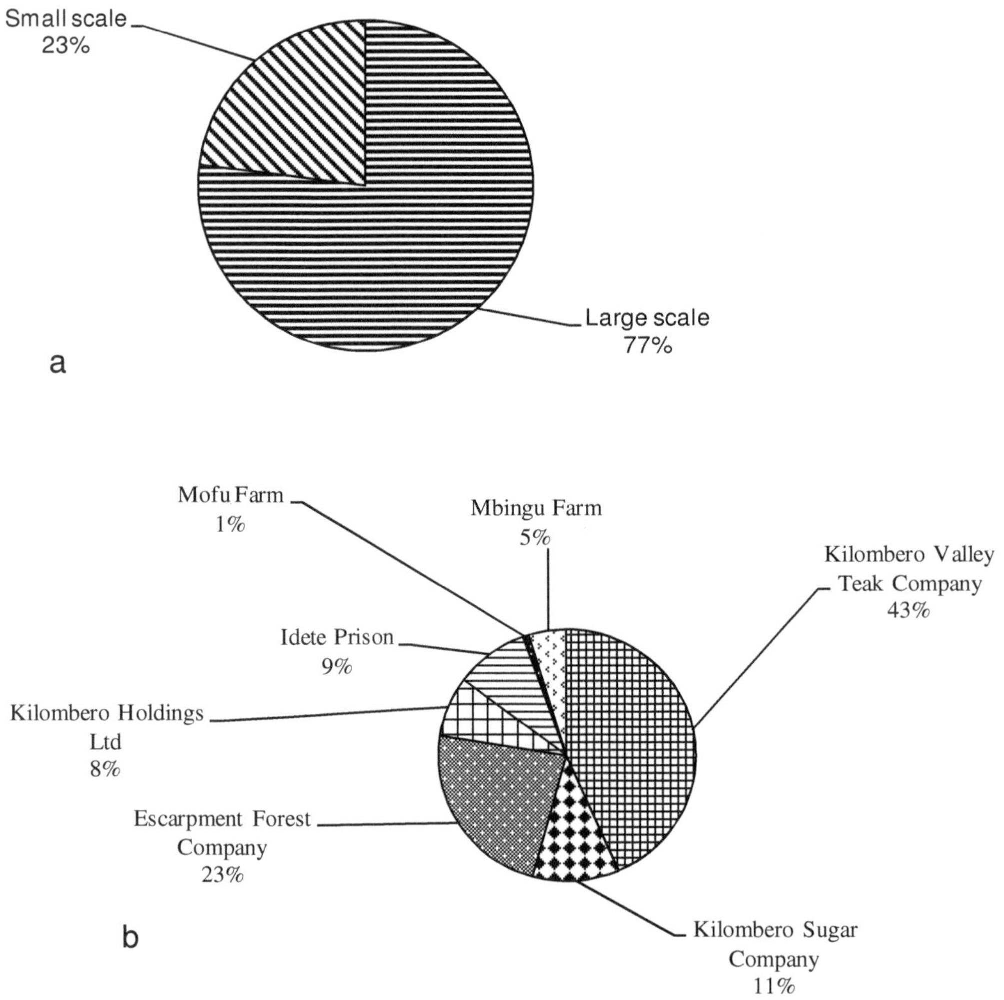

Figure 3. (a) Distribution of cultivated Land in Kilombero Valley, (b) Proportion of land holdings amongst large scale farmers in Kilombero Valley (Source: MNRT, 2007).

Table 1. Product/service rank by respondents in Kilombero Valley wetlands communities.

Rank	Product/service
1	Source of domestic water
2	Source of food and cash
3	Source of timber and energy
4	Grazing
5	Habitats
6	Cultural practices
7	Wild game
8	Source of non timber forest products
9	Provides burial grounds

governments, but sometimes it is also rented from owners.

The major food crops cultivated in the Kilombero Valley include maize, paddy rice, banana, and cassava. Cash crops include sugar cane, and paddy rice. Other minor crops also used as sources of food and supplemental income include sweet potatoes, corn, vegetables, peas, pigeon peas, groundnuts and sesame. Some figures on the yields and farm gate prices of selected crops in the harvest season are presented in Table 2

Due to the practice of shifting cultivation, the emergence of large scale farming and population growth, currently more than 50% of the flood plain is converted into agricultural land. At this moment, some people are even cultivating areas which were previously considered marginal, such as the river banks and around the swamps. As a result of this, there is increased soil erosion, which has resulted into siltation of downstream hydro power dams and also causing the drying up of streams and swamps. Several permanent streams, such as Kiberege, Idete, Kikwawila, Namawala and Idandu, in the past, are now only flowing seasonally. Furthermore there is speculation that water pollution and depletion are increasing due to the use of pesticides by the large scale farmers.

Table 2. The amount of agricultural crops obtained from wetlands and their prices.

Product (crop)	Amount per Season^{-1} Ha^{-1}	Selling price	
		(TAS Unit^{-1})^{b}	US$
Paddy rice	25 bags	30,000.00	23
Maize	10 bags	42,000.00	32
Sugarcane	75 tons	35,000.00	27
Banana	375 bunches	2,750.00	2

[a]TAS 1300 is equivalent to 1USD (2010), [b]1Bag is equivalent to 100 kgs.

Wetland as source of domestic and industrial water

Kilombero River is the main source of water for the wetland adjacent communities. The river connects Kilombero District and Ulanga District. The river is fed by 10 tributaries originating from the Udzungwa Mountains. It flows in the Rufiji River, which is the biggest river in the country. There are several sources of domestic water and it was revealed by the interviews that a household can get water from multiple sources. However, the majority (83%), of the local people use yard tap water for domestic purposes, whereas less than 20% of the sampled households have a private connection. These percentages vary depending on the season of the year since the water from tap is not always available especially in dry seasons. At these moments they use water from other nearby sources such as community wells and streams.

Productive water use in the area mainly constitutes irrigation in the large scale farming of sugar cane and rice. The estimated costs of irrigation water per cubic meter are 20 TAS (1TAS is equivalent to 1300 USD). This includes all expenses in irrigation such as water pumping, labour and electricity.

Wetlands as source of energy, timber and non – timber forest products

Most respondents claim to be using energy from various sources, depending on the availability and on the purchasing capacity of the household. Often different energy sources are used in combination, although the majority depends mainly on fire wood and charcoal for their cooking (Figure 4). Most of the people use kerosene for lighting and a negligible proportion (1%) uses electricity.

It is striking that the area which produces 70% of the countries' electricity only a small minority (1%) has access to electricity in their houses. In this way the local communities do not appear to have particular benefits from electricity production and hence the significance of the wetland in terms of power production.

Lusambo (2009) states that the households' overwhelming dependency on wood as energy source has negative environmental consequences. Based on the model developed by Lusambo (2009) the wood fuel consumption in the study area is unsustainable, estimated to be causing a net deforestation ranging between 3.37 and 21.59 ha day^{-1}. The most recent statistics on deforestation rate in Tanzania are those provided by Butler (2006) and UN (2007). In both cases deforestation is estimated to be about 410,000 ha year^{-1}. Household energy consumption is an important determinant of this trend but another important cause of deforestation is timber harvesting. To reverse this deforestation trend timber harvesting is prohibited in reserved parts of the wetlands and only firewood collection by women and children is allowed.

Furthermore, various programs on tree planting around homesteads have been introduced to reduce the pressure on the parks' resources. Notwithstanding official claims that these programs have been successful, we observed that except for fruit trees the amount of trees planted on the homesteads was limited. This observation is also supported by Kajembe et al. (2008) and Munishi and Mbeyale (2009) who specifically pointed out that while tree planting programs by the park authority stressed on planting the trees for fuel wood the needs of the local community was not solely for fuel wood but also for timber. In fact the fuel wood was seen to be a secondary need. This could be expected, because the local communities can still legally get fuel wood from the park through the existing arrangement and therefore due to tight regulations and probably high enforcement the need for timber and poles as raw materials for shelter is considered as a more pressing need by the local communities. Notwithstanding the increasing deforestation, still an extensive timber resource exists outside the forest reserve, although there had been illegal logging for many years which has resulted into removal of all important trees for timber. While illegal pit sawyers target individual and vulnerable trees leaving a good forest cover, the opening up of areas for agriculture is reducing the area of woodlands significantly (URT, 2005).

Finally, we have to note that there is also large scale production of timber which is mainly done by large companies. The data from MNRT (2004b) shows that the Kilombero Valley Teak Company (KVTC) owns 28,159 ha (Figure 3b) in the valley on a lease of 99 years for timber production, mainly through teak plantation. However, to establish the plantations the company cuts down

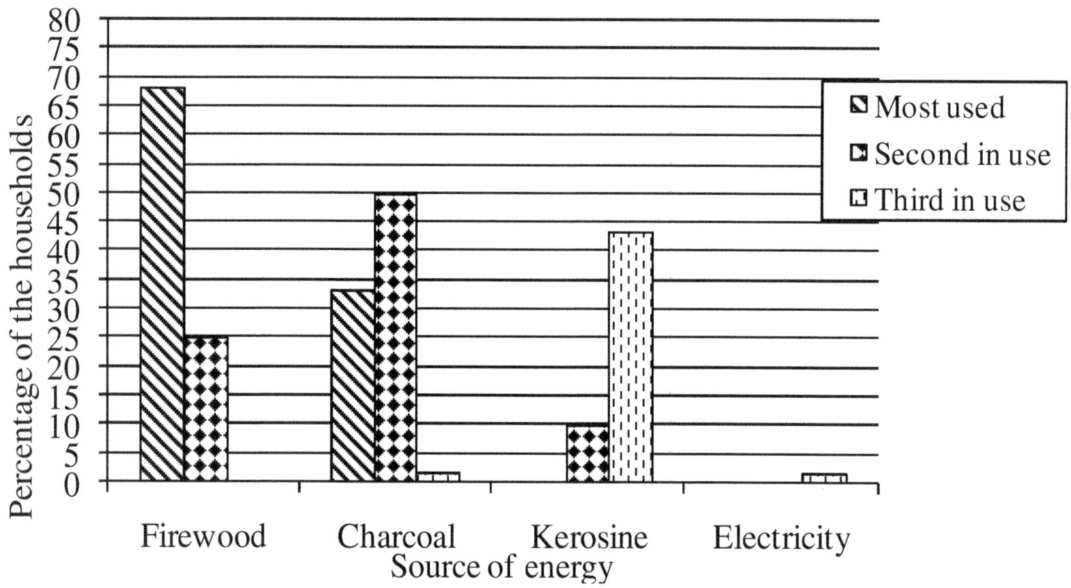

Figure 4. The ranking of energy used by the local communities in the wetlands.

indigenous woodland. This clearly has a negative effect on the wetlands' ecosystem.

Locals also obtain non-timber products such as bee products and reeds for weaving. The production volume for bee products is very low for example one of the associations in the area known as Miombo beekeepers Association with 300 beekeepers produces only an average of 5,000 Kg of honey and 500 Kg of bees wax from 613 beehives. Also the weaving is done on a very small scale (5%). Nevertheless such activities are important areas to look at as in most cases non-timber forest products are environmental friendly therefore an intervention on this as wise use (Ramsar, 1971) activity could be an important incentive for wetlands management if other users could contribute as compensation to the replacement of the activities which are very detrimental to the wetlands ecosystem.

Fishing activities

Fishing is one of the important economic activities in the study area. The fishing population depends on the location of the village with reference to the distance to the river and the flood plain. Following this, in some of the villages such as Mofu many people depend on fishing as their major economic activity. It is revealed by the study that fishing is negatively affecting the integrity of the wetlands in Kilombero Valley due to the way it is carried out. It was observed that due to occasional decrease in fish stocks (Figure 5), the fishermen in Kilombero Valley have developed techniques which destroy the integrity of the wetlands.

An example of a fishing technique is abstracting water from the swamps to reduce the water level which makes the fish easy to catch. This activity is carried out in the dry season and also involves planting reeds across the water inlets of the swamps to prevent too much water from entering in the swamps. Furthermore the fishermen also destroy the breeding sites of fish by uprooting the important vegetation called "Makongo" which are used by fish as their breeding sites. As a result to these destructtions, many swamps are claimed to dry up since the "Makongo" are also argued by the locals to prevent swamps from drying up during dry seasons. Furthermore this type of vegetation are claimed to prevent erosion along the river banks. Finally the locals argue that the vegetation has a kind of close relationship with the fish as its existence also depends on the availability of the fish in these water bodies. Although these claims need scientific verification the arguments are valid ecologically as this is also supported by other scientific findings as revealed by De Leo, and Levin (1997) and McClanahan (1999).

The destruction of breeding sites has resulted into the disappearance of a number of fish species and in the reduction of the quantities of others. The data with the trend of catch per species also suggests that fishing might be done in an unsustainable way as there is a shift after some time to a different species. These findings are also supported by claims from locals that some of the preferred species have decreased in availability forcing them to focus their efforts on other species (Figures 6 and 7). The species that appear to have disappeared are the *Heterobranchus longofilis*, *Citharinus congicus, Synodontis maculipiana* and *Labeo coubie*. Those that decreased in quantity are such as *Bagrus orientalis*, *Hydrocynus (*sp*), Alestes (*sp*), Distichodus fasciolatus* and an endemic species "Njogu" which we could not

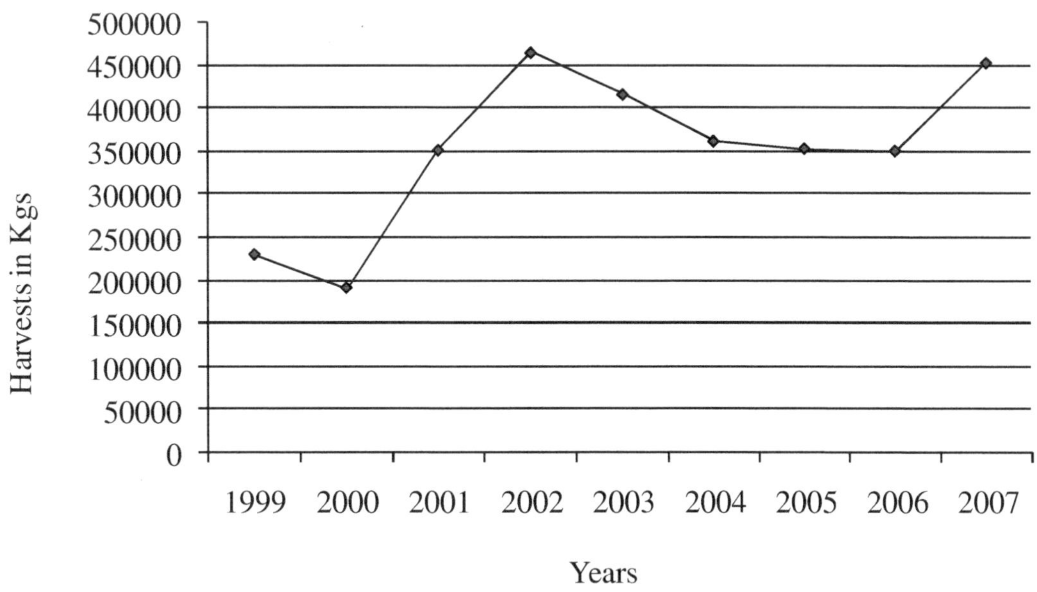

Figure 5. General trend of catch for the past nine years (1999-2007).

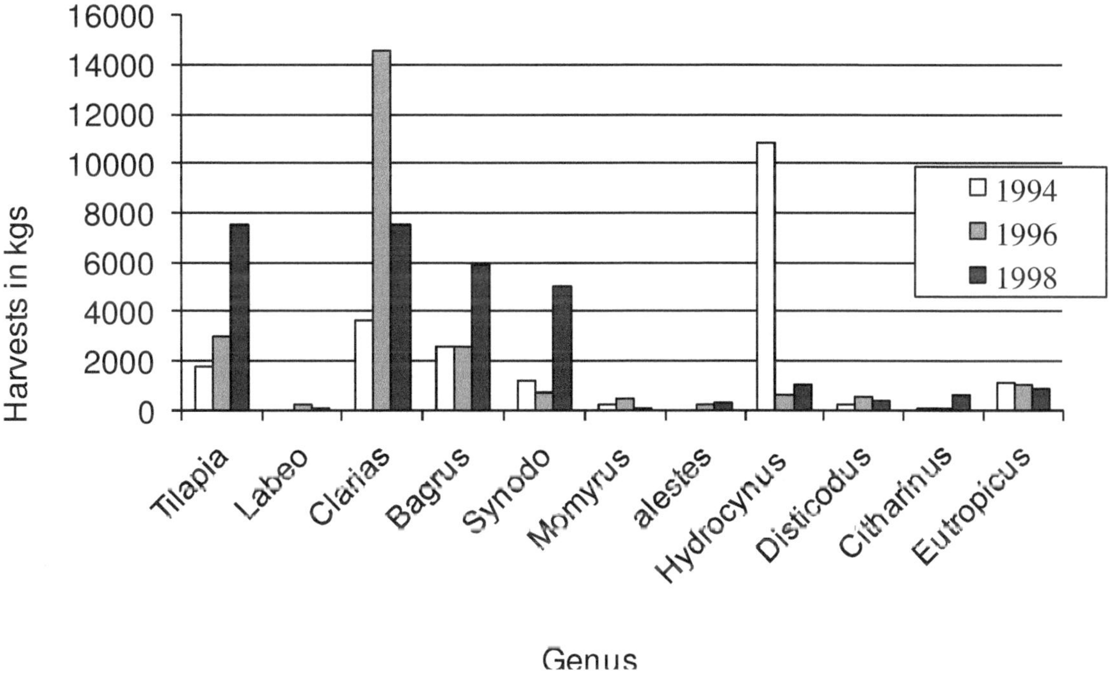

Figure 6. The catch per genus for 1994-1998 period.

identify.

Wetlands as source of grazing land and pasture

The wetland is also used as a grazing land for livestock. It is estimated that cattle stock constitutes about 300,000 cattle and that there are approximately 43,000 sheep and goats. The livestock population has been increasing during the past 10 to 15 years. This increasing population of livestock requires large land areas for grazing, which is not readily available in Kilombero Valley. This has resulted in a number of conflicts between livestock and crop farmers. Furthermore, in some parts of the flood plain the spongy character of the soils is changed due to livestock trampling. At these places soils are now more compacted.

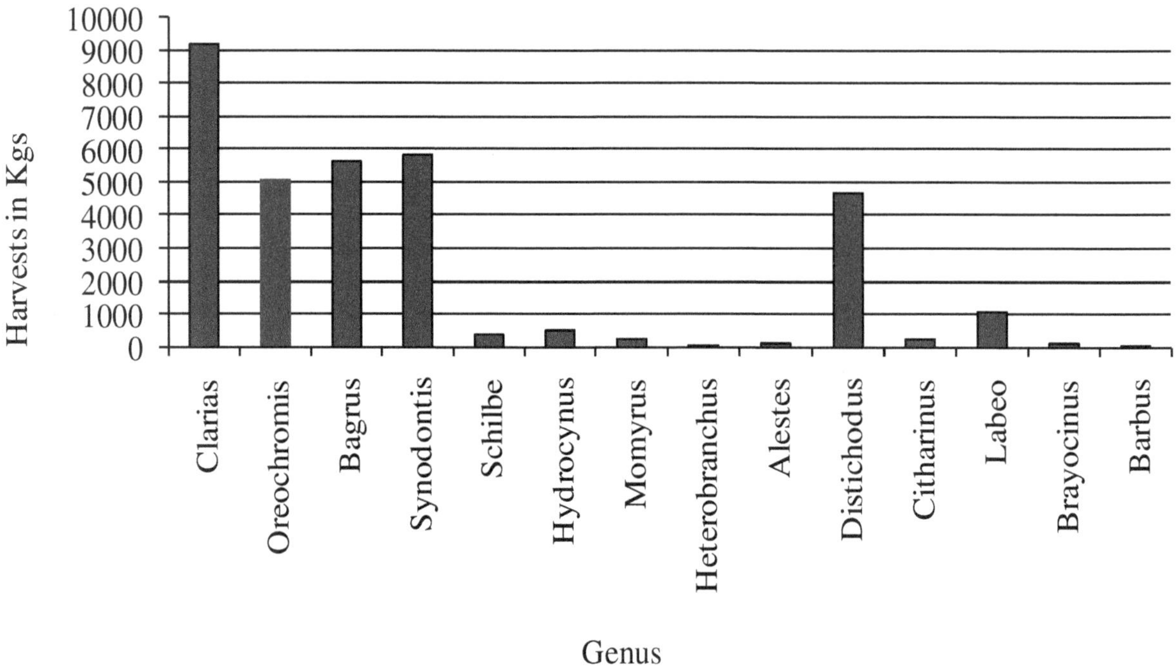

Figure 7. The catch by genus in 2008.

The livestock is also held responsible for the seasonally drying up of a number of water sources. Other negative effects of overstocking are overgrazing and water pollution with the chemicals used to treat the livestock against tick born diseases.

Community aspects

An important aspect drawing the relationship between the local communities and the wetlands is their income level. The majority of the households in the Kilombero wetlands are poor with an average consumption expenditure of 17,000 TAS annually which is significantly below the national average of 63,000 TAS (MNRT, 2004b). The majority of the households depend on agriculture as the main source of income, although the findings from this study show that there is a big percentage of households (30%) whose main income is derived from petty trading (Figures 7 and 8).

The figure clearly confirms that agriculture is a main source of livelihoods for the local communities. Following this it is evident that there is a big pressure on the wetlands and therefore a deliberate effort is needed to sustain the ongoing population of the area which is projected to be growing at the rate of 3.4 and 2% per annum for Kilombero and Ulanga Districts respectively. The population distribution is shown in Table 3.

Table 3 shows that Kilombero District is more densely populated compared to Ulanga District though care should be taken on how to interpret this distribution. It is equally true that a large part of Ulanga District falls in Selous Game Reserve, where people were relocated when the place was designated for the reserve and so most of the households are concentrated in some specific locations and not evenly distributed. This is the main problem with the population distribution in the valley as most of the area is taken from local communities and converged to either large scale farming or conservation reserves such as National Parks and game reserves. This pushes the rural poor to the marginal areas which in most cases are more vulnerable areas of wetlands.

In terms of production settings and technologies the region is underdeveloped. The local communities practice peasantry farming with hand hoe being the main tool of tillage. Although this is good for environment, it requires a lot of labor to put a large area into cultivation. Moreover, due to the poor technology the production is so low and limited to areas where there is enough moisture and where land is fertile (Table 3). Consequently since these areas are most of the time the important areas for ecosystem's health (De Leo and Levin, 1997; Robin et al., 2006) their use has a negative impact on the wetlands. More than that, the problem becomes so big when there are large companies which have good bargaining ability since these take good arable land and push the locals to very fragile and vulnerable areas.

Institutional arrangements and information sharing and their impacts on the management of the wetlands

The Kilombero Valley wetlands are recognized as Ramsar site and therefore are under the management of the

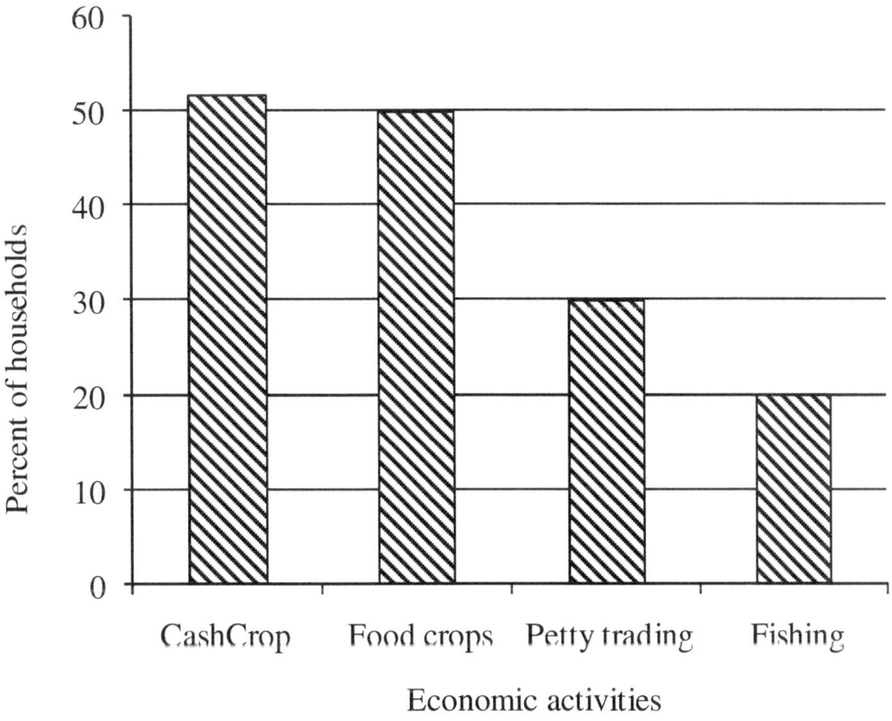

Figure 8. The distribution of household economic activities in Kilombero Valley wetlands.

Table 3. Population characteristics of Kilombero Valley wetlands communities (a map of population density would be better to use here).

District area	Total area (Km^2)	Land area (Km^2)	Population no			No. of households	Household size
			Men	Women	Total		
Kilombero	14,918	13,577	162,942	159,837	322,779	73,393	4.4
Ulanga	24,560	23,681	95,915	98,294	194,209	36,988	5.3
Total	39,478	37,258	258,857	258,131	516,888	110,381	4.8
Wetlands area	7,967	-	-	-			
Ulanga	-	-	56,519	58,668	115,187	20,953	
Kilombero	-	-	158,992	155956	314,948	-	
Total	7,967		215,511	214,624	430,135	-	

Source: MNRT (2004b) with minor modifications.

department of Wildlife Division in the Ministry of Natural Resources and Tourism. Under this division there is one unit which deals with wetlands in Wildlife protected areas (MNRT, 2009). According to an interview with key informants it was only recently that they started to have officials (Wetlands officers) in wetland protected areas which work on specific issues of wetlands. That the area is categorized as a Ramsar site implies that according to the Ramsar convention only wise use is allowed. The problem of the convention is that the term 'wise use' is not specific and therefore gives a loop hole of justification of uses especially for a poor country like Tanzania, which has to bargain between immediate developmental objecttives and conservation perspectives which will be beneficial for future generations.

Furthermore since the division is only supposed to interfere with the issues of wildlife protection it does not have enough power to protect the area from other types of mismanagement related to for example agriculture, water, forest and energy issues whose activities are supposed to be carried out by other ministries through sectoral policies (MNRT, 2004a; MoA, 2007; URT, 2005). There are advantages and disadvantages attached to segmentation of natural resources management compared to an integrated approach (Vatn, 2009; Vanhyulenbroeck, 2009). Yet it is important to have a detailed study to elaborate on how advantageous or disadvantageous as far as sustainable wetlands management is concerned. Important to note is there is an integrating organ under the same wetlands unit in the wildlife division. The problem arises when this organ is at the national level and therefore makes it difficult to solve problems which are in most cases important and/or detrimental at local level.

Not all wetlands in Tanzania (following the Ramsar definition of wetlands) are situated in Wildlife protected areas as is the case for Kilombero Valley. The Land Act and the Land Policy of 1999 even categorizes wetlands as hazardous areas and therefore these areas are not allowed by the law to be entitled to any individual or group of individuals for whatever use (URT, 1999).

Notwithstanding this legal statement many households live and earn their living in wetland areas. Thus there is a great contradiction to what is in the formal rules to the inform rules formed by the communities living in wetlands areas. Surprisingly even companies which are practicing large scale farming have legal occupancy to wetland areas. This however, raises the question as to whether what is defined to be wetlands by the Ramsar convention is the same as what is called wetlands in Tanzanian context let alone the local context. Following this, there is a need to thoroughly study this and come with appropriate suggestion as to what should be called wetlands and the reasons for that in Tanzanian's context so as to have a clear management area when it comes to sustainable wetland management.

Moreover, this problem of all wetlands not being protected by the Ramsar convention except those of international importance was once experienced by Canada when it first ratified the convention in 1981. Once the Ontario government realized that its 36 designated Ramsar sites represented only a small fraction of all wetlands in the country and that the clear majority of wetlands are not directly covered under the Convention, nor are they within the federal jurisdictional responsibility and since wetlands protection is a provincial and municipal governments responsibilities; the government decided to develop its own policy that would protect most of the wetlands which were by then fast degrading (Schulte-Hostedde et al., 2007). Tanzania does not have an independent policy on wetlands management however the new National wildlife policy has also incorporated wetlands management issues and by law they are mandated to protect all designated wetlands for protection (MNRT, 2009). In addition to this there is no legal instrument like a wetlands Act which would make the policy operational legally. From Canada's experience much could be learnt on sustainable management of wetlands in Tanzania.

Beside the above findings the survey in the area shows that most of local people (87%) do not know whether there is a specific law which prohibits them from performing specific activities in the wetlands areas beside that of cutting trees in the Udzungwa National Park. The people also do not know whether the area is designated as Ramsar site under the wildlife division which according to them owns Seleous Game Reserve. This shows how decision making is done by government elites living aside from the majority of the rural population of 87% unaware of ongoing conservation activities in the area despite many policies insisting on participatory approaches to natural resource management.

Moreover, the locals admit to be involved in conservation activities by Tanzania National Park Authority (TANAPA) and therefore have mentioned the institution as one of the forefront organs in conservation activities. This gives an encouragement that, if conservation programs are well structured like those of TANAPA by other institution present in the area with good coordinating unit then there would be a big impact in the sustainable management of the wetlands. However, all these conservation efforts by TANAPA are due to the fact that the community interactions with park resources were impacting the vital areas of attraction and therefore this was not an obligation of the park authorities to the environmental services provided by the whole wetland ecosystem. It is therefore important to have incentive mechanisms that would internalize the externalities created by the users of the wetlands. The internalization of externalities and synchronization of the rules governing the resource use in a Common Pool Resources (CPRs) such as wetlands would bring cooperation amongst the stakeholders involved in resources use and management (Turner et al., 2000; Gatzweiler and Hagedorn, 2001; Vanhuylenbroeck, 2002; Munishi et al., 2003; Ostrom, 2005; Hagedorn, 2008). As a result, the potentials of the stakeholders participation can be fully realized and utilized into bringing sustainable wetlands management.

CONCLUSION AND RECOMENDATIONS

The Kilombero Wetlands are an important source of livelihoods for the majority of dwellers in the area. However, currently these wetlands are threatened by ongoing use due to growing demands of the primary users which due to their poverty depend directly on the wetlands for their primary needs of energy, food, shelter and social well being. However, the present institutional setup amplifies the problem due to the fact that it encourages the ongoing lifestyle, which in the long run if not corrected would make the wetlands not able to sustain the needs of both primary and secondary users. Following this, it is perceived that the continuing management style is responsible for unsustainable management of the wetlands. It is now evident that although Kilombero is a Ramsar site there is still going on unsatisfactory use of products and services accrued from the wetlands a condition that threatens its sustainability. These findings suggest that ratification to the Ramsar convention may not necessarily be sufficient condition to sustainable wetlands management. Because of this, it is important to have appropriate incentive mechanisms that would be institutionalized to make both primary and secondary users of the wetlands obliged to internalize the externalities they create to their counterparts (other primary and secondary users) in order to have sustainable management of the wetlands. It is therefore recommended to:

(1) Evaluate the externalities (costs and benefits) created and find out the extent to which they are responsible for the resources miss-management.
(2) Examine how the created externalities can be internalized so as to have social optimal resources use in the wetlands which together with institutions rearrangement would ultimately lead to sustainable wetlands management.
(3) Evaluate the rules governing the use of the resources and their relationship to unsustainable wetlands management.

REFERENCES

Butler R (2006). Tanzania Forest Figures. http://Rainforests. Mongebay.com/20tanzania. html.

De Leo GA, Levin S (1997). The Multifaceted Aspects of Ecosystem Intogrity. Conservation Ecology. http://www.consecol.org/vol1/iss1/art3/

Environmental Support Programme (ESP) (2003). Wetlands Management (2004-2008) Tanzania Draft Component Document. Danish Ministry of Foreign Affairs. Denmark.

Fischer G (2002). Land and Water Use of Wetlands in Africa: Economic Values of African Wetlands. Interim Report No. IR-02-063. Luxemburg, Austria.

Gatzweiler F, Hagedorn K (2001). The evolution of institutions in transition. CEESA Discussion paper No. 4. Humboldt University of Berlin: Berlin.

Hagedorn K (2008). Particular requirements for institutional analysis in nature-related sectors. Eur. Rev. Agric. Econ. 35(4): 606-623.

Hanna S, Jansson A, Arrow K (1996). Rights to Nature: Ecological, Economic, Cultural and Political Principles of Institutions for the Environment. Washington (D.C). Island press.

Kajembe CG, Kimaro D, Kilahama F (2008). Participatory Mid-Term Review of the Project on Improving Natural Resources Use in Eastern Side of Udzungwa Mountain National Park, Tanzania.

Kilombero Valley Ramsar Site Project, (KVRSP) (2009). Aerial Census in the Kilombero Valley Flood Plains Ramsar Site. Kilombero, Tanzania.

Baron JS, Poff NL, Angermeier PL, Dahm CN, Gleick PH, Hairston NG, Jackson RB, Johnston CA, Richter BD, Steinman AD (2002). Meeting ecological and societal needs for freshwater. Ecol. Appl. 12(5): 1247-1260.

Lusambo LP (2009). Economics of Household Energy in Miombo Woodlands of Eastern and Southern Tanzania. Ph. D. dissertation. University of Wales. Bangor, UK.

McClanahan TR (2000). Recovery of a coral reef keystone predator, Ballstapus undulatus, In East African Marine Parks. Biol. Conserv. 94:191-198.

Ministry of Agriculture (MoA) (2007). Feasibility studies for Mngeta Farm Development. Dar es Salaam Tanzania.

Ministry of Natural Resources and Tourism (MNRT) (2004a). An Issue Paper for the Formulation of the National Wetlands Strategy: Dar es Salaam, Tanzania.

Ministry of Natural Resources and Tourism (MNRT) (2004b). The Development and Implementation of an Integrated Management Plan Of Kilombero Valley Flood Plain Ramsar. Site Identification Report: Dar es Salaam, Tanzania.

Ministry of Natural Resources and Tourism (MNRT) (2007). The National Wetlands Management Strategy. Draft Report. Dar es Salaam, Tanzania.

Ministry of Natural Resources and Tourism (MNRT) (2009). The New Wildlife Policy in Tanzania. The Draft Report. Dar es Salaam Tanzania.

Munishi PKT, Shear TH, Temu RPC (2003). Household level impacts on forest resources and the feasibility of using market based incentives for sustainable management of the forest resources of the Eastern Arc Mountains of Tanzania. Paper Presented at the Africa Mountains High Summit Conference. Nairobi, Kenya 6-10 May 2002.

Munishi PKT, Kilungu H (2004). Contribution of wetlands to household income and food security: A Case Study of 'Nyumba ya Mungu' Wetlands System in Northern Tanzania. Paper presented at the National Wetlands Working Group (NWWG) meeting, Dar es Salaam City, June 30.

Munishi PKT, Mbeyale G (2009). Lessons learnt for the WWF project on 'Improving Natural Resource Use on the Eastern Side of Udzungwa Mountains National Park, TanzaniaWWF TPO Dar es Salaam Tanzania.

Ostrom E, Schlager E (1996). The formation of Property Rights. In Ecological, Economic, Cultural and Political Principles of Institutions for the Environment. Washington (D.C). Island Press. pp. 127-156.

Ostrom E (2005). Understanding Institutional Diversity. United Kingdom.Prencenton University Press.pp. 141-64.

Ostrom E (2007). A diagnostic approach for going beyond panaceas. Proceedings of the National Academy of Sciences 104(39): 15181-15187.

Ramsar Convention (1971). The Ramsar Convention. Ramsar Center, Rue Mauverney 28, CH 1196, Gland, Switzerland.

Robin AJ, Allan DA, Lehner BI (2007). Unlocking the potential of protected areas for freshwaters. Biol. Conserv. 134(1): 48-63.

Schulte-Hostedde DW, Powell C, Shrubsole D (2007). Wetland management: An analysis of past practice and recent policy changes in Ontario. J. Environ. Manage. 82(1): 83-94.

Turner RK, van den Bergh JC, Soederqvist T, Barendregt A, Van der Straaten J, Maltby E, van Ierland E (2000). Ecological-economic analysis of wetlands: scientific integration for management and policy. Ecol. Econ. 35(1): 7-23.

United Nations (UN) (2007). The Millennium Development Goals Report. New York.

United Republic of Tanzania (URT) (1999). The National Land Policy. Dar es Salaam. Tanzania.

United Republic of Tanzania (URT) (2005). National Environmental policy. Dar es Salaam. Tanzania.

Van Huylenbroeck G (2002). Multifunctional Agriculture: How to provide Incentives to Farmers. Paper prepared for presentation at the 13th International Farm Management Congress, Wageningen, The Netherlands, July 7-12.

Van Huylenbroeck G, Vuylsteke A, Verbeke W (2009). Public Good markets: The Role of Hybrid Governance Structures in Institutions for Sustainability. In Beckmann V, Padmanabhan M (eds.), Institutions and Sustainability. Political Economy of Agriculture and Environment-Essays in Honour of Konrad Hagedorn. Springer, pp.175-191.

Vatn A (2009). Sustainability, Institutions and Behaviour. In Beckmann, V and Padmanabhan M (eds.), Institutions and Sustainability. Political Economy of Agriculture and Environment-Essays in Honour of Konrad Hagedorn. Springer, pp. 293-314.

Advanced evaluation of cowpea (*Vigna unguiculata*) accessions for fodder production in the central rift valley of Ethiopia

Ayana Etana[1], Estefanos Tadesse[2], Ashenafi Mengistu[3] and Abubeker Hassen[4]

[1]Forage Agronomy Research, Adami Tulu Agricultural Research Center, P. O. Box 35, Ziway, Ethiopia.
[2]Dairy Nutrition Research, Adami Tulu Agricultural Research Center, Ethiopia.
[3]Faculty of Veterinary Medicine, Addis Ababa University, Debrezeit, Ethiopia.
[4]Ruminant Nutrition and Pasture Science, Department of Animal and Wildlife Science, University of Pretoria, Pretoria 0002, Republic of South Africa.

The experiment was conducted at Adami Tulu Research Center in 2004 with 37 selected lines of grain type cowpea (*Vignia* species) germ-plasm introduced from Melkassa Agricultural Research Center, Lowland Pulse Improvement Program for the advanced nursery trial. Dolichos lablab (*Lablab purpureius* both var. Highworth and Rongai) and cowpea (*Vignia ungiculata* var. white worder training type) were included as a control for herbage yield and quality evaluation and ranking. The top ten accessions of cowpea (IT82D 889, TUX 1948-01F, TVU 1142 4, IT85F 2687, 82D 504-4, IT84D-448, IT93K2046-2, IT87D 551-1, IITAUK91-12, 87D -1802,) and Dolichos lablab var. Rongai was selected using the visual observation data taken during the 2004 growing season and promoted for further advanced evaluation. Advanced evaluation was carried out during the growth period, 2005 and 2006, using randomized complete block design (RCBD) of three replication of plot size 3 × 2 m. Fresh weight yield, hay yield, grain yield of both years showed statistically significant difference among accessions (p<0.005) during both growing periods. There was statistically significant difference in percentage dry matter (DM%), percentage organic matter (OM%), percentage ash (ASH%) and percentage crude protein (CP%) (P<0.0001) and the highest value of 88.94, 79.32, 12.09 and 22.3% were shown by accessions IITAUK91-12, IT87D 551-1, 87D -1802 and IT93K 2046-2, respectively. The different accessions responded to the different seasons differently based on the rainfall duration in their herbage and grain yield that needs further categorization.

Key words: Adami Tulu, Dolichos lablab, herbage var. Rongai yield.

INTRODUCTION

As cultivated land becomes scarcer and fertilizer prices continue to rise, the increasing demand for food and animal feed must be met by raising the productivity of land already under cultivation. Legumes are the most important forage plants that substantially improve the feed available for livestock as they can provide the essential protein for animals, improving soil fertility food crop production and household nutrition through a more reliable supply of milk and meat (Alemayehu, 1997).

*Corresponding author. E-mail: ayanaet@gmail.com

Adami Tulu is known by its agro pastoral farming system with maize as the major crop cultivated for food security while white haricot bean is cultivated as cash crop (ATRC, 1997). Cereals like maize, wheat, barley, tef and small plots of sorghum have been cultivated continuously on a single plot or one after the' other without integration of any legume crop for several years. As a result, the fertility and then the productivity of the land have been declining. In addition, poor families have been frequently seen victimized by protein deficiency (ESAP, 2005). Previous efforts in Adami Tulu agricultural research center have shown a good potential forage legume like Dolichos lablab (*Lablab purpureus*) that enhance the intake and utilization of poor quality roughage and subsequently improved livestock production and productivity. However, due to high seasonal variation of rainfall in the area, it is difficult to produce herbage dry matter yield of more than 2.5 t/ha from the improved forage. Hence there is a need to search for alternate species that could serve more purpose.

Cowpea (*Vigna unguiculata L. Walp.*), is a leguminous plant belonging to the Fabaceae family. Cowpea is largely produced in Africa (80%) with Nigeria and Niger predominating, but Brazil, West India, Myanmar, Sri Lanka, Australia, the United States, Bosnia and Herzegovina all have significant production (Quinn, 1999). Cowpea, like other grain legumes is an important foodstuff in tropical and subtropical countries (Chinma et al., 2008) because of its use mainly as a grain crop, a vegetable or fodder for animals. Although it occupies a smaller proportion of the crop area than cereals, cowpea contributes significantly to household food security in West and Central Africa (Langyintuo et al., 2004). It is grown under rain-fed conditions in the tropics (Sangakkara, 1998), for its high protein content and is consumed as dry seeds, green pods or leaves. The residues of the plant are used in animal feeding. Also, cowpea can be used as a summer fodder crop fed as fresh or hay. Nutrients provided by cowpea make it extremely valuable where many people cannot afford proteins from animal sources such as meat and fish (Akpapunam and Sefa-Dedeh, 1997). Nigeria is the largest producer and consumer of cowpeas, accounting for about 45% of the world's cowpea production. Cowpea is referred to as the "hungry-season crop" given that it is the first crop to be harvested before the cereal crops are ready. It is a crop that offers farmers great flexibility. They can choose to apply more inputs and pick more beans, or - if cash and inputs are scarce - they can pick fewer beans and allow the plant to produce more foliage. This means more fodder for livestock, so that lower bean yields are balanced by more livestock feed, which in turn translates into more meat and milk. This flexibility in use makes cowpea an excellent crop under the challenging climatic conditions faced by African farmers (Okike, 2000). Cowpea fixes atmospheric nitrogen up to 240 kg/ha and leaves about 60 to 70 kg nitrogen for succeeding crops (CRI, 2006). Cowpea is a valuable component of farming systems in many areas because of its ability to restore soil fertility for succeeding cereal crops grown in rotation with it (Carsky et al., 2002; Tarawali et al., 2002; Sanginga et al., 2003). Cowpea is highly valued for its ability to tolerate drought and the high protein content of about 25% (IITA, 2007). These qualities make it a choice crop for catering for the food security needs of societies. Cowpea plays a critical role in the lives of millions of people in Africa and other parts of the developing world, where it is a major source of dietary protein that nutritionally complements staple low-protein cereal and tuber crops, and is a valuable and dependable commodity that produces income for farmers and traders (Singh, 2002; Langyintuo et al., 2003). Integration of cowpea with the prevailing farming system could have significant importance in improving soil fertility and productivity, improving feed quality and withstand the effect impact of climate change. Therefore, this advanced evaluation was conducted with the objective of identifying productive dual purpose cowpea genotypes for the mid rift valley of Ethiopia.

MATERIALS AND METHODS

The study was conduct at Adami Tulu Research Centre located at 7º9'N and 38º7'E in semi-arid rift valley of Ethiopia. The mid-rift-valley of Ethiopia is the part of great rift-valley that runs from northeast to south of the country and separates the western and southeastern highlands. The highlands on each side of rift valley give way to extensive semi-arid lowlands to the east, south and west of the country. The area at an elevation of 1650 m asl receives variation in seasonal rainfall distribution with an average annual precipitation of less than 760.9 mm. The mean annual temperature is 26ºC with a mean minimum and maximum of 12.7 and 29.2ºC, respectively. The soil is fine sandy loam with sand silt and clay in the proportion of 34, 48 and 18%, respectively (ATARC, 1998).

The experiment was conducted with 37 selected lines of grain type cow pea (*Vignia species*) germ-plasm introduced from Melkasa Agricultural Research Center's low land pulse improvement program for the advanced nursery trial. Dolichos lablab *(L. Opurpureius* both var. Highworth and Rongai) and cowpea (*Vignia ungiculata* var. white worder trailing type) were included as a control for herbage yield and quality evaluation and ranking. The selected *V. ungiculata* species germ-plasm were grown as experimental treatment in a Randomized Complete Block Design(RCBD) of three replications and plot size 3 x 2m with an inter-row and intra-row spacing of 60 and 20 cm. Recommended fertilizer rate (100 kg DAP/ha) and weeding (two times before flowering) was practiced.

At 50% flowering stage, the middle row of each plot was harvested for biological yield estimation. The pods were harvested from the rest rows at optimum physiological maturity by hand picking for grain yield. Sub-samples of the above ground biomass was ground to pass through 0.2 mm sieve and maintained for chemical determination. Percentage dry matter (DM%), percentage ash (ASH%) percentage organic matter (OM%), and percentage crude protein (CP%) were determined using AOAC (Association of Official Analytical Chemists), (1990) procedure).

RESULTS

Forage yield

The herbage yield of cowpea accessions and Dolichos lablab (control) is presented in Table 1. The differences among accessions fresh weight yield during the first and the second growing periods were found highly significant ($p<0.0001$ and $p<0.005$), respectively. During the first year growing period, 29.09, 26.86 and 26.69 t/ha were obtained from accessions 87D -1802, IT82D 889 and TVU 1142 4, respectively. In the same way, during the second year growing period, highest fresh matter yield of 32.22, 29.16 and 26.60 t/ha were obtained from accessions 87D -1802, TVU 1142 4 and IT87D 551-1, respectively. From the two years, relatively high fresh

Table 1. Herbage and grain yield of different cowpea accessions and Dolichos lablab .

ATRC. acc no.	Accessions.	Fresh weight (t/ha)		Hay (t/ha)		Grain yield (Quantity/ha)	
		Year-1	Year-2	Year-1	Year-2	Year-1	Year-2
12	Dolichos lablab	12.37^{de}	14.57^{d}	5.1^{bc}	5.29	4.49^{ef}	5.98^{de}
64	87D -1802	29.09^{a}	32.22^{a}	7.67^{a}	6.98	10.12^{a}	11.47^{a}
65	IITAUK91-12	11.10^{e}	17.89^{cd}	3.12^{ef}	5.34	5.83^{cd}	4.47^{ef}
67	IT87D 551-1	16.16^{cde}	26.60^{abc}	3.31^{def}	6.35	4.93^{de}	6.05^{de}
69	IT93K2046-2	13.38^{de}	19.38^{cd}	4.65^{bcde}	4.89	7.63^{b}	9.17^{b}
74	IT84D-448	22.90^{abc}	22.94^{bcd}	4.29^{bcdef}	5.79	7.05^{bc}	9.07^{b}
75	82D 504-4	25.47^{ab}	26.31^{abc}	5.44^{b}	7.12	3.71^{f}	3.95^{f}
83	IT85F 2687	22.84^{abc}	24.13^{abc}	4.92^{bcd}	5.86	6.17^{cd}	6.57^{cd}
86	TVU 1142 4	26.69^{ab}	29.16^{abc}	3.62^{cdef}	5.98	5.02^{de}	5.69^{def}
89	TVX 1948-01F	19.68^{bcd}	17.86^{cd}	3.53^{cdef}	5.53	5.00^{de}	5.66^{def}
94	IT82D 889	26.86^{ab}	26.32^{abc}	2.78^{f}	5.87	6.73^{bc}	7.97^{bc}
	$\bar{x}$	20.59	23.4	4.4	6.91	6.06	6.91
	CV%	21.32	20.26	19.83	15.62	11.23	15.62
	P-level	0.0001	0.005	0.0001	NS	0.0001	0.0001

*Means in the same column followed by different letters are significantly different **at** œ = 0.05.

matter average mean yield (23.4 t/ha) was obtained during the second growing period.

Dry matter herbage yield was significantly different for first year (p<0.0001) and not for second growing year at œ = 0.05. During the first year growing period, highest hay yield of 7.67, 5.44, and 5.1 t/ha was obtained from accessions 87D-1802, 82D 504-4 and the control (Dolichos lablab), respectively. Although the hay yield data during the second year growing period did not show statistically significant difference, relative high hay yield of 7.12, 6.98, and 6.35 t/ha was obtained from accessions 82D 504-4, 87D-1802 and IT87D- 551-1, respectively. As of fresh matter yield, high hay mean yield (5.91 t/ha) was obtained during the second year growing period due to the same reason. Both the shortage and extension of rainfall brought yield difference among accessions (Figure 2).

Grain yield

The *Viginia* accessions considered in this study were found superior in their grain yield in comparison to Dolichos lablab. The two year data presented (Table 1) were found highly significant (p<0.0001). The highest grain yield of 10.12, 7.63 and 7.03 Qun/ha were obtained during first year growing period from accessions, 87D - 1802, IT84D-448 and IT93K2046-2, respectively (Table 1). Similarly during the second year growing period, highest yield of 11.47, 9.17 and 9.07 Qun/ha was obtained by the same accessions, respectively. The higher average mean grain yield (6.91 Qun/ha) was obtained during the second year growing period, due to fair distribution of rain during the growth months (Figure 1) that resulted in limited pest problem on majority of accessions.

Nutritional compositions

There were statistically significant difference among cowpea accessions and between accessions and Dolichos lablab (the control) in DM, ash, OM and CP% at (P< 0.0001) level of significance. Only one cowpea accession (acc. IITAUK91-12) was found outstanding (88.94%). The control (88.69%) and all the rest cowpea accessions were not significantly different (p>0.05) in their DM% with lablab and among each other. In their ash%, two of the cowpea accessions were underscored (Table 2) and the rest were superior to the control (Dolichos lablab). Accessions, 87D -1802, TUX 1948-01F and IITAUK91-12 were the top three accessions with 12.09, 11.99 and 11.73%, respectively (Table 2). OM% of only one accession of cowpea (IT87D 551-1) was found superior (79.32%) to the control (79.02%) and followed by accession IT85F 2687, (78.21%). The highest CP% was obtained by accession IT93K 2046-2 followed by accessions, IITAUK91-12 and 87D -1802 with corresponding values of 22.3, 22.17 and 21.13%, respectively. In this study, except one cowpea accession, all the rest were found superior to Dolichos lablab in their CP% showing that cowpea is more nutritious than Dolichos lablab.

Disease and pest

The disease observed on the crop during the experimental period was leaf spot. In comparison to lablab cowpea accessions were highly susceptible (Figure 3).

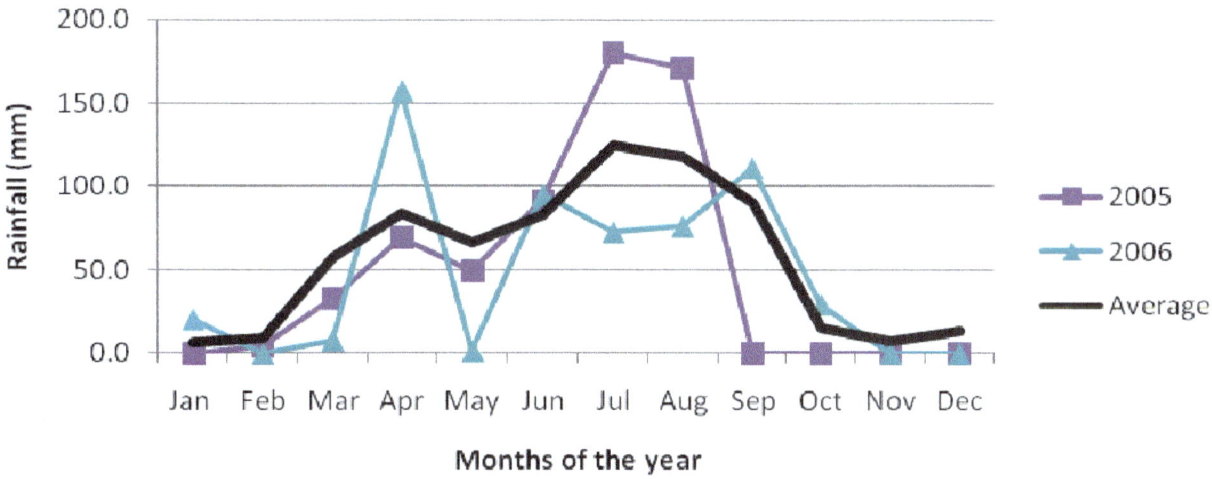

Figure 1. Monthly average rainfall distribution of the study site during the trial period.

Table 2. Mean Percentage dry matter (DM%), percentage ash (ASH%) percentage organic matter (OM%), and percentage crude protein (CP%) of composite sample of the two years (2005 and 2006) of cowpea accessions and Dolichos lablab.

ATRC acc. no.	Accession	DM%	ASH%	OM%	CP%
12	Dolichos lablab	88.69 [ab]	9.67 [d]	79.02 [d]	19.23 [d]
64	87D -1802	87.72 [ab]	12.09 [a]	75.75 [de]	21.13 [b]
65	IITAUK91-12	88.94 [a]	11.73 [ab]	77.24 [cd]	22.17 [a]
67	IT87D 551-1	88.31 [ab]	9.35 [d]	79.32 [a]	20.18 [c]
69	IT93K2046-2	87.88 [ab]	11.37[abc]	76.01 [de]	22.3 [a]
74	IT84D-448	87.16 [b]	9.30 [d]	77.86 [bc]	19.25[d]
75	82D 504-4	88.56 [ab]	11.3 [abc]	77.26 [cd]	20.85 [b]
83	IT85F 2687	88.74 [ab]	10.55 [c]	78.21 [abc]	20.18 [c]
86	TVU 1142 4	87.51 [ab]	11.16[bc]	76.35 [de]	19.43 [d]
89	TUX 1948-01F	87.33[ab]	11.99 [ab]	75.37 [e]	19.22 [d]
94	IT82D 889	87.42 [ab]	10.52 [c]	76.9 [cd]	19.76 [d]
Mean		88.02	10.82	77.21	20.33
CV%		0.95	4.32	1.03	1.9
P-level		0.0001	0.0001	0.0001	0.0001

*Means with the same letter in the same column are not significantly ($p<0.05$) different.

The common pests observed during the experimental period were cut worm and aphid. Cutworm mostly damaged when the seedlings were at their flag leaf stage and aphid's invasion occurred during early maturing on flag shoots and immature green pods.

DISCUSSION

The result is in agreement with the finding of Mohammad et al. (1993) who reported green fodder yield of cowpea of different accessions ranging between 16.49 to 26.02 t/ha in experiment conducted in Islamabad. When the rainfall was limited, those relatively late maturing accessions faced pest problem. Contrary to this, FAO (1988) indicated that during fair distribution of rainfall, the relatively late maturing accessions got advantage, while those relatively early maturing faced disease problem that led to final yield reduction and brought yield variation of the same accession between growing seasons (Table 1). The current result is in agreement with the report of Farming System Research and Extension Unit (1999) that stated that pests can severely reduce cowpea grain yield. Cowpea grain yield average only 200 to 400 kg/ha in Uganda (Sabiti et al., 1994; Omongo et al., 1997), and 200 to 300 kg/ha in Nigeria (Alghali, 1992) which was by far below the yield recorded in this study. Nelson et al. (2008) stated that cowpeas can have yield potential up to

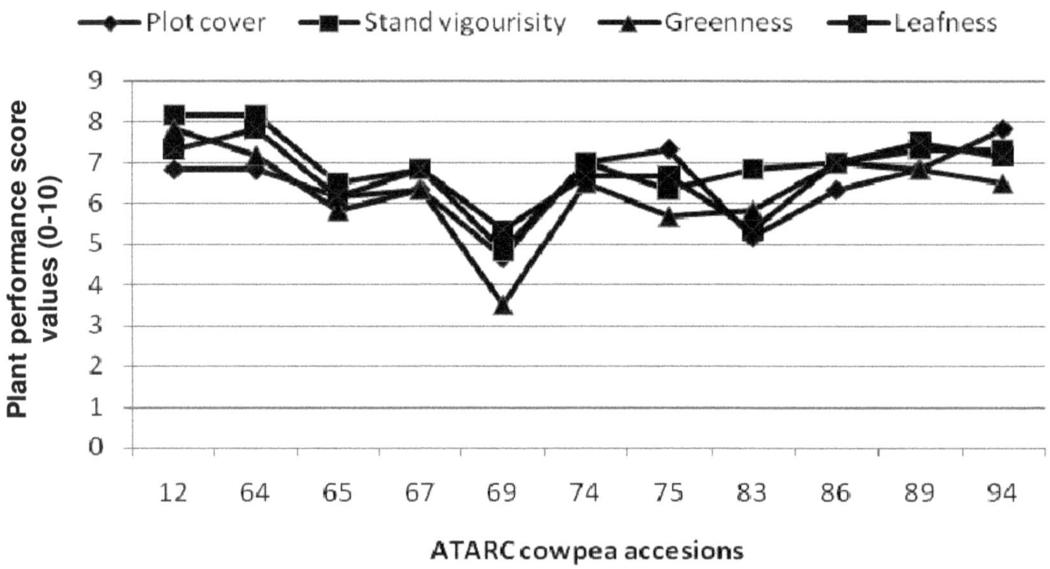

Figure 2. Growth characteristic of the accessions. Score: 0 = Poor condition; 10 = good condition.

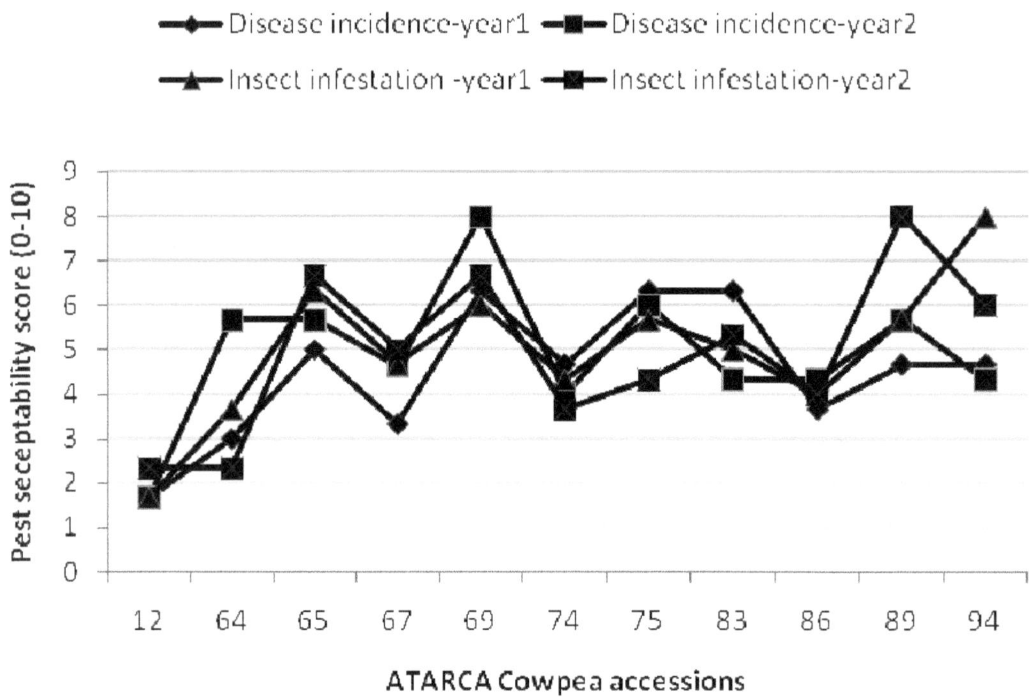

Figure 3. Pest and disease susceptibility level of the cowpea accessions and Dolichos lablab var. Rongai. Score: 0 = Resistant, 10 = highly susceptible.

2969 kg/ha in laboratory biosphere but most accessions grain yield under field condition ranges 6.38 to 11.21 Qunt/ha (Davis et al., 1991), which is almost similar to what was obtained in the present study. Similar finding was reported by Gwanzura et al. (2012) who showed that Dolichos lablab exceeds cowpea in its DM% and OM%, while these were high in cereals like sorghum. The OM and DM% demand of livestock in mid-rift-valley can be easily compensated by the major basal diets used viz. maize stover, tef straw and wheat straw. Based on their CP% content, Gwanzura et al. (2012) suggested that both cowpea and lablab have potential of being utilized as protein supplement for ruminants on low quality roughages. Ebro et al. (2004) envisaged in their study of legume supplementation that protein-N sources such as oilseed cakes and those of animal origin are produced in limited quantities and are beyond the economic reach of most of the small holder farmers in Ethiopia. The same authors realized that 0.5% level of supplementation of calves with cowpea hay stimulated the voluntary intake of tef-straw which as a result could bring an intermediate level of animal production. Gwanzura et al. (2012) supported this idea based on the high concentration of anti-nutritional factor (tannin) in Dolichos lablab (0.05%) than in cowpea (0.03%). Abule et al. (1995) confirmed that cowpea had similar degradation characteristics, rumen ammonia concentration, rate of degradation of teff straw, mean retention time, particulate passage rate and dry matter intake compared to lablab. Consumption of foods based on these cowpea varieties would be important step towards alleviating protein malnutrition (Appiah et al., 2011). Agfacts (2003) stated that lablab is relatively free of diseases such as root rots and foliar diseases while cowpea is susceptible to phytophthora stem and root rot that can be a serious disease and aggravated by poor drainage conditions. Although the incidence of cut worm on cowpea has been little stated, aphids are repeatedly reported as the common pest of cowpea all over the world during the growing season (Farming Systems Research and Extension Unit, 1999). This was observed during the first growing season when rain was stopped before the normal time. This was in agreement with Nabirye et al. (2003) investigation that aphid, pod borer and pod sucking bug are common pests of cowpea among which incidence of aphid is highly dependable on rainfall situation. Moreover, Nabirye et al. (2003) depicted that aphid incidence can be reduced by intercropping cowpea with cereal crops like sorghum.

Conclusion

Considerable variation exists among accessions, indicating the potential for selecting superior genotypes for both forage and grain (dual purpose). Some accessions showed variation in both herbage and grain yield with varying rainfall distribution and amount. High average herbage and grain yield was obtained not by high amount and short rain but with uniform distribution of rainfall during the growing period. The termination of rain earlier than the normal time exposes the crop to high infestation of pests, from which, aphid is of importance. A great variation existed among herbage and grain yield than the chemical composition (DM%, OM%, ASH%, and CP%) of the accessions. Cowpea is more nutritious than lablab and can supplement deficient roughage feeds. Dual-purpose cowpea species are the appropriate crop for the Adami Tulu area where the rainfall situation is erratic and irregular and agro-pastoral farming system is practiced. When the rain extends from the normal time, most species inclined to more of herbage yield.

RECOMMENDATIONS

Ranking and then promoting of cowpea in mid rift valley of Ethiopia should focused on herbage and grain yield. Based on that, accessions 87D -1802, IT82D 889 and TVU 1142 4, were ranked in decreasing order of importance and promoted for further agronomic evaluations for the study area and similar agro-ecologies. From the very nature, cowpea is a low yielder and its sole cultivation is not common and recommended to be sown mixed with cereals. So, it is recommendable that these accessions be evaluated for their intercropping potentials before promoting to end-users.

Cowpea is not a common crop even for its fodder value in rift valley of Ethiopia; however when it was taken to Mieso district through Integrated Project Managemebt Solutions (IPMS) project and Fentale district by Animal Feeds Research Team of Adami Tulu Research Center for demonstration at the end of this trial, people were found giving rather high value as a food crop (personal communication). In addition, giving due attention as food crops and less priority as forage crops by farmers limited extension of forage crops in most parts of Ethiopia. These all show that further research have to give due attention in identifying dual purpose than on sole purpose accessions in future animal feed improvement program.

REFERENCES

Abule E, Umunna NN, sahlai IV, Osuji PO, Alemu Y (1995). The effect of supplementing teff (Eragrostis tef) straw with graded levels of cowpea (*Vigna unguiculata*) and lablab (*Lablab purpureus*) hays on degradation, rumen particulate passage and intake by crossbred (Friesian × Boran (zebu)) calves. Livest. Prod. Sci. 44:221-228.

Agfacts (2003). Cowpea, lablab and pigeon pea (also available http://www.agric.nsw.gov.au). p 4.2.21.

Akpapunam MA, Sefa-Dedeh S (1997). Jack bean (*Canavalia ensiformis*) Nutrition related. Aspects and needed research. J. Plant Food Human Nutr. 10:123-127

Alemayehu M (1997). Conservation-Based Forage Development for Ethiopia. Addis Ababa, Ethiopia.

Alghali AM (1992). Insecticide application schedules to reduce grain yield losses caused by insects of cowpea in Nigeria. Ins. Sci. Appl. 13:725-730.

AOAC (1990). Official Methods of Analysis.15th. Ed. Association of Official Analytical Chemists,. Inc.,Virginia, USA. pp. 770-771

Appiah F, Asibuo JY, Kumah P (2011). Physicochemical and functional properties of bean flours of three cowpea (Vigna unguiculata L. Walp) varieties in Ghana. Afr. J. Food Sci. 2:100-104.

ATRC (Adami Tulu Research Center) (1997). Livestock production system survey in the Mid Rift Valley areas of Ethiopia. ATRC Bulletin 1. Adami Tulu, Ethiopia. p. 25.

ATARC (1998).ATARC (AdamiTullu Agricultural Research Center) thirty years of research experience (1968-1998) AdamiTullu agricultural research center , Bulletin NO 1. Oromiya agricultural development bureau fifnfinee Ethiopia.

Carsky RJ, Vanlauwe B, Lyasse O (2002). Cowpea rotation as a resource management technology for cereal-based systems in the savannas of West Africa. In: Fatokun CA, Tarawali SA, Singh BB, Kormawa PM, Tamo M (eds) Challenges and Opportunities for Enhancing Sustainable Cowpea Production. International Institute of Tropical Agriculture, Ibadan, Nigeria. pp. 252-266

Chinma CE, Alemede IC, Emelife IG (2008). Physicochemical and Functional Properties of Some Nigerian Cowpea Varieties. Pak. J.

Nutr.7(1):186-190.
CRI (2006). Cowpea Production Guide: Introduction to Cowpea Production. Available online: http://www.cropsresearch.org/publications/pdf/cowpea_Introduction.pdf
Davis DW, Oelke EA, Oplinger ES, Doll JD, Hanson CV, Putnam DH (1991). Alternative plant and animal products: Programs in information exchange and research. In: Alternative Field Crops Manual. Janick and J.E. Simon (eds). New crops. John Wiley and Sons, New York. pp. 133-143.
Ebro A, Nsahalai IV, Yami A, Umunna NN (2004). Effect of supplementing graded levels of forage legumes on performance of crossbred calves fed tef (*Eragrostis tef)* straw. J. Appl. Anim. Res. 26:107-102.
ESAP (Ethiopian Society of Animal Production) (2005). Participatory Innovation and Research: Lessons for Livestock Development. Asfaw Yimegnuhal and Tamrat Degefa (Eds). Proceedings of the 12th Annual conference of the Ethiopian Society of Animal Production (ESAP) held in Addis Ababa, Ethiopia, August 12-14, 2004. ESAP, Addis Ababa. p. 410.
FAO (1988). Traditional food plants. FAO Food and Nutrition, Paper 42. FAO, Rome.
Farming Systems Research & Extension Unit (1999). *Crop Pests in northern Namibia: cowpea bruchid.(Bruchid callosobruchus).* Internet ersion of the training manual. No. 62. Namibia. 2
Gwanzura T, Ng'ambi JW, Norris D (2012). Nutrient composition and tannin content of forage sorghum, cowpea, lablab and mucuna hays grown in Limpopo province of South Africa. Asian J. Anim. Sci. 5(12):1-7.
IITA (2007). Cowpea. www.iita.org/cms/details/cereal_legumes aspx?=a86&2=63
Langyintuo AS, Ntoukam G, Murdock L, Lowenberg-DeBoer J, Millera DJ (2004). Consumer preferences for cowpea in Cameroon and Ghana. Agric. Econ. 30:203-213.
Langyintuo AS, Lowenberg-DeBoer J, Faye M, Lambert D, Ibro G, Moussa B, Kergna A, Kushwaha S, Musa S and Ntoukam G (2003) Cowpea supply and demand in West Africa. Field Crops Res.82:215-231
Mohammad D, Hussain A, Khan S, Banaras B (1993). Variability for green fodder yield and quality in cowpea under rainfed condition.Pak. J. AgriC. Res.14(2,3):154-158.
Nabirye J, Nampala P, Ogenga-Latigo MW, Kyamanyawa S, Wilson H, Odeke V, Iceduna C, Adipala E (2003). Partipatory evaluation of cowpea integrated pest management (IPM) technologies in Eastern Uganda. Crop protect. 22:31-38.
Nelson M, Dempster WF, Allen JP, Silverstone S, Alling A, van Thillo M (2008). Cowpeas and pinto beans: Performance and yields of candidate space crops in the laboratory biosphere closed ecological system. Adv. Space. Res. 41:748-753.
Okike I (2000). An ealuation of potential adoption and diffusion of improed cowpea in the dry saannahs of Nigeria. IITA. Nigeria.
Omongo CA, Ogenga-Latigo MW, Kyamanywa S, Adipala E (1997). Effect of season and cropping systems on the occurrence of cowpea pests in Uganda. Afr. Crop. Sci. Soc. Proc. 3:1111-1116.
Quinn J (1999). *Cowpea, a versatile legume for hot, dry conditions.* Thomas Jefferson Institute. Columbia, USA. (www hort.purdue.edu/newcrop articles/ji-compea html; accessed 17/01/2013).
Sabiti A, Nsubuga ENB, Adipala E, Ngambeki DS (1994). Socioeconomic aspects of cowpea production in Uganda: A rapid rural appraisal. Uganda J. Agric. Sci. 2:59-99.
Sangakkara UR (1998). Growth and Yields of Cowpea (Vigna ungukulata (L.) Walp) as Influenced by Seed Characters, Soil Moisture and Season of Planting. J. Agron. Crop Sci. 180:137-142.
Sanginga N, Dashiell KE, Diels J, Vanlauwe B, Lyasse O, Carsky RJ, Tarawali S, Asafo-Adjei B, Menkir A, Schulz S, Singh BB, Chikoye D, Keatinge D,Ortiz R (2003). Sustainable resource management coupled to resilient germplasm to provide new intensive cereal–grain–legume–livestock systems in the dry savanna. Agric. Ecosyst. Environ. 100:305-314
Singh BB (2002). Recent genetic studies in cowpea. In: Fatokun CA, Tarawali SA, Singh BB, Kormawa PM and Tamo M (eds) Challenges and Opportunities for Enhancing Sustainable Cowpea Production. International Institute of Tropical Agriculture, Ibadan, Nigeria. pp. 3-13
Tarawali SA, Singh BB, Gupta SC, Tabo R, Harris F, Nokoe S, Fernández-Rivera S, Bationo A, Manyong VM, Makinde K, Odion EC (2002). Cowpea as a key factor for a new approach to integrated crop–livestock systems research in the dry savannas of West Africa. In: Fatokun CA, Tarawali SA, Singh BB, Kormawa PM and Tamo M (eds) Challenges and Opportunities for Enhancing Sustainable Cowpea Production. International Institute of Tropical Agriculture, Ibadan, Nigeria. pp. 233-251

6

Ash dieback associated with *Hymenoscyphus pseudoalbidus* in forest nurseries in Austria

Thomas Kirisits*, Peter Kritsch, Katharina Kräutler, Michaela Matlakova and Erhard Halmschlager

Institute of Forest Entomology, Forest Pathology and Forest Protection (IFFF), Department of Forest and Soil Sciences, University of Natural Resources and Life Sciences, Vienna (BOKU), Hasenauerstraße 38, A-1190 Vienna, Austria.

Dieback of *Fraxinus excelsior, Fraxinus angustifolia* and other *Fraxinus* species is an emerging infectious disease caused by the ascomycete fungus *Hymenoscyphus pseudoalbidus* (anamorph *Chalara fraxinea*). Investigations in five forest nurseries in Austria from 2008 to 2011 showed that ash dieback is a common and important disease of nursery seedlings. *H. pseudoalbidus* was consistently isolated at high frequencies from symptomatic *F. excelsior* and *F. angustifolia* seedlings, confirming that this fungus is associated with ash dieback. Symptom observations on potted seedlings revealed progression of the disease outside the vegetation period and thus a long incubation period in the disease cycle of ash dieback. Apothecia of *H. pseudoalbidus* were occasionally observed on dead ligneous parts of common ash, including stems of nursery seedlings. This suggests that a low portion of diseased seedlings can initiate new infections, which may be important for moving the pathogen to new areas. Based on our observations and studies in forest nurseries and the present knowledge on the disease cycle of ash dieback, recommendations for disease management in tree nurseries and concerning artificial regeneration are presented.

Key words: *Chalara fraxinea*, *Fraxinus excelsior*, *Fraxinus angustifolia*, emerging forest disease, disease management.

INTRODUCTION

During the last two decades, ash dieback caused by the ascomycete fungus *Hymenoscyphus pseudoalbidus* (Figure 1; anamorph *Chalara fraxinea*; Kowalski, 2006; Kowalski and Holdenrieder, 2009a, b; Queloz et al., 2011) has successively emerged in many European countries (Husson et al., 2011; Timmermann et al., 2011). Everywhere the disease appears it causes immense damage to common ash (*Fraxinus excelsior*) and in some parts of Europe also to narrow-leaved ash (*Fraxinus angustifolia*) (Kowalski and Łukomska, 2005; Kirisits et al., 2009, 2010; Schumacher et al., 2010; Schumacher, 2011). In addition, ash dieback has also been reported on a few ash species not native to Europe (Drenkhan and Hanso, 2010).

Its gradual spread and the high disease intensity may indicate that *H. pseudoalbidus* is an invasive alien organism (Husson et al., 2011; Queloz et al., 2011; Timmermann et al., 2011). Trees of all ages, in the forest, the landscape and in urban environments, both naturally regenerated and planted ones, are affected (Kowalski and Łukomska, 2005; Kirisits et al., 2009; Schumacher, 2011). Moreover, the disease is an economically important problem in shade tree and forest nurseries (Kowalski and Łukomska, 2005; Schumacher et al., 2010; Schumacher, 2011). Here, we report on observations and investigations on ash dieback in forest nurseries in Austria.

*Corresponding author. E-mail: thomas.kirisits@boku.ac.at.

ASSOCIATION OF *H. PSEUDOALBIDUS* WITH ASH DIEBACK IN NURSERIES

Investigations on ash dieback were conducted from 2008 to 2011 in five forest nurseries. Seedlings were inspected for disease symptoms on the stem and on side twigs. From seedlings showing early symptoms of disease (rela-

Figure 1. Apothecia of the ash dieback pathogen *H. pseudoalbidus* on ash leaf petioles and rachises from the previous year in the forest litter. Note black pseudosclerotial plates (see also inset) on petioles and rachises, from which apothecia emerge. Ash trees are infected by airborne ascospores produced in the apothecia.

tively small necrotic lesions in the bark), fungal isolation was attempted. After surface sterilization (1 min in 96% ethanol, 3 min in 4% NaOCl, 30 s in 96% ethanol) of stem segments, the outer bark was carefully peeled off and small discs containing wood and phloem were cut near the transition zone between necrotic and healthy phloem and placed on malt extract agar (MEA; 20 g/L malt extract, 16 g/L agar, 100 mg/L streptomycin sulphate). The primary isolation plates were at first incubated at room temperature in diffuse daylight (nursery 1, in 2008), but later (nurseries 1 to 5, in 2009 and 2011) at cool temperatures (between 4 to 10°C) in the dark. The latter was done in order to stimulate anamorph production of *H. pseudoalbidus* and to give it competitive advantage over other fungi, thereby increasing the likelihood to detect the ash dieback pathogen (Kirisits et al., 2009). *H. pseudoalbidus* was identified based on morphological characteristics of its *Chalara fraxinea* stage (colony morphology, phialophores and spores). Other fungi were not determined.

Typical symptoms of ash dieback, including necrotic lesions in the bark and phloem, wood discoloration as well as dieback of shoots and entire seedlings were frequently observed in all five nurseries (Figure 2), on one-year-old to three-year-old ash seedlings. *H. pseudoalbidus* was consistently isolated at high frequencies and often in pure culture from symptomatic *F. excelsior* seedlings in all five nurseries and in one nursery also from *F. angustifolia* seedlings. Overall, the ash dieback pathogen was obtained from 174 of the 192 (91%) investigated *F. excelsior* seedlings and from 14 of the 15 (93%) examined *F. angustifolia* plants. In agreement with a number of recent studies (for example Kowalski, 2006; Bakys et al., 2009; Kirisits et al., 2009, 2010; Kowalski and Holdenrieder, 2009a; Drenkhan and Hanso, 2010; Schumacher et al., 2010; Husson et al., 2011), the results clearly suggest that *H. pseudoalbidus* is associated with ash dieback and the cause of this disease.

DISEASE PROGRESSION OUTSIDE THE VEGETATION PERIOD

In mid-November 2010, three-year-old *F. excelsior* seedlings, rated as 'disease-free' based on external inspections were obtained from forest nursery 1. They were potted and placed in the garden of the institute. In spring 2011, the plants were again examined for the occurrence of ash dieback and isolations were done from diseased seedlings as described above. Upon inspection on 19 April 2011, 196 out of the 464 seedlings (42%) showed symptoms of ash dieback. The ash dieback pathogen was confirmed from all 41 plants (38 times in pure culture), from which isolations were made. As ascocarps of *H. pseudoalbidus* do not occur from October to May (Kirisits and Cech, 2009; Timmermann et al., 2011) and spores of *C. fraxinea* are most likely not infectious (Kirisits et al., 2009), the seedlings must have

Figure 2. Symptoms of ash dieback on seedlings of *F. excelsior*: (A) Necrotic bark lesion adjacent to a leaf scar, (B) Necrotic bark lesion with a small dead twig in the centre, (C and D) Extensive bark necrosis, (E) Package of ash seedlings prepared in autumn to be sold next year, containing diseased plants in spring.

been already infected by November, but were asymptomatic at that time and symptom progression took place outside the vegetation period. This is consistent with reports by forest nursery managers that considerable portions of apparently healthy ash seedlings, selected in autumn to be sold next year, were diseased when inspected again the following spring (Figure 2E). Disease progression during the cold period of the year is also consistent with the proposed disease cycle of ash dieback (Kirisits and Cech, 2009; Kirisits et al., 2009; Schumacher, 2011).

SPORULATION OF *H. PSEUDOALBIDUS* ON LIGNEOUS PARTS OF COMMON ASH

Structures resembling pseudosclerotial plates of *H. pseudoalbidus* were seen on stems, shoots and twigs of a few dead ash seedlings, in spring 2009 in nurseries 3 and 4 and in spring 2011 in nurseries 1 and 5, indicating that the fungus can probably form its apothecia on these seedlings later in the year. In summer 2011 stem segments of selected seedlings (from nursery 1) and natural regeneration as well as dead ash shoots and twigs with pseudosclerotial layers, collected from the forest litter at various localities, were placed in boxes filled with potting soil and incubated under moist conditions. Apothecia of *H. pseudoalbidus* subsequently developed, sometimes in high numbers (Figure 3), on a large portion of the incubated stems of nursery seedlings and naturally regenerated saplings as well as the ash shoots and twigs. This is consistent with previous reports on the occasional occurrence of apothecia on stems and shoots of one- to three-year-old dead nursery seedlings (Kowalski and Holdenrieder, 2009b) as well as on dead ash shoots and twigs in the forest litter (Kirisits and Cech, 2009). In contrast, teleomorph formation on ligneous tissues of ash is not known for *Hymenoscyphus albidus*, the long known and apparently native sibling species of the ash dieback pathogen (Kowalski and Holdenrieder, 2009b; Queloz et al., 2011).

H. pseudoalbidus is effectively disseminating via its airborne ascospores that are primarily produced in apothecia on leaf petioles and rachises from the previous year on the ground (Figure 1; Kirisits and Cech, 2009; Kowalski and Holdenrieder, 2009b; Timmermann et al., 2011). This mode of dispersal likely explains the rapid recent spread of the fungus in Europe (Kowalski and Holdenrieder, 2009b; Timmermann et al., 2011). However, trade with infected nursery seedlings could have

Figure 3. Apothecia of *H. pseudoalbidus* on *F. excelsior* twigs. Note black pseudosclerotial layers that had formed underneath the epidermis on each of the twigs. The size of the largest apothecial disc flats shown on each of the photos is about 5 mm.

been and still may be an important pathway to introduce the pathogen into new areas and to accelerate its spread (Kirisits et al., 2010; Timmermann et al., 2011). The occasional observation of *H. pseudoalbidus* apothecia on ligneous tissues of common ash indicates that nursery seedlings can indeed be infectious. It is likely that the movement of apparently and latently diseased seedlings, on which inoculum subsequently develops, can lead to the initiation of new disease centres, even far away from natural infection sources. Likewise, infected ash leaf petioles and rachises, including those with pseudosclerotial plates from the litter, may be moved together with bare-root and container-grown plants (Timmermann et al., 2011).

RECOMMENDATIONS FOR DISEASE MANAGEMENT

Although the origin of *H. pseudoalbidus* remains enigmatic, the pathogen behaves as an invasive tree pathogen and is a serious threat to *F. excelsior* and other

Fraxinus species (Kowalski, 2006; Bakys et al., 2009; Kirisits et al., 2009, 2010; Drenkhan and Hanso, 2010; Schumacher et al., 2010; Husson et al., 2011; Queloz et al., 2011; Timmermann et al., 2011). The disease cycle of ash dieback is now understood sufficiently well (Kirisits and Cech, 2009; Kirisits et al., 2009; Kowalski and Holdenrieder, 2009b; Schumacher, 2011; Timmermann et al., 2011) that recommendations for disease management concerning seedlings and artificial regeneration can be given. Measures are grouped according to the strategies presented by Tainter and Baker (1996). As knowledge on ash dieback is still limited amongst many practitioners and stakeholders, especially in countries and areas, where the disease presently does not occur, raising awareness on the disease forms the basis for implementing disease management measures.

Exclusion

i. Plant quarantine measures for nursery seedlings (import bans, imports from confirmed disease-free areas only, plant inspections, plant passports and certifications) may be effective to avoid or delay the movement of the ash dieback pathogen to geographically isolated parts of Europe such as the British Isles and to other continents. However, the long inclubation period in the disease cycle of ash dieback makes inspection of ash plants extremely difficult.
ii. Import bans for nursery seedlings and possibly other commodities (for example ash logs and timber) and thus closing potentially dangerous pathways would be the most effective measures to avoid the movement of *H. pseudoalbidus*, but are difficult to enforce politically.

Avoidance

i. Due to the risks posed by the disease, wide afforestation of susceptible ash species (especially *F. excelsior* and *F. angustifolia*) is no longer recommended. Site adapted alternative species should be given preference. If ash is used to some degree, it should be planted in mixture with other species.
ii. Observations in a number of countries (for example Schumacher, 2011) suggest that disease intensity is lower on drier sites within the ecological amplitude of *F. excelsior*. Still using ash on such sites may thus be justified, as well as in situations where wood production is not the main management goal, for example in shelterbelts, landscape and urban plantings.
iii. Flowering ash (*Fraxinus ornus*) that has so far not been affected by ash dieback may in some situations be an alternative.
iv. Little is known on the susceptibility of non-native *Fraxinus* species to the disease, but Drenkhan and Hanso (2010) reported ash dieback and confirmed the occurrence of *H. pseudoalbidus* on *F. americana*, *F. nigra* and *F. pennsylvanica* from North America and *F. mandshurica* from Asia. Exotic ash species should thus be used only with caution as alternative to native *Fraxinus* spp.
v. As sporulation of and infection by *H. pseudoalbidus* is favoured by high soil and air humidity, ash seedling beds should be established at sites with fairly dry microclimate, offering the prospect of lower disease levels.
vi. In order to decrease infections, seedling beds should be located large distances away from natural inoculum sources in the surroundings.
vii. Both producers and buyers should carefully inspect nursery plants for the occurrence of ash dieback. It is recommended that users continue the inspections after trees have been planted. If plants have been unambiguously infected in the nursery, buyers can claim for compensation.
viii. Many seedlings appear to be disease-free in autumn, but develop externally visible symptoms during winter and early spring. Purchasing trees in spring therefore increases the chance to obtain mostly healthy plants.

Inoculum reduction

i. The removal and destruction of shed ash leaves may in some situations be economically feasible to reduce inoculum and thus to decrease infections in nurseries. Leaves should be burnt, buried, ploughed into nursery beds or covered with soil.
ii. Routine removal of leaves in cities, in addition to dry climatic conditions in urban environments, not very conducive for inoculum production of and infection by *H. pseudoalbidus*, may explain why ash dieback is presently often of relatively minor importance on shade, avenue and park trees.

Protection

Fungicide treatments may be an option in nurseries, however, as the infectious period of *H. pseudoalbidus* is long (mid-June to early September in Central Europe) many applications are likely necessary. Overall, fungicide treatments will be of limited value to manage ash dieback, as seedlings remain susceptible when planted in the field.

ACKNOWLEDGEMENTS

The research leading to these results has received funding from the Austrian Federal Ministry of Agriculture, Forestry, Environment and Water Management (BMLFUW research project no. 100343, BMLFUW-LE.3.2.3/0001-IV/2/2008), the provincial governments of Lower Austria, Carinthia, Salzburg, Burgenland, Upper Austria, Styria, the Forest Office and Urban Agriculture

(MA 49) of the Vienna City Administration, the Austrian Federal Forests (ÖBf AG) as well as from the European Union's Seventh Framework Programme (FP7/2007-2013, KBBE 2009-3) under grant agreement no. 245268 (ISEFOR, Increasing Sustainability of European Forests: Modelling for Security Against Invasive Pests and Pathogens under Climate Change). We thank Susanne Mottinger-Kroupa, Rebecca Treitler and other colleagues at IFFF-BOKU for their technical assistance and the forest nursery owners and managers for their support.

REFERENCES

Bakys R, Vasaitis R, Barklund P, Ihrmark K, Stenlid J (2009). Investigations concerning the role of *Chalara fraxinea* in declining *Fraxinus excelsior.* Plant Pathol. 58: 284-292.

Drenkhan R, Hanso M (2010). New host species for *Chalara fraxinea.* New Disease Reports 22: 16. [doi: 10.5197/j.2044-0588.2010.022.016].

Husson C, Scala B, Caël O, Frey P, Feau N, Marçais B (2011). *Chalara fraxinea* is an invasive pathogen in France. Eur. J. Plant Pathol. 130: 311-324.

Kirisits T, Cech TL (2009). Beobachtungen zum sexuellen Stadium des Eschentriebsterben-Erregers *Chalara fraxinea* in Österreich [Observations on the sexual stage of the ash dieback pathogen *Chalara fraxinea* in Austria]. Forstschutz Aktuell 48: 21-25. Download: http://bfw.ac.at/400/pdf/fsaktuell_48_7.pdf.

Kirisits T, Matlakova M, Mottinger-Kroupa S, Cech TL, Halmschlager E (2009). The current situation of ash dieback caused by *Chalara fraxinea* in Austria. In: Doğmuş-Lehtijärvi T (Ed.), Proceedings of the conference of IUFRO working party 7.02.02, Eğirdir, Turkey, 11-16 May 2009. SDU Faculty of Forestry Journal, ISSN: 1302-7085, Serial: A, Special Issue: 97-119. Download: http://iufro.sdu.edu.tr/iufro_journal.pdf.

Kirisits T, Matlakova M, Mottinger-Kroupa S, Halmschlager E (2010). *Chalara fraxinea* associated with dieback of narrow-leafed ash (*Fraxinus angustifolia*). Plant Pathol. 59: 411. See also: http://www.ndrs.org.uk/article.php?id=019043.

Kowalski T (2006). *Chalara fraxinea* sp. nov. associated with dieback of ash (*Fraxinus excelsior*) in Poland. For. Pathol. 36: 264-270.

Kowalski T, Łukomska A (2005). Badania nad zamieraniem jesionu (*Fraxinus excelsior* L.) w drzewostanach Nadl. Włoszczow [Studies on *Fraxinus excelsior* L. dieback in Włoszczowa Forest Unit stands]. Acta Agrobot. 59: 429-440.

Kowalski T, Holdenrieder O (2009a). Pathogenicity of *Chalara fraxinea.* Forest Pathol. 39: 1-7.

Kowalski T, Holdenrieder O (2009b). The teleomorph of *Chalara fraxinea*, the causal agent of ash dieback. For. Pathol. 39: 304-308.

Queloz V, Grünig C, Berndt R, Kowalski T, Sieber TN, Holdenrieder O (2011). Cryptic speciation in *Hymenoscyphus albidus.* For. Pathol. 41: 133-142.

Schumacher J (2011). The general situation regarding ash dieback in Germany and investigations concerning the invasion and distribution strategies of *Chalara fraxinea* in woody tissue. EPPO Bull. 40: 7-10.

Schumacher J, Kehr R, Leonhard S (2010). Mycological and histological investigations of *Fraxinus excelsior* nursery saplings naturally infected by *Chalara fraxinea.* For. Pathol. 40: 419-429.

Tainter FH, Baker FA (1996). Principles of Forest Pathology. John Wiley & Sons, Inc., New York, USA, 805 p.

Timmermann V, Børja I, Hietala AM, Kirisits T, Solheim H (2011). Ash dieback: Pathogen spread and diurnal patterns of ascospore dispersal, with special emphasis on Norway. EPPO Bull. 40: 14-20.

The effect of malaria on staple food crop production in the Ejisu-Juaben Municipality of the Ashanti Region of Ghana

Asiamah Maxwell T., Patrick Dzadze and Bertha Gyasi

Department of Agricultural Economics, Agribusiness and Extension, Kwame Nkrumah University of Science and Technology, P. M. B, Kumasi, Ashanti Region, Ghana.

Malaria disease is a widespread problem in tropical countries. Recent publications have pointed to a correlation between enhanced agricultural activities and increasing prevalence of the disease. This paper therefore examined the relationship between food crop production and incidence of malaria in the Ejisu-Juaben Municipality of the Ashanti Region of Ghana. Ninety-two percent of sampled respondents had reported malaria cases in the 2010 major season. Out of these reported cases, 53.3% were female farmers. Chi-square tests of independence showed statistically significant association between education and measures adopted to control malaria. Self – medication was observed among respondents in the face of malaria symptoms. During periods of malaria incapacitation, 90.2% of affected farmers refrained from farm work. This threatened household food security and resulted in reduction in farm revenue. Average seasonal losses attributable to malaria incidence was estimated at GH¢126 ($66.67). The study recommended that efforts should be focused on malaria education during the second and third quarters of the year when malaria incidence is most prevalent.

Key words: Malaria, incapacitation, Kendall's coefficient, Ejisu-Juaben, Ghana.

INTRODUCTION

Malaria is a major contributor to poor public health and a leading cause of deaths and diseases in sub-Saharan Africa. At least 300 million cases of malaria are estimated to be reported each year globally resulting in over one million deaths. About 90% of malarial deaths occur in Africa with young children especially those under the age of five suffering the most mortality (WHO, 2008). Malaria also accounts for a high number of adult morbidity in endemic areas. Ghana's Ministry of Health and the national malaria control programme reported 2.8 million cases of malaria in 2002 and attributed 44% of out-patient clinical visits to malaria disease (Adams et al., 2004). The November 5th 2005 edition of the Daily Graphic also reported an estimated 17,000 deaths per year from malaria in Ghana.

Malaria has a negative impact on annual gross national product and up to 25% reduction in household income can be attributed to malaria control and treatment (Senzanje et al., 2002). Poor health as a result of malaria disease leads to incapacitation of the economically active population, reduces the capacity of labour force to work, affects the quality and quantity of available labour and decreases overall productivity (Asenso-Okyere et al., 2009). Thus, productive time lost and income lost to disease treatment and control, constitute some of the economically important effects of morbidity due to malaria disease.

Acquisition and diffusion of agricultural knowledge is also affected by malaria especially in cases where it leads to mortality. According to Asenso-Okyere et al., (2009), in cases of mortality, supply of labour is affected in addition to loss of farming knowledge.

Asenso-Okyere et al. (2009) further observed that risk of malaria transmission in malaria endemic areas increased with increasing levels of agricultural activities. Provision of irrigation water for agricultural activities through dams, reservoirs, bore-holes among others, creates favourable environments for mosquito activity all year long (Appawu et al., 2004; Yasuoka and Levins, 2007).

Studies in Eastern Ethiopia point to a correlation between type of crop grown and malaria incidence. Kebede et al. (2005) showed that maize pollen provides nutrition for larval mosquitoes. Dried leaves which fall and collect rain water, along with swampy areas reserved for rice cultivation have been shown to provide fertile grounds for mosquito breeding (Oladepo et al., 2010).

The correlation between agricultural activities and malaria incidence necessitated this study to identify the relationship between food crop production and malaria in the Ejisu-Juaben Municipality in the Ashanti Region. Specifically, the study sought to determine the period of the year when the prevalence rate of malaria amongst food crop farmers is highest, identify the effect of malaria incapacitation on farming activities, determine the coping strategies employed by farmers to reduce the effect of malaria and their relation with education, quantify seasonal losses attributable to malaria, and determine farm level activities most commonly affected by malaria incidence as perceived by farmers.

Hypothesis

H_0: Days of incapacitation, mode of diagnosis and coping strategies adopted are independent of education.
H_1: Days of incapacitation, mode of diagnosis and coping strategies adopted are dependent on education.

METHODOLOGY

Study area

Ejisu-Juaben Municipal is one of the 27 administrative and political Districts in the Ashanti Region of Ghana. The Municipal area covers 637.2 km^2 constituting about 10% of the Ashanti Region with Ejisu as its capital. It lies within Latitudes 1° 15' N and 1° 45' N and Longitudes 6° 15' W and 7° 00' W. Ejisu-Juaben Municipal shares boundaries with six other Districts in the Region. To the North East and North West of the Municipal are Sekyere East and Kwabre Districts respectively, to the South are Bosomtwe-Atwima-Kwanwoma and Asante-Akim South Districts, to the East is the Asante-Akim North Municipal and to the West is the Kumasi Metropolitan area.

As is the case for most parts of the middle belt in Ghana, the municipality experiences tropical rainfall (that is, bi-modal rainfall pattern and wet semi-equatorial climate). It is characterized by double maxima rainfall lasting from March to July and again from September and normally tapers off in the latter part of November. The mean annual rainfall is 1200 mm which is ideal for both major and minor season cropping. Temperatures range between 20°C in August and 32°C in March. Relative humidity is fairly moderate but quite high during rainy seasons and early mornings. The fair distribution of temperature and rainfall patterns enhances the cultivation of many food and cash crops throughout the district.

Types and sources of data

A multistage sampling technique was used to select 100 farmers for interviewing by purposively selecting four rural communities for ease of access to staple food crop farmers. Twenty-five farmers were then randomly selected using assigned random numbers from each of the selected communities. The towns selected were Essienimpong, Kwaso, Onwe and Besease.

Methods of analysis

Descriptive statistics including frequency distributions, means and percentages were employed in analysing the data obtained. Means and standard deviations were used in analysing the demographics. The Kendall's Coefficient of Concordance was employed to rank certain farm level activities/factors that were affected by malaria incidence as perceived by farmers.

Theoretical framework

Kendall's coefficient of concordance (*W*) is a measure of the level of agreement among independent Judges (*p*) assessing a given set of *n* objects. The approach of estimation employed in this study is adopted from Pierre (2005).

$$W = \frac{12S}{k^2\left(n^3 - n\right)}$$

$$S = \sum (SR)^2 - n(SR)^2$$

Where k = number of judges (farmers); n = number of problems ranked; S = sum of ranks; SR = mean of sum of ranks.

W ranges between 0 and 1, and indicates the strength of agreement; the closer to 1, the higher the level of agreement or concordance with *W* of zero signifying disagreement. For $n > 30$ the test of significance is achieved by computing the Friedman's chi-square X^2. Pierre (2005) cited Siegel and Castellan (1988: 270, 365) that $X^2 = k\ (n - 1)\ w$. He indicated that this quantity is asymptotically distributed like chi-square with (n−1) degrees of freedom and allows us to test *W* for statistical significance. *W* is significant if X^2cal > X^2crit at the prescribed level of significance.

The hypothesis was tested using chi-square test of independence.

RESULTS AND DISCUSSION

This study was conducted to identify the effects of malaria on activities of food crop farmers in the study area. Tables 1 and 2 show descriptive statistics of respondent farmers in the study area. The study sampled

Table 1. Descriptive statistics of respondents.

Variable	Frequency	Percentage
Gender		
Male	43	46.7
Female	49	53.3
Total	*92*	*100*
Educational level		
No formal education	28	30.4
Primary	22	23.9
Junior high school	11	12
Senior high school	28	30.4
Tertiary	3	3.3
Total	*92*	*100*

Source: Field Survey, January 2010.

Table 2. Summary statistics.

Variable	Min	Max	Mean	Std. dev.
Age(years)	27	66	46.32	9.69997
Household size	2	14	6.51	2.127
Farm size(Ha)	0.4	10	5.14	4.23
Farming experience (years)	2	20	13.76	5.106

Source: Field Survey, January 2010 (Std. dev.= Standard deviation).

Table 3. Effects of Malaria.

Variable	Frequency	Percentage
Prevalence periods		
1st quarter	16	17.4
2nd quarter	39	42.4
3rd quarter	29	31.5
4th quarter	8	8.7
Total	*92*	*100*
Mode of diagnosis		
Medical doctor	35	38
Traditional doctor	2	2.2
Self	55	59.8
Total	*92*	*100*

Source: Field Survey, January 2010.

100 farmers of which 92 had reported malaria cases in the 2010 major farming season, whist 8 had not report any malaria case indicating a high prevalence rate among food crop farmers in the district. The study therefore focused on those respondents who had reported malaria cases. About 53.3% of the reported cases were made by female farmers whilst 46.7% were by males. About seventy (69.6) percent of the respondents had received formal education with most (30.4%) attaining senior high school level of education. Respondents were on average 46 years old with about 14 years of farming experience and had an average of 7 persons per household. Majority of malaria cases during the 2010 major farming season, were reported in the 2nd and 3rd quarters of the year with the 2nd quarter recording the highest (42.4%) cases as shown in Table 3. This indicates a concentration of malaria cases during periods of peak agricultural activities under high rainfall conditions. It was also observed that respondents were well educated on the symptoms of the malaria disease. The number of respondents who diagnosed themselves of malaria and took remedial actions were 59.8%. Unfortunately, self-diagnosis and self-medication carry the risk of farmers misdiagnosing and treating themselves for malaria when they may be suffering from diseases like typhoid fever which have similar symptoms to malaria. Thirty-eight (38) percent sought qualified medical attention when they felt unwell and only 2.2% resorted to traditional healers for healthcare. These observations indicate a shift from traditional healthcare practices to modern medication.

Out of the total respondents who had malaria related problems, 90.2% refrained from farm work during the period of illness while the remaining 9.8% still carried on

Table 4. Period of staying away from farm work.

Period	Frequency	Percentage
3 days	19	20.7
1 week	21	22.8
2 weeks	26	28.3
1 month	17	18.5
Others	9	9.8
Total	*92*	*100.0*

Source: Field Survey, January 2010.

Table 5. Effect of illness on progress of farm work.

Effect on farm work	Frequency	Percentage
Ceases	45	48.9
Labourers work	35	38.0
Family works	3	3.3
Others	9	9.8
Total	*92*	*100.0*

Source: Field Survey, January 2010.

with their normal activities albeit with reduced vigour. Table 4 shows the distribution of the number of days that farmers had to stay away from work when ill. Also, 53.3% had to stay home to care for members of their household who had malaria whilst the remaining 46.7% still went about their normal activities when a household member had malaria. They attributed this to the fact that they had other family members who took care of those affected.

Effects of malaria on progress of on- farm activities

From Table 5, all but 9.8% of respondents refrained from active farm work during periods of malaria incapacitation. For most farmers (48.9%), farm work came to a total halt whilst the remaining 41.3% either used hired labour or relied on family labour for continuation of farming activities. Most of these (38%) relied on hired labour. Further Interviews revealed that each of the various alternatives had their shortfalls. Those who ceased from doing farm work altogether said they had to grapple with problems like emergence of weeds, late planting or delayed harvesting which resulted in low economic returns and household food insecurity. Those who hired labourers had to grapple with problems like labourers not working as expected and theft of farm produce. They also had to incur extra costs for labour. Those who had family labour as substitute did not have a lot of problems except that some of them did not have adequate farming knowledge. These observations confirm reports by Asenso-Okyere et al. (2009) that high rates of malaria transmission in rural farming communities usually coincided with the planting and harvesting seasons and so affect productivity.

Coping strategies adopted by farmers to reduce the effect of malaria and its relation with education

Farmers' views were sought on the strategies they used in the prevention of malaria. Identified strategies included visiting the hospital regularly, keeping clean surroundings (clearing bushes), using insecticide treated bed nets and clearing all choked gutters. Not all respondents demonstrated adequate knowledge on the subject, with responses like eating well, taking purgatives, and taking rest from work, among others. Chi-square test of independence showed that adoption of coping strategies was dependent on education and significant at 1% ($p> 0.001$). This meant that with higher levels of education, respondents were more inclined to adopt more effective or better coping strategies. This could be due to the fact that education gives a person a fair idea as to how to prevent or control malaria.

Losses due to malaria

Revenue losses due to incidence of malaria during the farming season ranged between GH¢0 and GH¢500 ($264.55) with the average loss being GH¢126 ($66.67) (std. dev. =52.428). Major factors contributing to farmers' losses in revenue were attributed to cost of medication, weed infestation of untended farms, delayed harvesting

Table 6. Perceived effect of malaria treatment on capital outlay.

Effect	Frequency	Percentage
Strongly agree	65	70.7
Agree	24	26.1
Undecided	3	3.3
Total	92	100.0

Source: Field Survey, January 2010.

Table 7. Ranking of activities mostly affected by malaria.

Farming activity	Rank total	Position	Comment
Labour	229	1	Activity most affected during incidence of malaria
Workdays	232	2	
Income	249	3	
Crop variety	471	4	
Household food security	472	5	
Labour intensive crops	519	6	
Area under cultivation	534	7	
Farm knowledge	607	8	Activity least affected during incidence of malaria

Source: Author's computations.

and cost of hired labour during periods of incapacitation. Despite the National Health Insurance Scheme (NHIS), farmers still incurred high medication bills because most preferred self-medication with off the counter drugs available at local pharmacies. A five point Likert scale was used to find out farmers' perception about the effect of malaria treatment on their capital. As shown in Table 6, majority (96.8%) of farmers agreed that cost of treatment affected their capital outlay adversely leading to reduction in total output for the season. None of the respondents disagreed with this statement.

Farm level activities/factors most commonly affected by malaria incidence as perceived by farmers

During the survey, some factors were listed for farmers to be ranked in order of those most affected with regards to malaria incidence as indicated in Table 7. From the table, the factor most affected is labour with a rank total of 229 and the factor least affected is farm knowledge. This is because according to the Kendall's Coefficient of Concordance, the factor with the least rank total is taken as the factor most affected.

Labour was perceived by farmers as the factor most affected in the incidence of malaria. From Table 5 above, it can be seen that 48.9% of the farmers said their work on the farm completely ceases when they were affected by malaria. Although 38.0% said labourers worked on the farm, they were not as productive as when the farmers themselves were around to supervise. Only 3.3% had family members working when they were taken ill. Second on the rankings was number of work days which is very obvious as the farmers were not able to go to work on those days due to incapacitation. The third was income. Combining the fact that the farmer was not able to go to the farm and spending his money on hired labour and hospital bills, income was likely to decrease. Fourth in the rankings was planting other crop varieties (mixed cropping). Farmers attributed this to the fact that they lose income and as such do not get enough money to invest in the production of different crops on the same field or plot of land. Also, they said they did not have ample energy to cultivate different crops. Household food security was perceived by the farmers as the fifth in the rank. This was attributed to the fact that input decreased in the case of malaria, resulting in low yields and as such farmers were not able to meet all their subsistence needs and were also not able to sell enough to cater for other crops they did not produce. Next were the labour intensive crops which farmers had to do away with in the case of malaria since they required a lot of energy and time input. Area under cultivation also decreased because malaria is mostly prevalent at that time of the year where the farmer is preparing his land and the farmer may either divert the resources he would use in preparing the land into treatment of malaria or may not be able to cultivate his normal land size due to incapacitation.

Table 8. Test of hypothesis.

Variable	Chi-square value	Degree of freedom	Significance level	Comments
Malaria diagnosis and education	36.041	12	0.000	Dependent
Days of incapacitation and education	106.8	16	0.000	Dependent
Coping strategies and education	9.530	4	0.049	Dependent

Lastly was loss of farm knowledge which resulted from the death of an experienced farmer or an experienced farmer who is incapacitated and had to stay away from work on the farm when he was supposed to give guidance on the farm as to how to go about work.

Assessing the degree of agreement of the farmers ranking

In order to assess whether farmers were in agreement with the order of the ranking computed above, the Kendall's Coefficient of Concordance was computed as:

$$W = \frac{S}{\frac{1}{12}k^2(n^3 - n)};$$

$$S = \sum(SR)^2 - n(\overline{SR})^2$$

$$S = (229)^2 + (534)^2 + (232)^2 + (471)^2 + (519)^2 +$$

S = 1535857-1371168

S = 164689

Therefore:

$$w = 164689 \Big/ \tfrac{1}{2} * 92^2(8^3 - 8).$$

$$w = \frac{164689}{355320}.$$

W=0.46

Therefore Kendall's coefficient of concordance is 0.46

The value of W shows that 46% of the crop farmers were in agreement with the order of ranking of these specific factors.

Test of hypothesis of agreement

H_0: *w=0* (No significance of the order of agreement by farmers).

H_A: $0<w\leq 1$ (Degree of agreement of the rankings by farmers is significant).

Employing the Friedman's Chi-square:

$$X^2 = k\,(n - 1)\,w$$

$$X^2 = 92(8-1)0.46$$

$$X^2 = 296$$

$$X^2_{crit} = 18.5$$

From the results above, the null hypothesis is rejected in favour of the alternative that the degree of agreement of the rankings by farmers is significant. From Table 8 we reject the null hypothesis which says that malaria diagnosis, days of incapacitation and coping strategies are independent of education and confirm the alternate hypothesis.

Malaria diagnosis and education

The results show that malaria diagnosis depends on the educational background of the farmer. Therefore, the higher the farmer's educational level, the better the methods of malaria diagnosis adopted and vice versa.

Days of incapacitation and education

As the more educated farmers seek better modes of diagnosis, they are likely to spend relatively short time at home to recover thus reducing the number of days they are incapacitated. An increase in education is likely to improve the standard of living of farmers and facilitate control of malaria, hence reduction in workdays lost (Awoyemi et al., 2009).

Coping strategies and education

The more educated farmers were the better coping strategies they were likely to adopt since they would have more knowledge on the source of illness and as such would direct strategies in that direction. Also, education is likely to improve the living standards of the individual, hence better, efficient and improved strategies were utilized.

Conclusion

Farming communities in the Ejisu-Juaben Municipality experienced high levels of malaria during the 2nd and 3rd quarters of the year. Out of a total of 100 respondents, 92% had been clinically diagnosed with malaria in the 2010 major season, showing that malaria is endemic in farming communities in the municipality. More females than males reported malaria cases. Respondents were on the average 46 years old, had farming experience of about 14 years with about 7 members per household and average farm sizes of 5 ha.

Most respondents had received some level of formal education. Farmers demonstrated adequate knowledge on malaria prevention with very few showing inadequate knowledge on the subject. There was statistically significant dependence between education and malaria coping strategies adopted. It was evident that most respondents relied on their knowledge of malarial symptoms for diagnosis and self-medication.

During periods of morbidity due to malaria disease, 90.2% of respondent farmers refrained from farm work, with the few that carried on with farm work doing so with reduced vigour. Abandonment of farms during periods of incapacitation threatened household food security, encouraged weed infestation of farms, delayed harvesting and in some cases hampered timely field preparation activities.

Losses in revenue attributable to malaria ranged between GH¢0 and GH¢500 ($264.55), with GH¢126 ($66.67) being the average loss per season. Major contributors to losses in revenue were costs of medication, cost of labour and losses due to delayed harvesting. Farmers ranked labour (quality and quantity) as the factor most affected by malaria incidence.

RECOMMENDATIONS

The study recommends intensification of malaria education during the second and third quarters of the year when malaria prevalence is at its highest. The use of insecticide-treated bed nets and insecticide sprays by farmers is also recommended.

Along with education on control and treatment of malaria, farmers should also be encouraged to promptly seek professional medical advice when they feel unwell instead of resorting to self-medication.

Farmers should be encouraged to register with the National Health Insurance Scheme to reduce costs incurred on medical treatment in times of morbidity.

REFERENCES

Adams I, Darko D, Accorsi S (2004). Malaria: A burden explored. Bulletin of health information.

Appawu M, Owusu-Ansah S, Dadzie S, Asoala V, Anfo F, Koram K (2004). Malaria transmission dynamics at a site in Northern Ghana proposed for testing malaria vaccines. Trop Med. Int. Health 9(1):164-170.

Asenso-Okyere K, Asante FA, Tarekegn J, Andam SK (2009, May). Linkages between malaria and agriculture: Issues for policy research and capacity strenghening. IFPRI discussion, P. 00861.

Awoyemi TT, Awoyinka YA, Babalola DA (2009). Effects of Malaria on Rural Farming Household Labour Supply: The Case of Ikenne Local Government, Ogun State, Nigeria. J. life. Phys. Sci. 1(3):20-24.

Kebede A, McCann JC, Kiszewski AE, Ye-Ebiyo Y (2005). New evidence of the effects of agro-ecologic change on malaria transmission. Am. J. Trop. Med. Hygiene 73(4):676-680.

Oladepo O, Tona GO, Oshiname FO, Titiloye MA (2010). Malaria knowledge and agricultural practices that promote mosquito breeding in two rural farming communities in Oyo State, Nigeria, Malaria J. 9(91), April 9.

Pierre L (2005). Species Associations: The Kendall's Cofficient of Concordance Revisited. J. Agric. Biol. Environ. Statist. 10(2):226-245.

Senzanje A, Hackenitz E, Chitima M (2002). Malaria and irrigated agriculture in zimbabwe: Impact assessment, costing and quantification under field conditions. (E. Boelee, F. Konradsen, & W. Van der Hoek, Eds.) Malaria in Irrigated Agriculture, pp. 15-23.

World Health Organization (2008). World Malaria Report. Geneva: WHO.

Yasuoka J, Levins R (2007). Impact of deforestation and agricultural development on Anopheline ecology and malaria epidemiology. Am. J. Trop. Med. Hygiene 76(3):450-460.

8

Adoption of improved agricultural technologies among smallholder farm households in Nakuru District, Kenya

Immaculate Njuthe Maina[1]*, Ingrid-Ute Leonhäuser[2] and Siegfried Bauer[3]

[1]Kenya Agricultural Research Institute (KARI), Socioeconomics and Applied Statistics Division, Kenya.
[2]Justus-Liebig-University Giessen, Nutrition Education and Consumer Behaviour, Germany.
[3]Justus-Liebig-University Giessen, Project and Regional Planning, Germany.

Households are institutions that nurture, develop and sustain capabilities, material and social resources and activities necessary for their members to sustain livelihoods. This is possible through socialisation, communication, diffusion and adoption of improved technologies. In developing countries like Kenya, adoption of improved agricultural technologies is critical in facilitating households' productive and consumptive capabilities and functions for better livelihoods. Yet, the adoption of such technologies remains low. Households are made up of male and female members whose roles, responsibilities, rights and entitlements often differ. Therefore, there is need to understand how the conditions supporting the adoption process vary across male and female farmers and within households. This paper adopts the livelihood systems framework in analysing the influence of socio-economic characteristics that influence men and women to adopt improved agricultural technologies. The field survey involved individual interviews with 190 randomly selected rural households from Nakuru District, Kenya. Data analysis procedures included descriptive statistics, factor analysis and a binary logistic regression model. Results indicate that men are more likely than women to adopt improved technologies while the propensity to adopt increases with increasing distance to market. Adoption is supported by social inclusion and peace and is more likely to occur among poorer, younger farmers, without skills for off-farm employment. Given that a wide range of male and female farmers' socio-economic characteristics influence adoption, careful analysis should always precede all efforts aimed at encouraging adoption of new and improved agricultural technologies.

Key words: Nakuru, Kenya, sustainable livelihoods framework, agricultural technology, adoption, gender, sustainable livelihoods.

INTRODUCTION

Kenyan economy is largely agro-based, with the agricultural sector accounting for 26% of the gross domestic product (GDP), 60% of the export earnings, over 80% of employment and 57% of national income (GoK, 2004). Agriculture also links with manufacturing, distribution and service related sectors to make an indirect contribution of 27% to the gross domestic product (GDP) of the country. There is therefore a direct relationship between growth in the agricultural sector and that of the entire economy (Figure 1).

In developing countries like Kenya, where most of the rural people are involved in agricultural production, adoption of improved agricultural technologies is an important strategy for facilitating households' productive and consumptive capabilities for better livelihoods. Households are institutions that act individually and in social context to nurture, develop and sustain capabilities, material and social resources so as to develop human potential and shape everyday culture (Schweitzer, 2006). They are not homogenous units but instead are made up of male and female members whose spheres of influence, roles and responsibilities and

*Corresponding author. E-mail: inmaina@kari.org, Maina@gmail.com.

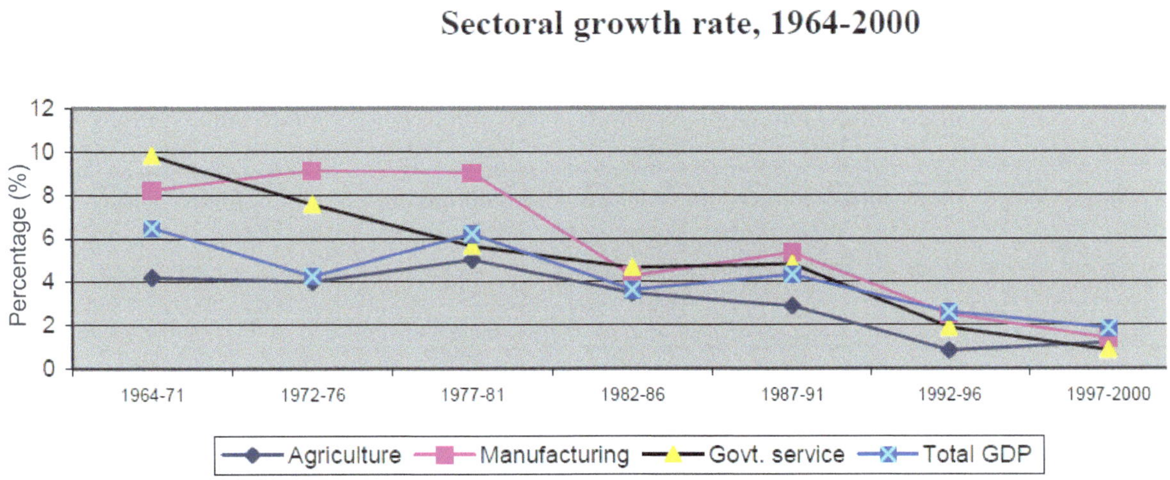

Figure 1. The relationship between growth in the agricultural sector and that of the entire economy.

entitlements differ. Thus, there is need to understand, the influence of individual characteristics of the female and male farmers and those of their households on adoption decisions.

Adoption of innovations has received great attention from scholars. The subject has been studied from the view of active knowledge accumulation (Feder and Slade, 1994); adoption of single farm inputs such as fertilizers or herbicides (Mbata, 1994; Omamo et al., 2002; Dadi et al., 2004; Freeman and Omiti, 2003); from the view of the household level characteristics including gender (Marenya and Barrett, 2007; Franzel et al., 2003), from the platform of broad policy issues (Lee, 2005; Smale, 2005; Anandajayasekeram et al., 1997); as a social learning process (Conley and Udry, 2001; Isham, 2000); as a method for improving research priority–setting (Batz et al., 2003; Elliott, 2004); it has been integrated into geographic information systems (GIS)-measures for understanding and differentiating locational effects such as market access, demographics and agro-climatic zones (Staal et al., 2002); it has been assessed as an interactive process between public investments and community health (Ersado et al., 2004) and also as a part of international development efforts for providing development support to developing countries (Baker and Edmonds, 2004). These studies emphasize the socio-economic environment within which the new technology is adopted and applied. They neglect to consider the factors that drive the innovation-decision process. This paper seeks to fill this gap by considering the factors that influence the way that men and women in small farm holdings make adoption decisions.

Adoption decisions have been shown to occur following the innovation-decision process (Rogers, 1995). The innovation-decision process consists of a series of actions and choices over time through which individuals or other decision-making unit evaluates a new idea and decides whether or not to include the innovation into existing practices. The process underscores demand for information about new ideas while creating opportunities for farmer empowerment through their active involvement in decision-making. Empowered clients are better able to articulate their needs, can negotiate effectively with public and private sector service providers, command recognition in collaborative efforts and can bear the consequences of their involvement in collaborative efforts (Lightfoot, 2004). The innovation-decision process (Figure 2) evolves through the following stages: the knowledge stage; the persuasion stage; the decision stage; implementation stage; and the confirmation stage (Rogers, 1995).

In the knowledge stage, farmers are exposed to innovations in existence within the agricultural knowledge system and they gain some understanding of their functioning. This occurs in an active process of communication between farmers and various external actors including agricultural researchers and extension agents. This stage is a social process in which men and women may have different entitlements, priorities, leadership styles or socio-economic characteristics all affecting how knowledge is sought and utilized (Rogers, 1995). For farmers to make the decision to invest their limited resources in new technologies, these technologies must possess certain attributes and components that appeal to the farmer. The technologies should also fulfil the farmers' varied needs. Perceptions of farmers towards technologies have been found to affect adoption of technologies (Rogers, 1995; Heffernan et al., 2008). At the persuasion stage, farmers' groups actively seek information about technologies that are deemed to have the greatest potential in satisfying their needs and discriminate against those that do not. Information plays an important role in mediating the understanding and effective use of the technology. This information base is

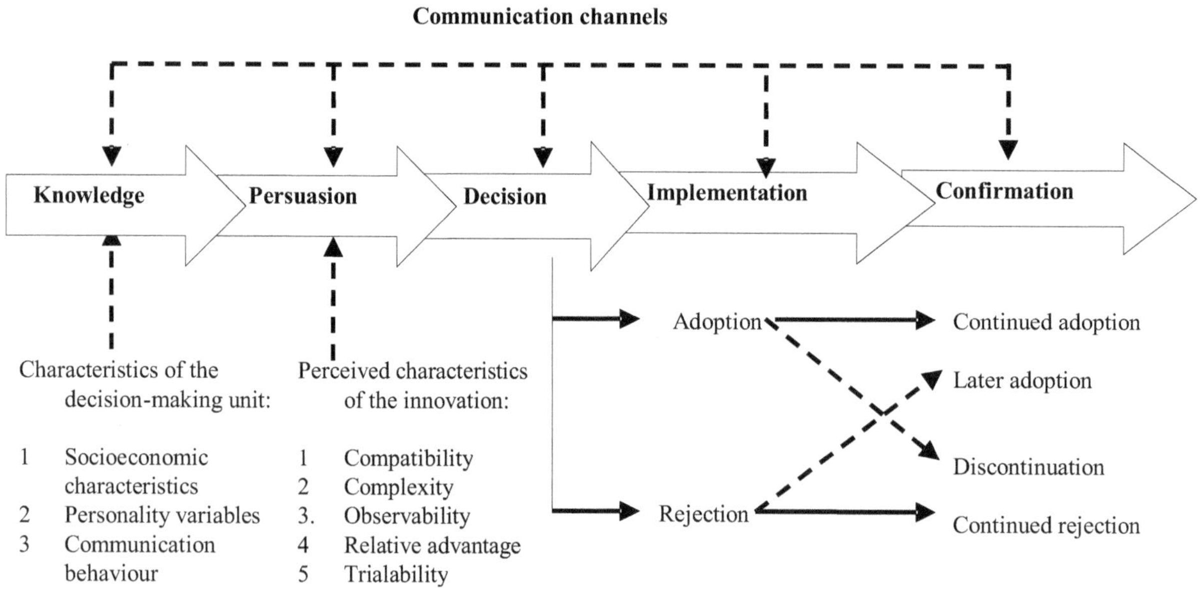

Figure 2. The decision-innovation process. Source: Rogers (1995).

intrinsic and not supplementary to any technology because a focus on the product without knowledge of its use and application renders the physical entity useless (Bozeman, 2000). Rogers (1995) argues that it is at the decision stage that a technology is fully accepted (continued adoption) or is rejected. However, sometimes a technology may be used for a time and then abandoned (discontinuance) and at other times the rejected technology may be adopted later (later adoption) or it may be subjected to continued rejection. Implementation of the demanded technology occurs when the decision-making unit puts the innovation to use (Rogers, 1995). For effective application of new knowledge farmers require guidance and support. Confirmation occurs when the innovation-decision is reinforced (Rogers, 1995).

Furthermore, most of these adoption studies use the female-headed households and male-headed households' dichotomy to establish the determinants of technology adoption in each gender. The female-headed, male-headed households dichotomy may mislead because such a dichotomy assumes that adoption decisions, made by the household head apply broadly within the household and overrides those of individual (male and female) farmers within the same household (Morrison et al., 2007). Yet, empirical evidence shows that adoption decisions within households are undertaken in decentralized processes (Ashaw and Admassie, 2004). Many economic decisions made within households are dependent upon the characteristics of both men and women members (Doss and Morris, 2001; Doss et al., 2003). This study analyzes the gender differences in factors that constitute the innovation-decision process using the dichotomy of male and female farmers, instead of female-headed households and male-headed households. This is intended to establish whether, or not, gender is a statistically and practically significant determinant of technology adoption, and whether, or not, women are systematically disadvantaged relative to men with respect to the factors under study. The acknowledgement of gender as a policy issue that has far reaching implications in society can be found in the third Millennium Development goal (MDG): promote gender equality and empower women (UN Millennium Project, 2005).

Theoretical framework

In discussions regarding rural development, poverty reduction and environmental management, the term sustainability is linked to sustainable rural livelihoods (Scoones, 1998; Chambers and Conway, 1991; Ellis and Alison, 2004). The concept of sustainable rural livelihoods seeks to understand the context (including the politics settings, politics, history, agro-ecology and socioeconomic conditions) and the livelihood resources available to rural populations and which determine their livelihood strategies. Rural persons pursue different livelihood strategies in support of their families. Some of the strategies include agricultural intensification, livelihoods diversification and also migration. Analysis of sustainable rural livelihood is ultimately interested in the outcomes realised by rural households and the roles played by the formal and informal institutions upon which rural households depend on for their everyday living conditions.

For ease of analysis the aforementioned concepts are

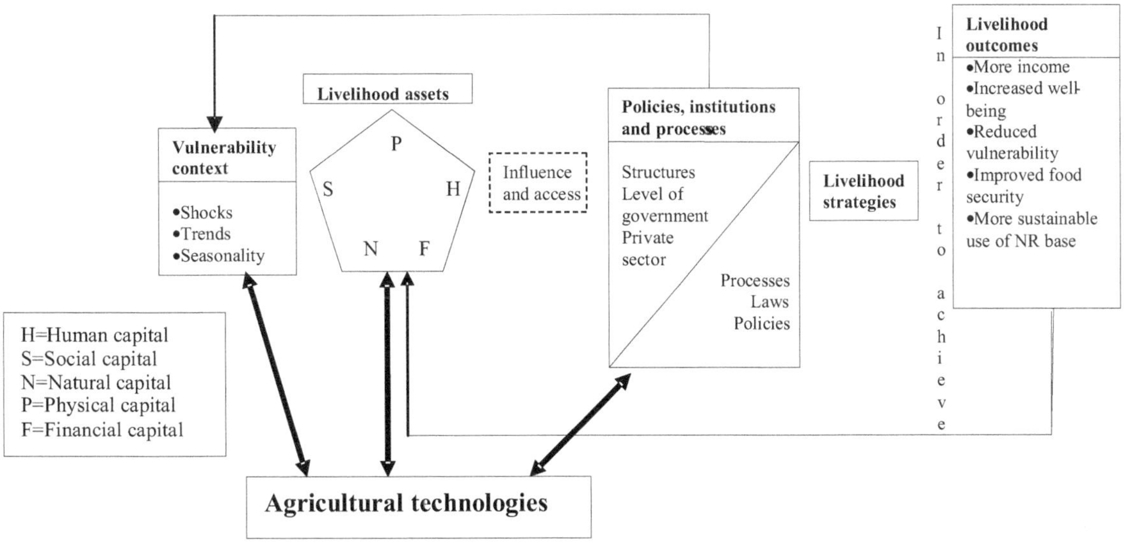

Figure 3. Sustainable livelihoods framework. Source: Adato and Meinzen-Dick (2002).

often organized into a framework; the sustainable livelihoods framework (Figure 3). The sustainable livelihoods framework is used to analyse these cross-sectional and multi-occupational characters of contemporary rural livelihoods (especially in low-income countries), as well to understand the complexity of livelihood strategies needed to reduce rural poverty. The framework provides a checklist of issues and their linkages; shows schematically the main influences and processes and depicts the multiple interactions that exist between the different factors that affect rural livelihoods. The five main features of the framework include the vulnerability context, the livelihood assets, the transforming structures and processes, livelihood strategies and the livelihood outcomes.

In the framework, sustainable livelihoods, are defined in relation to livelihood resources (natural, economic, financial, human and social capitals) which are combined in the pursuit of different livelihood strategies and also to the range of formal and informal organisational and institutional factors that influence sustainable livelihood outcomes (Scoones, 1998). Another important feature of the sustainable livelihoods framework is that it recognizes people, whether poor or not, as actors with assets and capabilities with which they actively pursue their own livelihood goals (Adato and Meinzen-Dick, 2002). This paper specifically assesses the five livelihood capital assets (human; social natural, physical, financial) and their influence on adoption of improved agricultural technologies.

Assets matter because assets beget assets; they influence earnings; they are required for production; they smooth consumption; buffer risk; they are required to enable households make additional purchases and their accumulation is instrumental to households' escape from poverty (Schreiner, 2005). Assets also influence behaviour, so that, the better off have positive attitudes towards their future (Sherraden, 1991). Wealth indicator variables have been linked to adoption of technologies (Asfaw et al., 2011) and to reduction in risks associated with adoption of technologies (Mariam et al., 1993).

MATERIALS AND METHODS

Study area

This study was carried out in Nakuru district (Figure 4), which holds 4% of Kenya's total population at a population density of 164 persons per square kilometre and at a growth rate of 3.4%. According to the 1999 National Population and Housing Census, the gender division of this population was 588,336 (or 49.6%) females and 598,708 (or 50.4%) males (GOK, 2001).

Sampling and data collection

The sampling frame was constituted from lists of men and women, members of farmers' groups that were implementing technologies listed in Table 1. Farmer groups are organizations for collective action often formed among neighbours and peers based on the principle of free membership, to pursue specific common interests of their members (Hussein, 2001). Farmers' groups, especially those formed by smallholders, are important institutions for agricultural technology uptake in Kenya (Place et al., 2002; Davies et al., 2004).

The simple random sampling procedure was used to select men and women from 20 farmers' groups. The total membership of these groups was 494, made up of 200 men and 294 women. To ensure representation, the size of the sample selected for the study followed the procedure for "estimation of a population proportion" outlined by Scheaffer et al. (2006). Following the procedure: $y_i = 0$ if the ith element sampled does not possess the character of interest, (non-adopter) and $y_i = 1$ if it does (adopter). Then the total number of elements in the sample of size n possessing the specified

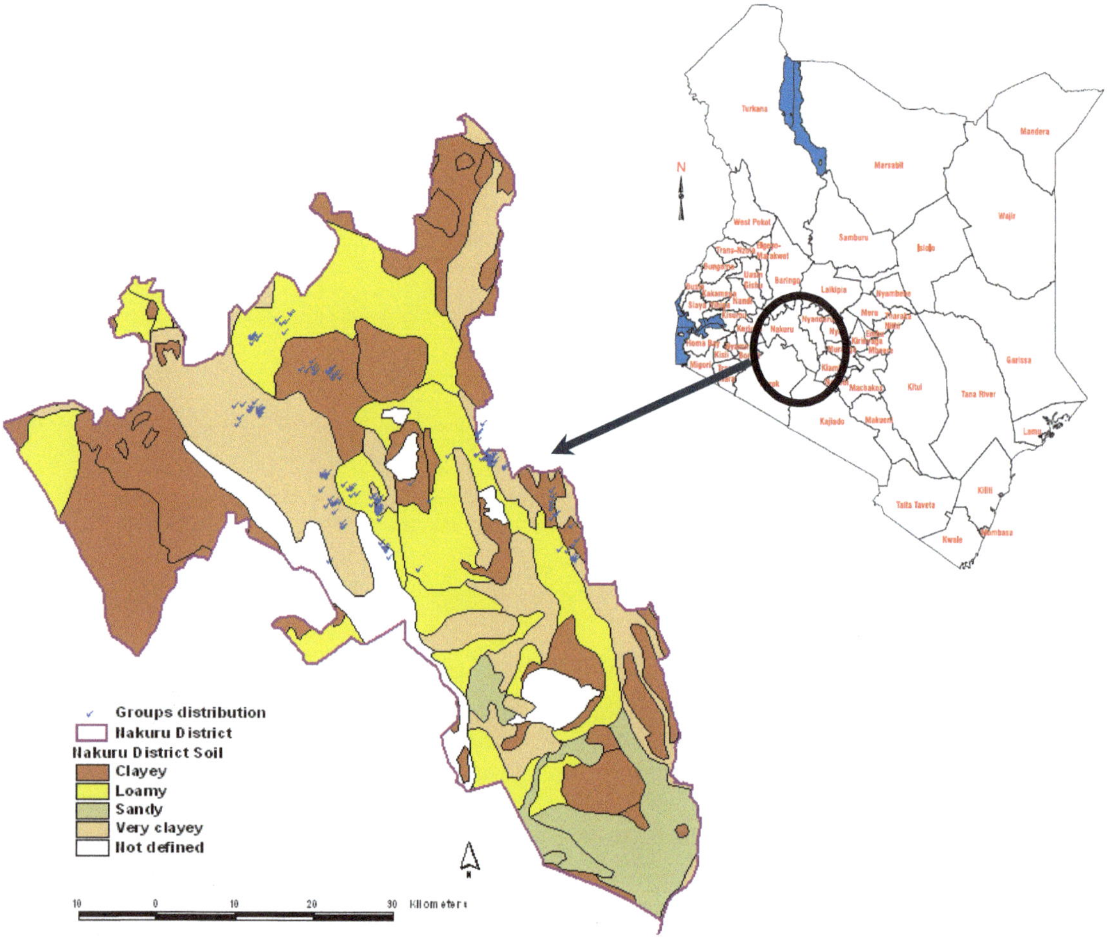

Figure 4. Map of Nakuru District showing the soil types in the district and the location of sampled households Source: International Livestock Research Institute (ILRI); geographic information systems (GIS) laboratory with distribution of group members' households from the field data.

Table 1. Types of technologies implemented by the farmers groups.

Technology class	Type of technology	Number of farmers' groups
Livestock production	Rearing dairy goats	6
	Rearing sheep	1
	Rearing indigenous chickens	1
	Beekeeping	2
Farm mechanization	Use of motor chaff cutters	1
Soil and water management	Drip irrigation	2
	Water harvesting technology	2
	Tree nursery	1
Crop management	Sunflower	1
	Tissue culture banana	1
	Seed potato production	1
	Fruit production	1
Total		20

characteristic is:

$$\sum_{i=1}^{n} yi \quad (1)$$

When a simple random sample of size n is drawn, the sample proportion, $\hat{p}$, is the fraction of the elements in the sample that possess the characteristic of interest. In this case, $\hat{p}$ of adopting farmers in Nakuru District is:

$\hat{p}$ = Number of adopting farmers sampled or:

$$\hat{p} = \frac{\sum_{i=1}^{n} yi}{n} = \bar{y} \quad (2)$$

Number of farmers sampled:

$\hat{p}$ is the average of the 0 and 1 values from the sample. Similarly, the population proportion is the average of the 0 and 1 values for the entire population, hence, $p = \mu$. The estimator of the population proportion p is:

The estimated variance of $\hat{p}$ is:

$$\hat{p} \cdot \bar{y} = \frac{\sum_{i=1}^{n} yi}{n} \quad (3)$$

Where $\hat{q} = 1 - \hat{p}$

$$\hat{V}\left(\hat{p}\right) = \frac{\hat{p}\hat{q}}{n-1}\left(\frac{N-n}{N}\right) \quad (4)$$

The bound on the error of estimation is:

$$\sqrt[2]{\hat{V}\left(\hat{p}\right)} = \sqrt[2]{\frac{\hat{p}\hat{q}}{n-1}\left(\frac{N-n}{N}\right)} \quad (5)$$

Sample size required to estimate p with a bound on the error of Estimation B is:

$$n = \frac{Npq}{(N-1)D + pq} \quad (6)$$

When the value of p is unknown, as was the case in this study, (since there exists no other past surveys for the population of interest; that is, the farmers adopting KARI/ATIRI technologies in Nakuru District), in such circumstances, substitution into the equation with p = 0.5 is done (Scheaffer et al., 2006: 95). Hence:

$$n = \frac{Npq}{(N-1)D + pq}$$

$$= \frac{(494)(0.5)(0.5)}{(493)(0.000625)+(0.5)(0.5)} = \frac{123.5}{0.558125} = 221 \quad (7)$$

Therefore, the required sample size was thus determined to be 221 respondents.

Procedure for determining proportionate samples of men and women

The population, from which the sample of men and women was drawn, was considered as two strata made up of the gender. Sampling at the level of the population and also at the level of the group was stratified between the two genders. In order to establish proportional numbers of men and women in the total population for all the groups the following formula was applied:

$$s = \left(\frac{N}{M}\right)n \quad (8)$$

Where:

s = Strata, N = sample size, M = total population, n = total number in the strata.

Therefore, the required sample for men was:

$$\left(\frac{221}{494}\right)200 = 89$$

and for women:

$$\left(\frac{221}{494}\right)294 = 131 \quad (9)$$

Procedures for data collection

Data was collected in a field survey through structured individual interviews. A formal survey using a structured questionnaire was administered to individual men and women, members of farmers' groups, by the researcher with the help of enumerators. The questionnaire included on-site recordings of the exact location of the household of the individual group member by global positioning system (GPS) and physical verification of the adopted technology. At the end of data collection and after data cleaning and entry, 190 questionnaires (69 questionnaires from male respondents and 121 questionnaires from female respondents), were admissible for further data analysis.

Operationalisation of variables

Human capital refers to the knowledge, skills and capabilities that people need for life and work, including the education and health levels of people as they affect economic productivity (World Bank, 2007). The factor of human capital is represented by education

Table 2. Rotated varimax factor loadings for social capital.

Statements describing social capital	Principal components		
	Factor 1	Factor 2	Factor 3
	Social inclusion	Trust	Social cohesion
Most people in this neighbourhood can be trusted	0.006	0.811	0.101
In this neighbourhood one has to be alert or someone is likely to take advantage of you	0.010	-0.760	0.029
Most people in this neighbourhood are willing to help if you need it	0.334	0.404	0.263
In this neighbourhood, people generally do not trust each other in matters of lending and borrowing money	-0.571	-0.336	0.283
If I need immediate help, members of my group would come to my help	0.789	-0.160	0.194
People in my group are always interested in each others welfare	0.740	0.067	0.351
I feel accepted as a member of my group	0.108	0.164	0.771
Our group contributes to the welfare of the community	0.153	-0.036	0.798
Eigenvalues	1.643	1.569	1.553
% of variance explained (59.567)	20.538	19.616	19.412

Extraction method: Principal component analysis with Kaiser Normalization.

levels of the respondents and their children, their personal skills and the occupations that employ them to earn additional off-farm income, as well as the amount household labour available to them. Thus, human capital is hypothetically mediated by gender, age and household size.

Social capital refers to social resources upon which people draw in pursuit of their livelihood objectives. This includes networks and connectedness, membership to formalised groups and relationships of trust, reciprocity and exchanges including common rules, norms and sanctions in society (Woolcock and Narayan, 2000; Sobel, 2002). To assess levels of social capital, this study focused on the perceptions that farmers hold towards eight statements describing social capital (Table 2). Since social capital is unlikely to pervade situations of conflict and despondency, it was also assessed through five ranks of peace in the neighbourhoods and perceptions towards levels of social deprivation. Deprivation of five public services including education, water resources, health services, transportation and justice were used as a proxy for social exclusion.

Natural capital refers to the natural resources and ecosystem services for livelihoods including soils, vegetation, forests, water and natural habitats that people depend on in support of their livelihoods. In this study, natural capital was taken to refer to all activities that support the sustainable use of soil and water resources, as the principle natural elements supporting livelihoods in any farming community. Ten technologies for soil and water conservation were considered together to provide an appropriate proxy for the natural capital availbe to the respondent by virtue of belonging to the household.

Physical capital refers to the basic infrastructure and producer goods needed to support livelihoods. In this study, the two aspects of physical capital were considered: affordability of transport from the village to the market and the physical structure of the family house. The costs one pays to and from the market, travelling by public means (matatu), were considered as transaction costs that affects accessibility of the market by respondents and hence, as a proxy for economic isolation. The variable was divided into two levels: low level of economic isolation refers to paying Kenya shilling (KShs) 40 to 80, while and high level of economic isolation means paying more than KShs. 80. These costs were approximated as almost equivalent to paying less than or more than 1 US dollar, the monetary poverty line, that is often used. The mean annual exchange rates of the Kenyan shillings to the US dollar according to the Central Bureau of Statistics were 72.101 (2006) and 75.554 (2005).

The physical quality of the house was measured as a factor of the quality of walls, floor and roof of the main family house. An additive index for 'quality of housing' was constructed in which scores were assigned to each type of roof, floor and wall according to durability and quality of material used.

Financial capital as used in the sustainable livelihoods framework denotes availability of cash and equivalents that support livelihood decisions and actions. The types of financial resources considered in this study were household income, land resources and livestock holding. Household income refers to all sources of income whether from paid or self-employment, involvement in primary products, rents, interest and dividends received, patents, current transfers (such as social security benefits, pensions and life insurance annuity benefits, alimonies), etc; and any other benefits flowing to the household (GoK, 2005). When land has secure tenure, through a title deed, it can readily be exchanged in the market for money or used as collateral to obtain money from a financial institution. For this reason, land is considered a financial good. Livestock is important as a means of asset accumulation and as a saving and as insurance when there is need for immediate cash (Christoplos et al., 2001). Such cash may influence the investment in new improved agricultural technologies. The individual man or woman was assumed to be a beneficially of financial resources administered at the level of the household rather than at the individual level. This is because the assumption was made that resources for investment in new technologies are most likely negotiated at household level.

In examining adoption of technologies, a distinction is often made between the proportion of farmers that adopt a given technology, regardless of the level of use (rate of adoption) and the intensity of adoption in which consideration of levels of use of the technology is made (Doss and Morris, 2001). In this study, the proportion of farmers that used a given technology was considered adopters of the technologies. This was regardless of the level of use. The level of use refers to the self-assessment designation of use of a technology that progresses from non-use; orientation; preparation; mechanical use; routine; refinement; integration to renewal (Griffin and Christensen, 1999). A similar understanding of adoption has been asserted by Karugu (2006), where a farmer is presumed to have adopted a technology if he/she uses it to any extent. During data collection, each farmer who claimed to have adopted a technology was also expected to show evidence of that technology within their farm. This helped to verify their adopter status.

Data analysis

Descriptive statistics that included percentages, frequencies, means, and standard deviations provided the basic scenarios of the various aspects of the study. Factor analysis based on the principal component analysis (PCA) was used in data reduction to identify a small number of factors that explained most of the variance that was observed in the statements describing social capital.

To determine the influence of capital assets on adoption, a binary logistic regression was applied. Binary logistic regression was adopted because of the qualitative nature of the binary dependent variable, y =1 for adoption and y = 0 otherwise. The objective was to find out the probability of occurrence of the event:

$$P(Y) = \frac{1}{1 + e^{-(b_0 + b_1 X_1 + b_2 X_2 + \ldots + b_n X_n) + \varepsilon_i}}$$

Following Field (2005: 220), the binary logistic regression model was specified as:

Where:

1) $P(Y)$ is the probability of Y occurring, 2) e is the base of natural logarithms, 3) b_0 is a constant, 4) $b_1..b_n$ are the coefficients associated with each of the explanatory variables, 5) X_1 - X_n are predictor variables and 6) ε_i is the error term.

RESULTS AND DISCUSSION

The respondents were 63% female and 37% male. Of these, 80.5% were married, 3.7% were single and 15.8% were widowed. Male-headed households were 81.1 and 18.9% were female-headed households. They ranged in age from 24 to 90 years old, with a mean of 51 and a standard deviation of 12. Most of the households were medium in size with the number of children in each family varying between 0 and 12 with a mode of 5. The minimum number of persons providing labour to the households was 1 and the maximum was 14. External labour through employment was available to 37% of the households. There existed a significant association in the number of persons available to provide labour to male-headed and female-headed households (Pearson chi-square 32.155; df = 13; $p<0.005$).

Labour is a major variable input into farming, whose scarcity limits production. Farmers seek higher returns to labour by adopting new crops, production methods and production combinations deemed to improve returns to work (Ruthenberg, 1976). In Kenya the bulk of farm labour is provided by the household members. In cases where the household is short of its own labour and does not have the means to hire external labour, its farm productivity is highly compromised. New and improved technologies and innovations often indicate new labour expenditures; hence, the amount of household labour available influences adoption decisions. Marenya and Barrett (2007) found that the number of adults per household was statistically significant and positively associated with the adoption of natural resources management practices in Western Kenya. It is thus expected that, the more the number of persons in a household, the higher is its labour force. This factor should therefore elicit positive affects on adoption decisions since available labour can be used in implementation of the new technology.

It was found that a high number of women respondents were uneducated or had primary school level education. Few respondents had attended technical schools or undertaken apprenticeship courses. Levels of education between female and male-headed households differed significantly ($p<0.05$). Female-headed households recorded lower levels of education. Among the 975 children (48.1% female and 51.9% male), and a significant difference ($p<0.05$) between the education levels for male and female children was noted. More female children had lower levels of education than the male children. The results are consistent with the overall estimates of literacy in Nakuru district where literacy levels for males and females, are 52.9 and 47.1%, respectively (GoK, 2002).

New or improved technologies are often knowledge intensive, and hence the target population is best able to benefit from them when there exists within such communities or organizations, a basic level of formal education. Where this is not the case, extension and agricultural development projects present an opportunity through which low levels of educational attainment can be improved through management training and skills building (Marenya and Barrett, 2007). Higher levels of education also increase ones potential to earn a higher income which could be invested in acquisition of improved technologies, especially those technologies that require an initial investment. However, in Kenya, schooling attainment has been found to have no impact on technology adoption (Place et al., 2007). Thus, education could either encourage greater adoption of technologies, or it may have no implication on adoption decisions.

Most respondents (68.4%) did not have skills with which to earn off-farm income, hence majority (80.2%) was engaged full-time as small scale farmers. The respondents (19.8%) who had ventured out to engage in non-farm activities, were employed both in formal and informal settings. A chi-square test showed that there was no significant gender association of men and women engaging in non-farm livelihood diversification strategies (Chi square = 1.146, p = 0.284).

It is well established that due to high levels of material uncertainty and risk, rural populations often diversify their livelihoods through occupational flexibility, mobility, including rural-urban migration and dependence on rural non-agricultural income generating activities, even if farming continues to play a major role in their lives (Baumann, 2004). Livelihood diversification can assist households to shield themselves from environmental and economic shocks, trends and seasonality hence reducing their vulnerability. For a person to engage in off-farm activities, they require skills that can support an occupation other than agriculture. The earnings from off-farm activities may be invested in new agricultural technologies to increase household productivity. Or, the non-farm income diversification activities may be in competition for time and energy with implementation of new technologies. This could discourage the skilled persons engaged in a non-agricultural occupation from adopting new technologies.

Results of factor analysis (Table 2), provide three main factors of social capital. Factor 1 also called 'social cohesion', which explained 20.5% of the total variance, shows that social capital is associated with 'help' and 'concern' from the members of the group. Factor 2, which explained 19.616% of variance is related to 'trust'. Factor 3 also referred to as 'social inclusion', explained 19.4% of variance, and was associated with ones 'acceptance within the group' and also with the 'contribution that the farmers' groups made to community welfare'. Up to 75% of the respondents perceived their neighbourhoods as peaceful. There was no significant difference between the views of men and women towards the level of peace in their neighbourhoods (Pearson chi-square = 7.239; df = 4; p-value = 0.124). This could be an indicator that gender-based violence is not widespread in the study area.

Social capital describes circumstances in which individuals use membership in groups and networks to secure benefits (Grootaert, 2001). It has been found that communities with high levels and diverse supply of social capital are better able to deal with poverty, vulnerability, disputes and also to take advantage of new opportunities (Woolcock and Narayan, 2000). Social cohesion ensures that the community will pull together to initiate mechanisms for self support and in provision of necessary resources and services (Kummssa and Mbeche, 2004; Amudavi, 2005). Such cohesion is favourable to peaceful co-existence. However, social capital is double edged; it has positive and negative aspects to it. On the positive side it helps the poor manage risk and vulnerability. Its manifestations are negative in isolated and parochial situations such as within organized crime syndicates or when it supports caste inequality, ethnic exclusion or gender discrimination (Woolcock and Narayan, 2000). The former is called productive social capital while the latter has been referred to as perverse social capital. Therefore, social capital can positively or negatively influence adoption depending on the existing social networks and social relationships.

The percentage of respondents identifying deprivation in public services in their neighbourhoods are shown in Figure 5. Overall, the community under this study faces institutional barriers, that could inhibit access to developmental and welfare opportunities including attainment of the Millenium Development Goals (MDGs). A high number of respondents perceived deprivation in education (facilities and opportunities) (64.7%), water resources (59.5%), health services (47.4%) and means of transportation including poor roads (43.9%). Deprivation in education (facilities and opportunities) has implications for human resources and institutions, which are important predictors of growth in an economy, so much so that, poor technological flow to poor countries has been linked to poor human capital endowment (Lucas, 1990).

Water is a critical commodity without which agriculture and daily living are highly compromised. Deprivation of water sources means that a lot of time is wasted in collecting the commodity from far off sources which increases the burden of domestic tasks, especially for women (Morrison et al., 2007). Where such sources are of low quality, water-borne diseases compromise the health status of household members, increase household expenditures due to increased payments for medical care and also affect labour allocation decisions seriously compromising the likelihood of adoption of improved agricultural technologies (Ersado et al., 2004). The situation could be highly compounded by the lack of means of transport and health facilities which are important institutions in support of the general well-being of rural communities.

Figure 6 shows the frequency of use of each of the soil and water conservation technologies. The frequencies of use of technologies, give credence to the fact that rural people depend on natural resources and ecosystem services for their livelihoods and often contribute to their improvement by adoption of improved technologies and management practices (Bewket, 2007).

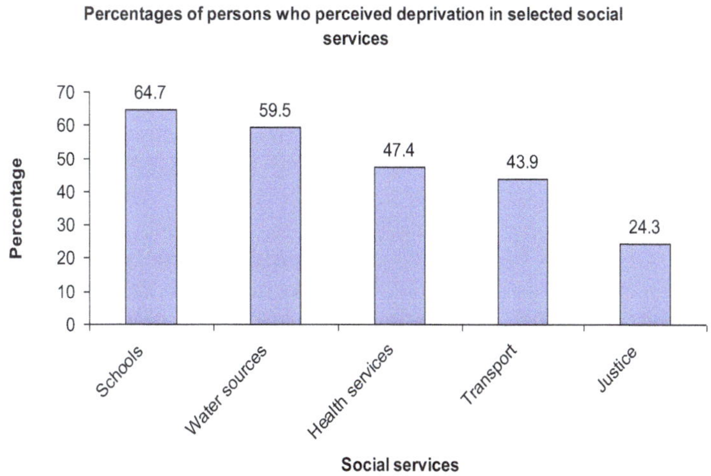

Figure 5. Deprivation within the neighbourhoods.

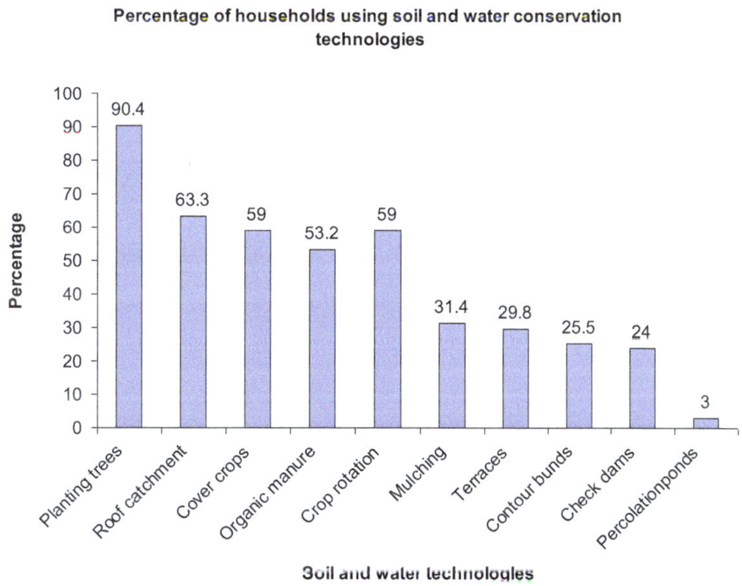

Figure 6. Technologies for soil and water conservation.

Table 3. Affordable transport as a measure of level of economic isolation.

Level of economic isolation	Frequency	Percent
Low level of economic isolation – paying up to KShs. 80 to travel to and from market (=$1)	84	44.2
High level of economic isolation – paying up more than KShs. 80 to travel to and from market (>$1)	106	55.8
Total	190	100

More than half of the study population (56%) faced a high level of economic isolation as shown in Table 3. This could increase their costs of operations and constrain their active participation in the market. In Kenya, smallholder farmers living far from the market face constraints in accessing service providers and consumers of their produce, as well as high exchange costs related to poor transportation systems occasioned by lack of information, and poor access to assets (Alene et al., 2007).

Shelter in the study site can generally be considered as good, because durable roofs made of iron sheets or clay tiles were noted in 97% of the households, 31% of the households had houses with durable walls made of stone or bricks, while 48% of the households had good quality cemented or wooden floors. According to Kenyan participatory poverty assessment studies, poor shelter is among the leading manifestations of poverty (Kimalu et al., 2002; Nafula et al., 2005). Quality of housing was used as a proxy for wealth status. It is expected that wealthier households own better quality housing made up of durable roofs and walls as well as clean floors. Such households are more likely to have surplus income to invest in acquisition of new technologies. These households are also likely to be involved in non-farm activities for income generation, a fact that may remove their participation in agriculture.

Of the total number of respondents, 75% recorded total monthly incomes of up to Kshs. 5000.00. Considering a mean dollar rate of KShs. 80, this translates to an income of about $62.5 a month, which means a daily income of approximately $2 per day for a 30-day month. This means that most households in the study area are relatively poor. Most of the respondents had up to 5 acres of land as shown in Table 4. An acre of land is equal to 4,046.85 m^2. The respondents with more than 10 acres of land were 6.1%. Of all the land owners, 86% had title deeds to their land while 14% did not have title deeds. Livestock ownership including livestock breeds and their average farm-gate prices, as well as total value, were as shown in the Table 5. The farm gate price was the price that the farmer was willing to offer for each animal at the time of data collection.

Financial capital denotes financial resources that people use to achieve their livelihood objectives. The three forms of financial resources considered in this model are: household livestock holding, household monthly income and household land resources. Household livestock holdings and household land resources are expected to have a positive influence on adoption of technologies. Livestock can be sold easily to raise cash. Land can be used as collateral to acquire money from financial institutions to invest in new technologies. Availability of land also directly supports the adoption of new technologies by availing the necessary physical space to do so. Monthly income can positively influence adoption where such non-farm income is invested in new technologies.

The study found that a large percentage of respondents were adopters of the technologies that had been demanded by their groups (Figure 7). This high number of adopters may be related to the integration of the individuals within farmers' groups. A similarly high number of adopters were noted in Brazil by Filho et al. (1999) who used a dynamic econometric framework (duration analysis) to analyze the determinants of farmers' decisions on whether or not to adopt low-external-input and sustainable agriculture (LEISA) technology. After considering a wide range of potential determinants (both economic and non-economic) they suggest that the probability of a farmer adopting increased if the farmer was more integrated within farmers' organizations.

Factors influencing adoption of technologies: Model explanatory variables and expected signs

The wide range of factors, each related to the five livelihood assets, is assessed in relation to their influence on adoption decisions. These variables and their expected signs are summarised in Table 6.

Results of the binary logistic model

Results of the model (Table 7) show significance coefficients for 'age', 'gender of respondent', 'availability of a skill', 'cost of travelling to and from the market', 'social inclusion', 'social cohesion', 'level of peace in the community and monthly income'.

Age was significant ($p<0.05$) but negatively related to the dependent variable. This suggests that younger farmers are more likely to adopt new technologies compared to the older ones. Tauer (1995) found that generally, farmer's efficiency increases and then decreases with age. This has implications for the efficiency of farming and decisions to adopt new or improved technologies. In Kenya older farmers are often the owners of the land, the patriarchs of their families when they are male and repositories of farming knowledge, especially indigenous knowledge. They may be more unwilling to try out new ideas than the younger more innovative farmers.

Gender was positively related to the adoption of technologies ($p<0.05$). The positive sign suggests that men are more likely to adopt new technologies compared to women. Gender is an important consideration of human capital since in many studies, equality and equity in the determinants of technology adoption including, access to production resources such as land, labour, credit and information, between men and women, is shown to be elusive (Doss and Morris, 2001; GoK, 2004; Akinboade, 2005; Morrison et al., 2007). In Kenya, women and girls bear most of the direct toll of this disparity: they work longer hours contributing between 60 and 80% of labour in household, reproductive activities and in agricultural production, are poorer in health, are less educated, have lower nutritional status and higher mortality rates (GoK, 2004). In spite of this overwhelming evidence to the contrary, in Kenya, the policy environment considers gender a social issue. In the stratgey for revitalizing agriculture (SRA) (GoK, 2004), gender is considered a part of the social development

Table 4. Classes of land ownership.

	Frequency	Percent
Landless	2	1.1
Landsize between 0.1 - 2.5 acres	56	30.9
Landsize between 2.6 - 5.0 acres	66	36.5
Landsize between 5.1 - 7.5 acres	26	14.4
Landsize between 7.6 - 10 acres	20	11.0
Landsize greater than 10 acres	11	6.1
Total	181	100.0

Table 5. Livestock breeds and their value.

	Number of owning households	Number of animals	Average Unit cost (KShs.)	Value (KShs.)
Zebu cattle	9	34	16,000	544,000
Cross breed/exotic cattle	146	433	20,000	8,660,000
Indigenous goats	46	222	2,000	444,000
Dairy goats	47	85	8,000	680,000
Indigenous sheep	127	759	2,000	1,518,000
Cross breed/exotic sheep	6	43	3,500	150,500
Poultry layers	3	387	250	96,750
Poultry broilers	1	200	300	60,000
Indigenous chicken	163	2675	200	535,000
Beehives	36	172	2,100	361,200

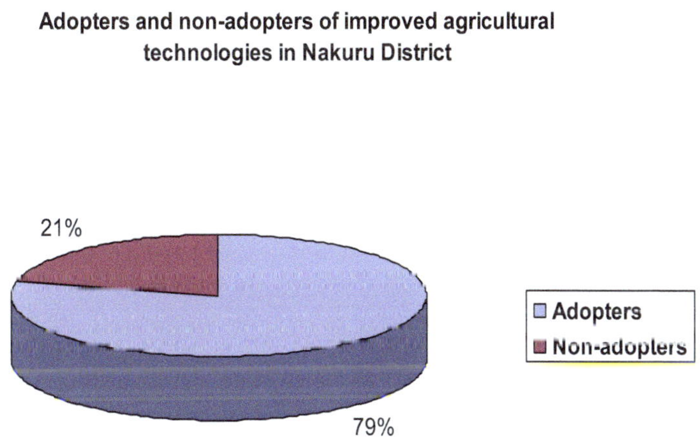

Figure 7. The number of respondents reporting adoption of improved agricultural technologies.

issues that are necessary to coordinate agricultural development with other sectors. Furthermore, although existing laws in Kenya provide for equal rights and privileges for both men and women, cultural and social interpretation of these laws, especially through common laws and social conventions often compromise them (GoK, 2004). Thus, in Kenyan society, the existing legal systems may lead men and women to experience either equalities or inequalities in opportunities depending on the interpretation adopted.

The 'skill' variable was significant and negative in the model results ($p<0.05$). The negative sign indicates that those without skills were more likely to adopt the new technologies as compared to the skilled. This is acceptable because non-farm income diversification activities may be in competition for time and energy with

Table 6. Socio-economic characteristics as explanatory variables and their expected signs.

Description of the explanatory variables	Expected signs of the model coefficients
Human capital	
Age of respondent in years	+/-
Gender of the respondent (1 = male; 0 = female)	+/-
Number of persons in the household	+
Number of persons aged 15-64 in the household	+
Dummy for respondent with a skill (1 = yes; 0 = no)	+/-
Dummy for income diversification (1 = yes; 0 = no)	+/-
Respondents' level of education (1>Secondary school;0 = otherwise)	+
Natural capital	
Score for soil and water conservation measures	+
Physical capital	
Cost of travelling to and from the market ($)	+/-
Score for quality of main family house	+/-
Social capital	
Social inclusion as a component score from PCA	+
Trust as a component score from PCA	+
Social cohesion as component score from PCA	+/-
Score for perceived social exclusion	+
Score for perceived level of peace in the community	+
Financial capital	
Total value of household livestock holding in KShs. (10,000s)	+
Estimated monthly income in KShs. (1000s)	+/-
Total household land holding in acres	+

implementation of new technologies and could discourage the skilled persons, who are engaged in a non-agricultural occupation, from adopting new technologies. In addition, for a person to engage in off-farm activities, they require skills that can support an occupation, away from farming. However, as noted earlier, a large percentage of the respondents did not possess employable skills.

'Economic isolation' as a factor of cost of travelling, to and from the market, was positively and significantly related to the dependent variable ($p<0.1$). This suggests that the propensity to adopt technologies increased with longer distance to the market. In Kenya, smallholder farmers living far from the market face constraints in accessing service providers and consumers of their produce (Alene et al., 2007). Therefore, this result goes against the norm that increased costs of operations constrain active participation in the market and negatively affect adoption of improved technologies. This would suggest that availing improved technologies to people living far from the market provides them an impetus to seek means for active participation in the market in spite of the constraints.

Social inclusion was positively and significantly related to adoption ($p<0.05$). The positive relationship exhibited in the variable 'social inclusion' suggests that where farmers trust and share their knowledge with others, adoption levels are better.

Social cohesion shows a significant ($p<0.05$) but negative relationship to adoption of technologies. Social cohesion deals with connections and relations between individuals and groups (Berger-Schmitt, 2000). Loyalty and solidarity in a group serves to cement relations and connections between individuals. However, the negative relation between adoption and social cohesion might be evidence of negative social capital. This is exhibited when members of farmers groups exclude other members of their communities. This may occur in the form of restrictive networks in which information flow is restricted so as to reduce competition by excluding the participation of others in the innovation-decision process. When social capital is used to exclude others it may lead to social exclusion, and increase inequalities and disparities in access to information and services, hence limiting adoption to a select few. It would seem that some of the farmers' groups in the study area exhibit this form

Table 7. Socio-economic characteristics that influence adoption of technologies.

Explanatory variable	Dependent variable: Adoption of technologies with y = 1 for adoption and y = 0, otherwise				
	Variable description	**B**	**S.E.**	**Sig.**	**Exp(B)**
Age	Age of respondent in years	-0.072	0.026	0.005**	0.930
Genders	Gender of the respondent (1 = male; 0 = female)	1.323	0.589	0.025**	3.754
HH Size	Number of persons in the household	-0.014	0.104	0.892	0.986
TTHHL abourforce	Number of persons aged 15 to 64 in the household	0.154	0.095	0.104	1.167
Skillres	Dummy for respondent with a skill (1 = yes; 0 = no)	-1.259	0.613	0.040**	0.284
Occupationres	Dummy for income diversification (1 = yes; 0 = no)	-0.979	0.628	0.119	0.376
Educ Level Dummy	Respondents' level of education (1 > secondary school;0 = otherwise)	-0.888	1.136	0.434	0.411
Soil water con	Score for soil and water conservation measures	-0.831	1.365	0.543	0.435
Cost travel mrkt$	Cost of travelling to and from the market ($)	1.216	0.691	0.078*	3.375
Hse quality	Score for quality of main family house	0.053	0.122	0.664	1.054
S inclusion	Social inclusion as a component score from PCA	1.187	0.439	0.007**	3.276
Trust	Trust as a component score from PCA	-0.111	0.260	0.669	0.895
S cohesion	Social cohesion as component score from PCA	-0.775	0.235	0.001**	0.461
S exclusion	Score for perceived social exclusion	0.322	0.206	0.118	1.379
Peace comm	Score for perceived level of peace in the community	3.932	1.654	0.017**	50.984
Value total livestock	Total value of household livestock holding in KShs. (10,000s)	0.025	0.043	0.568	1.025
M monthly income thou	Estimated monthly income in KShs. (1000s)	-0.168	0.077	0.028**	0.845
HH landholding	Total household land holding in acres	-0.007	0.061	0.905	0.993
Constant		2.607	2.060	0.206	13.560

Number of observations: 162; percentage of correct predictions: 84%; omnibus test of model coefficients: Chi-Square = 56.254; df = 18; significance = 0.000***; model summary: -2 log-likelihood = 110.215; Cox and Snell R Square = 0.293; NagelKerke R Square = 0.457; Hosmer and Leme show test: Chi-Square = 5.660; df = 8; Significance = 0.685 (>0.05)
*, **, *** denote significance at 10 5 and 1% respectively.

of negative social capital.

Peace in the community was significant (p<0.05) with a positive coefficient suggesting that adoption is most likely to take place more effectively within peaceful communities. Obviously, peaceful conditions support exchange of information and allow free movement of individuals including external agents involved in agricultural research and extension. This is unlike in situations of conflict where such movement is curtailed and fear of attacks, limits the necessary contacts between the farming communities and external agents.

Estimated mean monthly income in thousands of Kenya shillings has a negative coefficient but significant relationship (p<0.05) with adoption of technologies. This means that adoption increased with reducing income levels. This inverse relation implies that poorer households were more likely to adopt improved technologies than well-off households. This may be so because the pooling of meagre resources amongmembers of farmers' groups enables poorer group members, to take the collective risk associated with new technologies. This ensures the participation of poorer households in the innovation-decision process and gives credence to the importance of farmers' groups as institutions that support the poor. This underscores the importance of supporting agricultural production for improved household food security and increased household incomes.

Conclusions

In Nakuru district, adoption of technologies occurs in an environment that is conditioned by the socioeconomic characteristics of the adopting household and the demographic characteristics of the individuals therein. Important factors include age and gender. Younger farmers and men were more likely to adopt new technologies. Thus, dissemination of technologies cannot be conceived as affecting all persons in a similar manner and adoption behaviour cannot also be assumed to have similar effects on men and women. Extension workers working in similar rural systems must endeavour to understand the gender relations inherent within specific study areas. Additionally, they must be deliberately sensitive to existing social inequities in order for men and women to benefit from research and extension services.

The lack of employable skills could affect adoption of new technologies because the unskilled may sell their labour to earn an income at the expense of spending time on their farms to improve their production systems using the new technologies. Thus technologies targeted at unskilled need to incorporate a 'money-earning' component or should realize monetary benefits in the short term to enable unskilled persons deal with immediate daily consumption needs unlike a focus on technologies whose benefits are in the long term.

Targeting households who are off the main infrastructural developments such as roads is beneficial first because their physical isolation often translates to less contact with extension agents and hence with new technologies. Thus, they would be most needy of such services. Secondly, lack of many alternative technologies may make them enthusiastic learners of the new technology which they may consider a means of greater interaction with markets.

Social inclusion and peace are positive factors of social capital that are beneficial of any innovation-decision process. These two factors underlie collective action an approach that allows poorer farmers to pool resources to undertake activities they may consider risky like adopting new technologies. These factors require promotion by all actors working within rural systems.

Social cohesion, another factor of social capital showed a negative relationship to adoption. This could be indicative of exclusiveness, gate-keeping tendencies and lack of community representation among farmers' groups. It could imply that groups that work directly with extension agents 'hoard' information from the rest of the community. This indicates that tapping into existing social networks requires careful consideration to ensure that existing social capital remains intact and that such engagement should seek to remove barriers to self-organization.

RECOMMENDATIONS

Any external agent, either in agricultural research or extension, should make the effort to understand the dynamics within which men and women function and their livelihood asset endowments while promoting any agricultural technology. Policies for rural development stand to have greater impacts if they focus on building rural assets such as schools, roads, security infrastructure and the natural resource base while encouraging self-organization and social networking by tapping into, building and supporting existing networks to deliver relevant services.

ACKNOWLEDGEMENTS

This work was carried out as part of a doctoral research supported by the German Academic Exchange Service (DAAD). We acknowledge and are thankful for the funds to undertake the study. Many thanks to the Director KARI for granting the first author a study leave from work to undertake the study and for logistical support during the field work. We thank the many colleagues and friends in Kenya and at the Justus Liebig University who have supported this work through discussions, critical reviews and consultations. The farmers of Nakuru District, who participated in this study, are gladly acknowledged for their patience and co-operation.

REFERENCES

Adato M, Meinzen-Dick R (2002). Assessing the Impact of Agricultural Research on Poverty Using the Sustainable Livelihoods Framework: Concepts and Methods. EPTD discussion paper 89. IFPRI, Washington, D.C., USA. http://www.ifpri.org/divs/eptd/dp/papers/eptdp89.pdf

Akinboade OA (2005). A review of women, poverty and informal trade issues in East and Southern Africa. UNESCO, ISSJ, 184: 255-275.

Alene AD, Manyong VM, Omanya G, Mignouna HD, Bokanga M, Odhiambo G (2008). Smallholder market participation under transactions costs: Maize supply and fertilizer demand in Kenya. Food Policy, 33: 318–328.

Amudavi D (2005). Exploring the effects of farmer community group participation on rural livelihoods. Strategies Analysis for Growth and Access (SAGA). Policy Brief.

Anandajayasekeram P, Masters WA, Oehmke JF (1997). Impact assessment of African agricultural technology development and transfer: Synthesis of finding s and lessons learned. Policy synthesis for USAID- Bureau for Africa Office of Sustainable Development. Number 28. Michigan State University, USA.

Asfaw S, Shiferaw B, Simtowe F, Hagos M (2011). Agricultural technology adoption, seed access constraints and commercialization in Ethiopia. J. Dev. Agric. Econ., 3(9): 436-447. http://www.academicjournals.org/JDA

Ashaw A, Admassie A (2004). The role of education on the adoption of chemical fertilizer under different socioeconomic environments in Ethiopia. Agric. Econ., 30(3): 215-228.

Baker KM, Edmonds RL (2004). Transfer of Taiwanese ideas and technology to the Gambia, West Africa: A viable approach to rural

development? The Geo. J., 170(3): 189-211.
Batz Franz-J, Janssen W, Peters KJ (2003). Predicting technology adoption to improve research priority setting. Agri. Econ., 28(2): 151-164.
Baumann P (2004). Improving Access to Natural Resources for the Rural Poor: A critical analysis of central concepts and emerging trends from a sustainable livelihoods perspective. FAO, Livelihood Support Programme (LSP) WP 1, Access to Natural Resources Sub-Programme.
Berger-Schmitt R (2000). Social cohesion as an aspect of the quality of societies: Concept and measurement. EU reporting working paper No. 14. Towards a European system of social reporting and welfare measurement. Centre for survey research and methodology (ZUMA), Social Indicators Department, Mannheim. [WWW document] http://www.gesis.org/fileadmin/upload/dienstleistung/daten/soz_indikatoren/eusi/paper14.pdf
Bewket W (2007). Soil and water conservation intervention with conventional technologies in northwestern highlands of Ethiopia: Acceptance and adoption by farmers. Land Use Policy, 4(2): 404-416.
Bozeman B (2000) Technology transfer and public policy: a review of research and theory. Res. Policy, 29: 627–655.
Chambers R, Conway GR (1991). Sustainable rural livelihoods: Practical concepts for the 21st Century. IDS Discussion paper 296.
Christoplos I, Farrington J, Kidd AD (2001). 'Extension, Poverty and Vulnerability: Inception Report of a Study for the Neuchâtel Initiative', Working Paper 144, ODI, London. [WWW document] http://www.odi.org.uk/resources/odi-publications/working-papers/144-extension-poverty-vulnerability-neuch%C3%A2tel-initiative.pdf
Conley T, Udry C (2001). Social learning through networks: The adoption of new agricultural technologies in Ghana. Am. J. Agric. Econ., 83(3): 668-673.
Dadi L, Burton M, Ozanne A. (2004) Duration analysis of technological adoption in Ethiopian Agriculture. J. Agric. Econ., 55(3): 613-631.
Davies K, Franzel S, Hildebrand P, Place N (2004). Extending technologies among small-scale farmers in Meru: Ingredients for success in farmer groups. In proceedings of the 20th Annual Conference of the International Association for International Agricultural and Extension Education (AIAEE), "Education and Extension for Multi-functional Agriculture", 23-29, May 2004. Dublin, Ireland. http://www.aiaee.org/2004/ Accepted/016.pdf.
Doss CR, Morris ML (2001). How does gender affect the adoption of agricultural innovations? The case of improved maize technology in Ghana. Agric. Econ., 25(1): 27–39.
Doss CR, Mwangi W, Verkuijl H, De Groote H (2003). Adoption of Maize and Wheat Technologies in Eastern Africa: A Synthesis of the Findings of 22 Case Studies. CIMMYT Economics Working Paper 03-06. Mexico, D.F.: CIMMYT http://www.cimmyt.org/english/docs/eco wpaper/ewp03-06.pdf
Ellis F, Allison E (2004). Livelihood diversification and natural resource access. FAO, Livelihood Support Programme (LSP), Overseas Development Group, University of East Anglia, UK.
Elloitt H (2004). Sharing experiences on reform of agricultural technology development and dissemination systems in Eastern Africa. ASARECA strategic Planning Paper No. 1. Report of a practitioners retreat, Addis Ababa.
Ersado L, Amacher G, Alwang J (2004). Productivity and land enhancing technologies in northern Ethiopia: Health, public investments, and sequential adoption. Am. J. Agric. Econ., 86(2): 321-331.
Feder G, Slade R (1984). The acquisition of information and the adoption of new technology. Am. J. Agric. Econ., 66: 312-320.
Field A (2005). Discovering statistics using SPSS. Second Edition. Sage publications limited. ISBN 0 7619 4452 4.
Filho HMS, Young T, Burton, MP (1999). Factors Influencing the Adoption of Sustainable Agricultural Technologies Evidence from the State of Espírito Santo, Brazil. Technological Forecasting Soc. Change, 60(2): 97-112.
Franzel S, Phiri D, Mafongoya P, Jere I, Katanga R, Phiri S (2003). The blending of participatory research and quantitative methods: wealth status, gender and the adoption of improved fallows in Zambia. In the proceedings of the 25th International Conference of Agricultural Economists (IAAE), 16-22 August 2003. Durban, South Africa. pp. 700-708. http://www.iaae-agecon.org/conf/durban_papers/papers/021.pdf
Freeman HA, Omiti JM (2003) Fertilizer use in semi-arid areas of Kenya: Analysis of smallholder farmers' adoption behavior under liberalized markets. Nutr. Cycling Agroecosystems, 66: 23–31. The Netherlands.
GoK (2001). The 1999 Population and Housing Census. Counting Our People for Development, Volume I. Ministry of Planning and National Development. Government Printer, Nairobi.
GoK (2002). Nakuru district Development Plan (2002-2008). Effective management for sustainable economic development and poverty reduction. Ministry of Planning and National Development. Government Printer, Nairobi.
GoK (2004). Strategy for revitalizing agriculture, 2004-2014. Ministry of Livestock and Fisheries Development. Nairobi, Kenya.
GoK (2005). Interviewers Manual for MSE Survey. Ministry of Planning and National Development, Central Bureau of Statistics, Nairobi, Kenya.
Griffin D, Christensen R (1999). Concerns-Based Adoption Model Levels of Use of an Innovation (CBAM-LOU). Adapted from Hall, Loucks, Rutherford, & Newlove (1975). Denton, Texas: Institute for the Integration of Technology into Teaching and Learning.
Grootaert C (2001). Does social capital help the poor? A synthesis of findings from the local level institutions studies in Bolivia, Burkina Faso and Indonesia. World Bank Local Level Institutions Working Paper No. 10. World Bank, Washington, DC.
Heffernan C, Thomson K, Nielsen L (2008). Livestock vaccine adoption among poor farmers in Bolivia: Remembering innovation diffusion theory. Vaccine, 26: 2433-2442.
Hussein K (2001). 'Farmers' Organisations and Agricultural Technology: Institutions that Give Farmers a Voice. Development, 44(4): 61-66.
Isham J (2000). The effect of social capital on technology adoption: Evidence from Rural Tanzania. A paper prepared for the conference on opportunities in Africa: Micro-evidence on firms and Households. The Centre for the study of African Economies, 9-10 April 2000. St. Catherine's College, Oxford, USA.
Karugu WN (2006). An assessment of the effects of technology transfer on gender roles within a community: The development of tea and coffee production among smallholder farmers in Kiambu District, Central Province Kenya. ATPS Working Paper Series No. 44. African Technology Policy Studies Network, (ATPS). Nairobi, Kenya. ISBN: 9966-916-76-8.
Kimalu P, Nafula N, Manda DK, Mwabu G, Kimenyi MS (2002). A situational analysis of poverty in Kenya. KIPPRA Working paper No. 6.
Kummssa A, Mbeche IM (2004). The role of institutions in the development process of African countries. Int. J. Social Econ, 31(9): 840-854.
Lee DR (2005). Agricultural sustainability and technology adoption: Issues and policies for developing countries. Am. J. Agric. Econ., 87 (Number 5, 2005).1325-1334 http://www.blackwell-synergy.com/doi/full/10.1111/j.1467-8276.2005.00826.x
Lightfoot C (2004). East Africa: Catalytic Action for the Emergence of Farmer Empowerment for "Demand-Driven Extension In: Rivera, W. and Alex, G (Eds.) Demand-Driven Approaches to Agriculture Extension Volume 3. Case Studies of International Initiatives. Agriculture and Rural Development Discussion Paper 10 Extension Reform for Rural Development. The International Bank for Reconstruction and Development / The World Bank.
Lucas RE (1990). Why doesn't capital flow from rich to poor countries? Am. Econ. Review, 80: 92-96.
Marenya PP, Barrett CB (2007). Household level determinants of adoption on improved natural resources management practices among smallholder farmers in western Kenya. Food Policy, 32: 515-536.
Mariam Y, Galaty J, Coffin G (1993). Strategic Decision-Making: Adoption of agricultural technologies and risk in a peasant economy. http://mpra.ub.uni-muenchen.de/387/1/MPRA_paper_387.pdf
Mbata JN (1994). Fertilizer adoption by small-scale farmers in Nakuru District, Kenya. Fertilizer Res., 38: 141-150.
Morrison A, Raju D, Sinha N (2007). Gender equality, poverty and

economic growth. Policy research working paper 4349. Poverty Reduction and Economic Management Network, Gender and Development Group. The World Bank, Washington, DC. [WWW document] http://www-wds.worldbank.org/external/default/WDSContentServer/IW3P/IB/2007/09/11/000158349_20070911132056/Rendered/PDF/wps4349.pdf

Nafula NN, Onsomu EN, Mwabu G, Muiruri S (2005) Review of policy options for poverty reduction in Kenya. Kenya Institute for Public Policy Research and Analysis, (KIPPRA), Social Sector Division, Discussion paper No. 49.

Omamo SW, Williams JC, Obare GA, Ndiwa NN (2002). Soil fertility management on small farms in Africa: evidence from Nakuru District, Kenya. Food Policy, 27(2002) 159-170.

Place F, Adato M, Hebinck P, and Omosa M (2007). Impacts of agricultural research on Poverty: Synthesis of findings and implications for future directions in MEINZEN-DICK, R. AND ADATO, M. (2007), Eds: Agricultural Research, livelihoods and poverty: Studies of economic and social impacts in six countries. pp. 149 –197. IFPRI.

Place F, Kariuki G, Wangila J, Kristjanson P, Makauki A, Ndubi J (2002). Assessing the factors underlying differences in group performance: Methodological issues and empirical findings from the highlands of central Kenya. Paper presented at the CaPRi workshop on Methodologies for studying Collective Action, Nyeri, Kenya. [WWW document] http://www.capri.cgiar.org/pdf/capriwp25.pdf

Rogers EM (1995). Diffusion of innovations. Fourth edition. The free press, New York. ISBN 0-02-926671-8.

Ruthenberg H (1976). Farming Systems in the Tropics, Oxford University Press, Oxford.

Scheaffer RL, Mendenhall W, Ott L (2006). Elementary survey sampling. Sixth edition. Thomson Higher Education, Belmont, USA. ISBN 0-534-41805-8.

Schreiner M (2005). Asset-Building in rural Areas. Center for Social Development. CSD Working Paper, George Warren Brown School of Social Work. Washington University. http://csd.wustl.edu/Publications/Documents/WP05-55.pdf

Schweitzer R (2006). Home economics Science and Arts: Managing sustainable everyday life. Schriften zur Internationalen Entwicklungs und Umweltforschung. Zentrum für internationale Entwicklungs-und Umweltforschung der Justus-Liebig-Universität Gießen. Peter Lang GmbH. ISBN 3-631-54986-5.

Scoones I (1998). Sustainable Rural Livelihoods: A Framework for Analysis. IDS Working Paper 72, Sustainable Livelihoods Programme (SLP). http://www.ids.ac.uk/ids/bookshop/wp/wp72.pdf.

Sherraden M (1991). Assets and the Poor, Armonk, NY: M.E. Sharpe, 1991.

Smale M (2005). Issues facing agricultural technology adoption in developing countries: Discussion. Am. J. Agric. Econ., 87(5): 1335-1336.

Sobel J (2002). Can We Trust Social Capital? J. Econ. Lit., 4: 139-154.

Staal SJ, Baltenweck I, Waithaka MM, deWolff T, Njoroge L. (2002). Location and uptake: Integrated household and GIS analysis of technology adoption and land use, with application to smallholder dairy farms in Kenya. Agric. Econ., 27(3): 295-315.

Tauer L (1995). Age and Farmer Productivity. Review Agric. Econ., 17(1): 63-69.

UN Millennium Project (2005). Investing in development: A practical plan to achieve the Millennium Development Goals. New York.

Woolcock M, Narayan D (2000). Social Capital: Implications for Development Theory, Research, and Policy. World Bank Res. Observer, 15(2): 225-250.

World Bank (2007). World Development Report 2007: Development and the Next Generation. The World Bank, Washington, D.C. USA.

Impact of youth rural-urban migration on household economy and crop production: A case study of Sokoto metropolitan areas, Sokoto State, North-Western Nigeria

Ango A. K.[1], Ibrahim S. A.[1], A. A. Yakubu[2] and A. S. Alhaji[3]

[1]Department of Agricultural Extension and Rural Development, Usmanu Danfodiyo University, Sokoto, Nigeria.
[2]Department of Agricultural Economics, Usmanu Danfodiyo University, Sokoto, Nigeria.
[3]Department of Liberal Studies, Waziri Umaru Federal Polytechnic, B/Kebbi, Nigeria.

Over the past decades, rural-urban migration research has focused on the hypothesis that the migration of household population, especially youths, has significant influence on the economy and crop production of households. This study aimed at examining the impact of youth rural-urban migration on household economy and crop production in Sokoto metropolitan area of Sokoto State, Nigeria. A multistage sampling technique was used to draw the sample of the study. Descriptive and inferential statistical tools were used to analyze the data obtained. The results indicated that all the migrant household heads were males, married and most of them fell within the ages of 45 to 54 years, while the majority of the migrants were within the ages of 18 to 27 years, single, males and literate. The common place migrated to by the migrants was Kaduna State due to lack of social amenities in their original place of residence. Majority of the migrants practiced temporary migration by coming back home at the onset of the rainy season to assist their parents on the farm, resulting to minimal losses in agricultural production for the household and at the same time remitting to their parents a certain amount of money which was used to cater for the welfare of the family. A non significant relationship was realized between the age of the migrants and some reasons behind their migration to the urban centres. It is recommended that social amenities should be provided by the government, nongovernmental organizations and the rural people with the aid of community driven development to the rural areas.

Key words: Impact, youth, rural-urban, migration, household income, crop production, Sokoto.

INTRODUCTION

In the past decades, youth labour migration and household agricultural economic research has focused on the hypothesis that the migration of the household population, especially youth, has significant influence on the economy and crop production of the households. The family labour which the rural farmers depends solely upon was reduced drastically due to migration of the youth to the rural centres. This phenomenon consequently resulted to high cost of production, low productivity, and reduction in annual income and a fall in standard of living of the rural populace (Akangbe et al., 2006). In a related study by Zimmerer (2004) reported that increased migration and transnational, as well as growth of forest product based handicraft industries, have led to rural households and communities abandoning agricultural lands, resulting to the growth of the imported agricultural packages. Farm labour provided by active and energetic youth is considered as an essential component of agricultural productivity in rural areas, because agriculture in isolated areas of an open country with low population density solely depends on family labour. Rural farmers, due to peasantry nature of the farm business and low income status, mostly depend on family labour, which is mostly provided by the youth. Despite the importance of youth migration to the urban centres due to, most especially, lack of social infrastructure in the rural settings, and lack of rural job opportunities during the dry season of the year, and its repercussions resulted to low yield and high cost of farm labour. Farm labour seasonal migration is often tremendous in magnitude and is widespread throughout the nation of Nigeria. Its net result has been described as having negative impact on the local development and production due to the reduction in human resources (Ray, 2001).

Migration, whether at the international or local level may be a deliberate decision or attempt by the migrant to reap social or economic benefit associated with changing locations. There are different types of migration. Internal migration is termed as the movement of people to a new home within a state, country or continent while external migration is the movement to a new home in a different state, country or continent (Adepoju, 2003; Adamu, 2009; Agbonlahor and Enilolobo, 2013). According to Chandna and Bala (1994) migration is the movement that involves a permanent or semi-permanent change in residence from one settlement to another. Rural - urban migration or labour migration on the other hand is defined as the movement of able bodied individuals from rural villages of origin to cities to earn a labour wage (Hunnes, 2012). Migrants are people who have left their homes to a new location, either temporarily or permanently in order to reap private social or economic gains (Adepoju, 2003). Youth rural urban migration was defined as the movement of youth from one geographical region to another, which may be on temporary or permanent basis (Adewale, 2005). Youth tend to move away from one place to another due to need to escape violence, political instability, congestion, drought in various dimensions and suspected or real persecution (Fadayomi, 1998). The level of poverty, lack of job opportunities and gross inadequacy of social infrastructures was found to be one of the reasons behind youth rural-urban migration (Aworemi et al., 2011). Also the seasonality of the primary job or the agricultural activities which the rural populace engaged in, results in seasonal unemployment and in addition, small scale business opportunities in rural areas are being wiped out (black smith, pot making, processing of agricultural produce, fishing industry etc) by the supply of technological products from the urban centers thus, causing structural unemployment for parts of the year (Akinyele, 2005).

The movement of youth from rural to urban areas is a common occurrence in Nigeria where most of the rural areas are ignored by the government despite the fact that majority (60 to 70%) of the country population resides there. Studies by Ehirim et al. (2000), Adesiji et al. (1998) and Akinyele (2005) on rural-urban migration revealed negative effects of the migration on both the rural areas and the urban settings in Nigeria. According to Adesiji et al. (1998) when the energetic and productive members of the rural populace migrated to the cities, the original place of residence experience low food production, and high cost of labour, while the new location on the other hand may be faced with over population, resulting in unemployment, high rates of crime, prostitution, outbreak of diseases etc.

Due to inadequate or lack of scientific studies on impact of youth rural-urban migration on household economy and crop production in the study area, there is therefore a need to conduct research on the field in order to create a proper understanding of its effects on the households of the study area.

The specific objectives of the study are to describe the socio-economic characteristics of the migrants' household members and the migrants themselves; determine the nature of youth rural-urban migration and its relationship to the household heads farm yield; determine the reasons leading to youth rural-urban migration; find out the businesses engaged in by the migrants and their perceived income; determine the food coping strategies adopted by the household heads

METHODOLOGY

The study on the impact of youth rural-urban migration on household crop production and economy was conducted in some metropolitan areas of Sokoto city. The areas are Kware, Wamakko and Dange/Shuni. Kware is approximately 12 km east of Sokoto city, Wamakko is 10 km west of Sokoto city and Dange/Shuni is approximately 10 km north of the city. Wamakko Local Government Area (LGAs) was purposively selected for the study. The LGA was carved out of Sokoto Local Government Area in 1991. The LGA has four (4) districts which are Dundaye; Wamakko; Gumbi and Gumburuwa. The study area is located on the longitude 3 and 9° East and latitude 10 and 14° North. It is bordered to the North by Tangaza Local Government, South by Bodinga Local Government and Yabo Local Government Areas, West by Silame Local Government and to the East by Sokoto and Kware Local Government Areas. The area has a population of 176,619 (NPC, 2006). The major occupation of the people is farming, fishing and trading. The major tribes of the area are Hausa/Fulani and many other Nigeria minor and major tribes are also found.

The climate of the study area is characterized by a long dry season (October/November-April/May) with a short rainy season (May-September/October), (Singh and Babaji, 1989). Rainfall starts

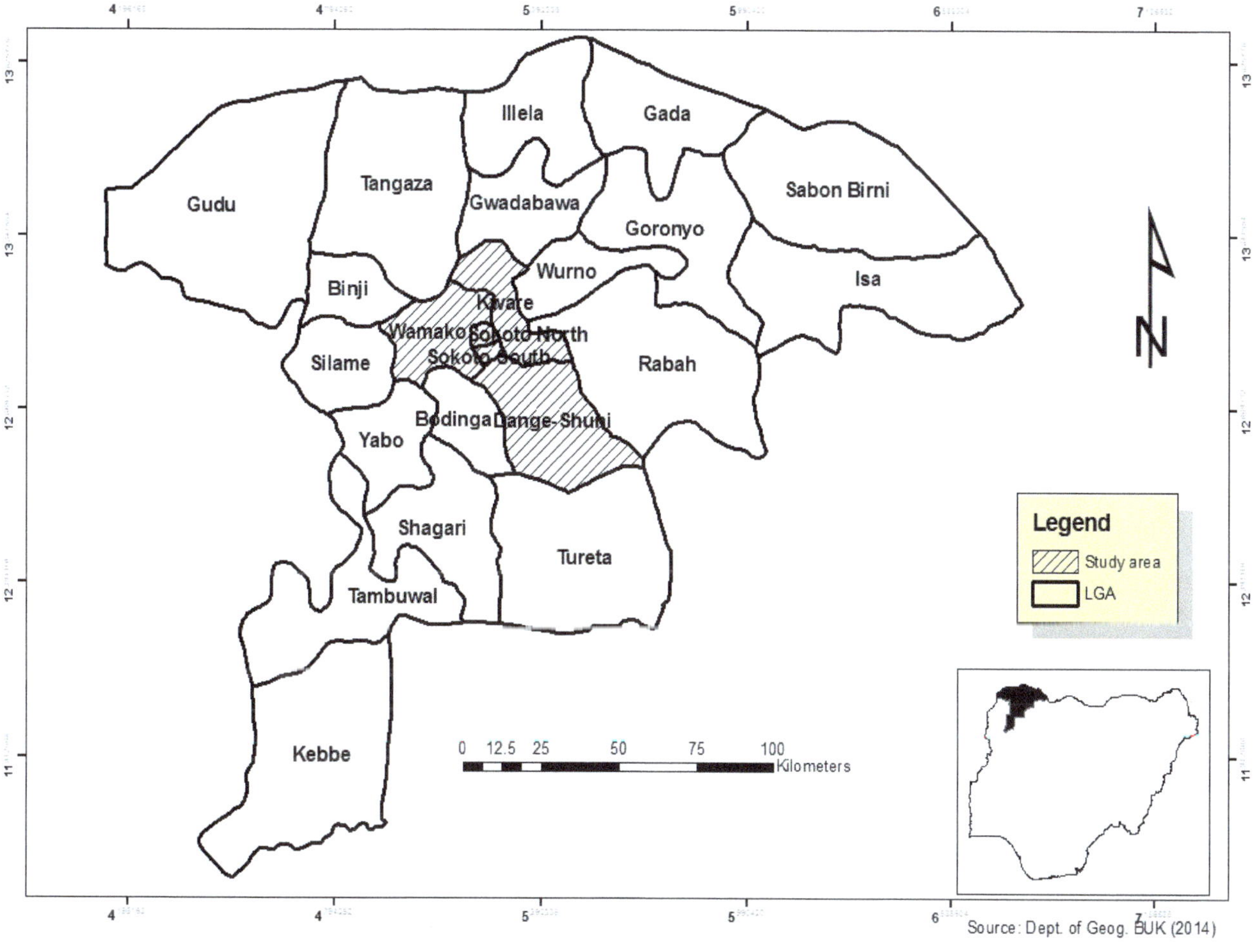

Figure 1. Map of Sokoto State showing the study area.

in late May and ends in late September or early October with annual rainfall ranging from 400 to 700 mm (Singh, 1995). The minimum and maximum temperatures are 19 and 34°C respectively with mean annual temperature of 21.5°C with relative humidity of 52 to 66%. The study area experiences harmattan wind (N-E Trade wind) which is a dry cold dusty wind blowing between the months of November to February. The soil of study area is predominantly sandy to sandy-loamy with low fertility level particularly poor in primary nutrients like nitrogen, phosphorus and potassium. The vegetation of the area falls within the Sudan Savannah agro-ecological zone charactorized by sandy soil, loamy soil and some patches of Fadama land. An assortment of various species of grasses and legumes, patches of bushes and sparsely distributed indigenous tree species majority of which are thorny tree species are also found. Such trees include *Acacia spp* and *Adansonia digitata* (Figure 1).

Multistage sampling technique was employed to arrive at the sample size of the study. The first stage was the purposive selection of Wamakko Local Government Area out of the three Local Government Areas that constitute Sokoto metropolis because of the high numbers of youth known to have migrated to the urban centres. The third sampling technique involved the random selection of two districts out of the four districts in the Local Government Area, and the fourth technique involved the purposive selection of two villages from each of the selected districts due to the concentration of migrants in the communities. Lastly, the sample size of the study constitutes 98 comprising 58 households and 40 migrants.

The primary data of the study was solicited from the respondents through the use of two sets of structured questionnaires; one set was administered to the households and the other to the migrants while the secondary data was restricted to published documents such as textbooks, journals, seminar papers, internet sources and past student projects. The data obtained were subjected to both descriptive statistics (percentages and frequency distribution) and inferential statistics (Chi-square) tests. A test of null hypothesis on the significant relationship between the some selected socio-economic characteristics (age, occupation, marital status) of the migrants and the reasons for their migration was tested using Chi-square analysis.

The study considered two sets of variables; dependent variable which was socio-economic factors influencing migration and independent variables which include the reasons for migration, type of business engaged in by the migrants in cities and the estimated income of the migrants. Age of the migrants was measured in years; marital status of the migrants was measured as single, married, divorced and widow. Educational attainment of the migrants was measured based on Qur'anic education, primary school education, secondary school education, tertiary education, adult education and never attended school. Household size of the head of household and migrants was measured based on the number of people in the house; and the youth rural-urban migration

Table 1. Distribution of household heads and their wards socio-economic characteristics (n=98).

Variable	Household heads (58)		Migrants (58)	
Age (years)	**Frequency**	**Percentage**	**Frequency**	**Percentage**
18 - 22	0	00.0	29	72.5
23 - 27	0	00.0	9	22.5
28 - 34	0	00.0	2	05.0
35 - 44	15	25.9	0	00.0
45 - 54	30	51.7	0	00.0
55 - 64	12	20.7	0	00.0
>64	1	01.7	0	00.0
Age of wards at first migration (year)				
10-14			6	10.3
15-19			41	70.7
> 19			11	19.0
Marital status				
Married	58	100.0	29	72.9
Single	0	00.0	11	27.5
Educational attainment				
Qur´anic education	34	58.6	8	20.0
Primary education	8	13.8	28	60.0
Secondary education	6	10.3	8	20.0
Adult education	10	17.3	0	00.0
Household size (people)				
9 - 13	31.0	53.4	22.0	55.0
14 -18	22.0	37.9	14.0	35.0
>18	05.0	08.6	04.0	10.0
Occupation of household heads	**Frequency**		**Percentage**	
Farming	49.0		84.5	
Trading	02.0		03.4	
Farming and trading	07.0		12.1	
Mode of land acquisition				
Inheritance	40.0		69.0	
Inherited and purchased	18.0		31.0	
Farm size (ha)				
0.1 – 0.9	40.0		69.0	
>1	18.0		31.0	

Source: field survey, 2012.

was measured based on reasons for migration such as lack of basic social amenities, search for job, household food security, looking for better education and the perceived income of the migrants was measured in Naira.

RESULTS AND DISCUSSION

Most (51.7%) of the household heads fell within the age range of 45-54 years, 25.9% falls within the age bracket of 35-44 years, and 20.75% of the household heads fall within the age ranges of 55-64years while only (1.7%) of the household heads were 64 years or older (Table 1).

The findings of the study imply that most of the household heads were within the age bracket of 45-54 years, indicating old ages and as such not active in terms of agricultural productivity. This could be due to the fact that the energetic youth belonging to the household have migrated to the cities which may lead to trend of agricultural productivity trend in the study area (Akangbe, 2006).

Majority (72.5%) of the migrants fell within the age ranges of 18 - 23 years, 22.5% were within the age brackets of 24 - 29 years while only (5%) of the wards (migrants) were within the ages of 30 -35 years (Table1).

The findings shows that majority of the migrants were within their youthful stage of development, a stage when a youth could take decision in conjunction with their parents to travel outside the domains of their parents. The implication of this finding is that the young migrants have the strength and risks bearing ability associated with such population movement. This finding is line with Gimba and Kumshe (2000) that majority of the migrants from the villages to cities are young and energetic to cope with the hurdles of activities faced in cities.

The majority (70.7%) of the migrants surveyed had migrated at the age range of 15-19 years, 19.0% of the migrants migrated at the ages of 19 years and above, 10.3% migrated at the ages of 10-14 years (Table 1). Although the majority of migrants were only within the ages 15-19 years, and in child development stage, it appears they were able to identify the benefits accrued to migrants and were therefore influenced to migrate to the cities.

All (100%) of the household heads were married men while the majority (72.5%) of the migrants were single; 27.5% were married (Table 1). The finding that the majority of the migrants were single suggests it is easier for them to travel than married men, proceeds obtained from labour in the city are then sent to their parents at home. This contradicts the findings of Agbonlahor and Enilolobo (2013) in the South-western Nigeria where majority of the migrants were married, which indicates that they are likely to pay a frequent visit to their families at regular intervals to minimize social impact of separation from place of origin or residency.

The finding that all household heads were males could be due to the fact that in this part of the country, based on their custom, tradition and culture only the males are involved in rigorous agricultural activities (farming). This finding is in agreement with Ango et al. (2011) who reported that majority of the male of the rural populace in the northern part of the country engages in farming whereas the female folks partake only in rearing of children, domestic, other household chores and processing of agricultural produce.

None of the household heads and the migrants was illiterate as they have attained one form of education or the other. Most (58.6%) of the household heads had attained Qur'anic education, 17.3% of the household heads attained adult education, 13.8% of the household heads attained primary education, while only (10.3%) of the household heads attained secondary education, 60% of the migrants had attained primary education while 20% of them attained Qur'anic education and secondary school education respectively (Table 1). It is a normal trend or practice among the people of northern part of Nigeria, to first enroll children into Qur'anic education before being engaged in to any type of education (Chaudna and Bala, 1994).

Both the household heads and the migrants were from moderate to large family size. As shown in the findings in Table 1, most (53.4%) of the migrants household heads had a family size of 9 to 13 people, 37.9% had a family size of 14 to 18 people while only 8.6% of the household heads had a family size of 18 peoples and above. While in the case of the migrants 55% of them were from a family size of 9 to 13 people, 35% were from family size of 14 to 18 people while only (10%) of the migrants were from a family size of 18 people and above. This finding is in agreement with Kamaldeen (2003) and Ejeoma (2000) that majority of the households in the rural settings exceed 15 people due to the nature of the way they live together, and dependence on family as a source of farm labour, and that household size of a rural people is always much greater than their counterparts living in the city.

Based on the findings it was also observed that families with a greater number of people living in their households tended to have their children migrating to the cities purposely to ensure that the rest of the family members left at home were able to sustain themselves with the little that the farm produced. This finding is in accordance with Waliu (2003) that rural household with large family size produced food that is insufficient for their living, thus, sending their children to migrate to urban areas so as to reduce the amount of food consumed by the family.

Socio-economic characteristics of household heads

Majorities (84.5%) of the household heads were farmers, 12.1% were engaged in farming and trading while only (3.4%) of the household heads occupation was solely traders (Table 1). This implies that the rural populace who mostly are farmers engaged mostly in agricultural activities and their end-produce are either meant for family consumption and little for sale to provide those items not produced in the farm. The rural areas are known to have limited or non availability of social amenities which if present as in the urban areas, can serve as an opportunity for varying job opportunities. The finding of the study concurs with United Nations (2001) and Aworemi et al. (2011) that majority of the villagers engaged in farming since they lack social and basic amenities that will provide them with better opportunity to engage in other occupation similar to those that are available in the urban areas.

Majority (69%) of the household heads inherited their pieces of land, 31% inherited and bought the land used for agricultural activities while none of the household heads were found barrowing and leasing the land used for agricultural activities (Table 1). The finding reveals that majority of the household heads inherited the piece

Table 2. Distribution of migrants based on reasons for migration (n = 40).

Reasons for migration	Frequency/percentage			
	Yes	%	No	%
Search for better employment	32.0	80.0	08.0	20.0
Change environment	15.0	37.5	25.0	62.5
Further education	21.0	52.5	19.0	47.5
To learn trade	16.0	40.0	24.0	60.0
Look for money through labour	13.0	32.5	17.0	67.5
Join family members in the city	02.0	05.0	38.0	95.0
Better housing in the city	04.0	10.0	36.0	90.0
To improve welfare	18.0	45.0	22.0	55.0
Better transport in the urban areas	13.0	32.5	27.0	67.5
Social status of one's parent	00.0	00.0	40.0	100
Escape from punishment	00.0	00.0	40.0	100
Free in terms of movement	00.0	00.0	40.0	100
Crop failure and famine	05.0	12.5	35.0	87.5
Lack of social facilities	37.0	92.5	03.0	07.5
	176*		**374***	

Source: Field survey, 2012; *Multiple response.

of land from their parents; because it is known to be part of the tradition and culture of the people of the study area that land belonging to deceased parent is shared to his heirs.

The findings showed that 69% of the household heads had a farm size ranging from 0.1-0.9 ha and 31% of them had a farm size of 1 ha and above (Table 2). This implies that majority of the respondents had small plots of land for their agricultural activities. The finding concurs with Adamu (1997) the plot of land which the rural populace posses are mostly small in nature because it has to be shared among the heir of deceased owner.

Factors responsible for rural – urban migration

Majority (92.5%) of the migrants migrated to the cities because they lack social amenities and infrastructures in their places of residence 80.0% migrated for better employment, 52.5% migrated to further education, 45% of the migrants migrated to improve present welfare, 40.0% migrated to learn trade, 37.5% migrated to change environment, 32.5% migrated to look for money through labour as well as better transport in the urban areasrespectively (Table 2). The findings also reveals that 12.5% of the migrants travel to the cities because of crop failure and famine, 10% travelled to cities for better housing, while only (5%) migrated to join family members in the city. The findings showed that majority of the migrants leave their villages to the cities because they lack social amenities which are found in the cities which if present in the rural areas will provide them with many job opportunities that would deter them from travelling to the cities. This finding corroborates the works of Adepoju (2003), Martin and Taylor (2003), Adamu (2009), Aworemi et al. (2011) and Hunnes (2012).

Places (Cities) migrated to and nature of migration practiced

The findings reveal that 27.5% of the migrants migrated to Kaduna State, 22.5% travel to Kano State, 12.5% travel to Lagos State, 10.5% migrated to Zamfara State, 7.5% travel to Sokoto city, Abuja and Niger States respectively while 5.0% of the migrants travel to Kebbi State (Table 3). Kaduna State having the higher percentages migrants from this study may be due to presence social infrastructure, high rate of job opportunities, better educational facilities as well as its serene environment. According to most of the migrants, they migrated to Kaduna State due to its closeness to Sokoto State in terms of distance and were found to be better than their original place of residence in terms of social infrastructures and amenities.

Nature of migration, jobs engaged in, and amount (naira) earned by the migrants

As indicated in the findings of Table 3, the majority (87.9%) of the household heads agreed that their wards return home after a while and 12.1% were of the view that their wards stay where they migrated to for life. The findings imply that majority of the household wards who migrated returned home on the onset of the rainy season, did so to assist their parents in farming activities and thus, these group of migrants' practices temporary

Table 3. Migrants distribution based on places migrated to, nature of migration, jobs engaged in, and amount earned/month (n = 40).

Places migrated to	Frequency	Percentage
Niger State	3	07.5
Kano State	9	22.5
Sokoto city	3	07.5
Kaduna State	11	27.5
Zamfara State	4	10.5
Lagos State	5	12.5
Abuja	3	07.5
Kebbi State	2	05.0
Nature of migration		
Temporary migration	7	12.1
Permanent migration	51	87.9
Nature of Labour engaged in		
Labour works	4	10.0
Okada/Kabu-kabu riding	7	17.5
Trading	21	36.2
Taxi driving	3	07.5
Studies	5	12.5
Amount earned monthly(#)		
Less than or equal to #10,000	1	2.5
#10,100 - 20,000	5	12.5
#20,100 - 30,000	4	10.0
#30,100 - 40,000	8	20.0
#40,100 - 50,000	5	12.5
> #50,000	10	30.0
Nothing remitted home	5	12.5

Source: Field survey, 2012.

migration. This type of migration practiced by the respondents goes contrary with the type practiced by the southern Nigerians who mostly returned home only during the festive periods of the year. This finding concur with Adewale (2005) and Olorunshogo (2007) that migrants from the village to the cities either stay in the cities to practice permanent migration or often go home frequently to pay visit to their people thus practicing temporary migration.

As shown in Table 3, that 36.2% of the migrants engaged in petty trading, 17.5% were Okada (Motorcycle) riders, 12.5% went for studies, and 10.0% were labourers while only 7.5% of the migrants were taxi drivers. The findings indicate that majority of the migrants engaged in one form of business or the other that provides them income. No migrants were involved in white collar job that generates appreciable amount of money due to lack of good educational background and capital to invest in the cities. Based on the findings, majority of the migrants were of the opinion that despite not engaged in white collar, migration to the cities is better than remaining idle at home. This finding is in contradiction with Okpara (1983) that the rural people in the cities only end up in petty businesses and unskilled vocation jobs in the urban areas that attracts no much income.

The findings depicts that 30% of the migrants earned more than #50,000 (Naira) monthly, 20% earned between #30,100 and #40,000 monthly, 12.5% earned between #10,100 and #20,000 and between #40,100 and #50,000 respectively while only 2.5% of the migrants earned below or up to #10,000 per month. Part of the income generated / gained was used for self-keeping while the remaining is remitted home to family (Table 3).

Impact of youth rural – urban migration on household food crop production

As shown in the findings 36.2% of the household heads stated that two of their wards had migrated, 34.5% stated that only one of their wards migrated, and 27.6% of the

Table 4. Distribution of household heads based on number of migrated wards and their yield/year (n=58).

Variable	Frequency	Percentage
Migrated wards		
One child	2	34.5
Two children	21	36.2
Three children	16	27.6
Four children	1	01.7
Harvest per year (bags)		
4 - 8	32	55.2
9 - 13	18	31.0
>13	8	13.8

Source: Field survey, 2012.

household heads stated that three of their wards migrated and only (1.7%) of the household heads stated that four of their wards ever migrated (Table 4). The finding of the study implies that the higher percentage of the household heads had only two of their children that have ever migrated; or it could be due to the fact that there is enough agricultural harvest by the family. This result is in contradiction with Kamaldeen (2003) who reported that from every household there always found some wards migrating and their number ranges from 4 and above this is because the agricultural produce harvested was small and it could not be enough to feed the family for the whole year. Migration of the wards according to him is the next alternative to purchase of seed and other farm inputs that could be used next year and is therefore considered as an act of supplementing the family income or food stock.

The result in Table 4 reveals that most (55.2%) of the household heads had 4 to 8 bags of assorted grains/year, 31% 9 to 13 bags of assorted grains/year and 13.8% of the household heads had realized a yield of 13 bags and above of assorted grains/year. The findings imply that most of the household heads produced between 4 to 8 bags of assorted grains/year, an indication of the peasantry nature of the agricultural activities where farmers cultivate mostly for household consumption and selling some of the farm produce to provide essential commodities not produce in the farm. This type of agricultural activities made farmer to operate on a small piece of land due to lack of capital and government incentives.

Food copping strategies adopted by the heads of the household in the community

As indicated in Table 5, majorities (93.1%) of the household heads received assistance from their migrating wards in form of monies sent while the remainder did not. This finding implies that majority of the migrants engages in one form of business venture or the other from which some incomes were generated. It's out of these incomes that some were sent home for up keep of the family. The findings in Table 5 also revealed that 46.6% of the migrants remitted between **#**10,000 and **#**20,000 to the family at home, 17.2% of the migrants remitted less than #10,000 and #20,000 – #30,000 respectively to their family at home while only 8.6% of the migrants did not remit any amount of money to their family. The finding shows that majority of the migrants remitted certain amount of money to their family at home.

Majority (72.4%) of the household heads were of the opinion that the money remitted home by their wards was not sufficient for home demands, 19% of the household heads agreed that the money remitted was enough while only (8.6%) household heads said that their wards that migrated did not remit money home (Table 5).

As shown in the findings in Table 5, majority (70.7%) of the household heads used the money remitted in purchasing food stuff, 19% paid labourers for services rendered with the money remitted and 8.6% of the household heads bought domestic animals with the money remitted while only (1.7%) household heads engaged in business with the money remitted by their wards that migrated. The result indicated that most of the money remitted home by their migrating wards was used for the purchase of food stuff because the majority of the farmers practiced subsistence farming out of which the yield obtained could not sustain the family.

The findings in Table 5 reveal that the majority (79.3%) of the migrants' absence was not felt by their family while only (10%) of the migrants' family felt the absence of their wards due to the separation with members of their family. The reason behind majority of the families of the migrants not feeling the absence of their wards could be due to temporary migration practiced, and some amount of money remitted home. In addition, the use of information

Table 5. Distribution of Heads of Household Based on copping strategies adopted, amount remitted by the wards and the sufficiency of the remitted amount (n=58).

Variable	Frequency	Percentage
Assistance received from wards		
Received assistance from wards	54	93.1
No any assistance from wards	4	06.9
Amount remitted (#)		
<#10,000	10	17.2
#10,000 - 20,000	27	46.6
#20,100 - 30,000	10	17.2
#30,100 - 40,000	5	08.6
>#40,000	1	01.7
Nothing remitted home	5	08.6
Sufficiency of allowance (Remittance)		
Sufficient	11	19
Not sufficient	42	72.4
No remitting of money home	5	08.6
Utilization of money sent		
Purchasing of food stuff	41.0	70.7
Buying domestic animals	05.0	08.6
Paying labourers	11.0	19.0
Engage in business	01.0	1.7
Feeling of wards absence by the families		
Yes	12	20.7
No	46	79.3

Source: Field survey, 2012.

technology such telephone to interact with the family members also helps in reducing the long silence between the households and their migrated wards.

Testing of research hypothesis

There is no significant relationship between reasons for migration and food provision coping strategies of the heads of the household.

The Chi-square analysis result in Table 6 revealed a significant relationship between migrants search for better employment in the urban centres and the money remitted home (X^2 = 0.36, P-values = 0.55). This finding implies that the migrants migrated to the cities due to the presence of job opportunities that yields some money out of which some were remitted home for the upkeep of the family.

The Chi – square analysis also revealed that there is significant relationship between lack of social amenities and infrastructures in the rural areas and migrating of the youth to the cities (X^2 =0.26, P-values = 0.61). The finding implies that the migrants migrated from their place of origin which is characterized with low availabilities of social infrastructure to urban areas where there is more opportunities for venturing into businesses that attracts some income, thereby remitting to the family at home to purchase food stuffs and other needs of the family at home (Table 6).

CONCLUSIONS AND RECOMMENDATIONS

Youth rural urban migration directly impacts the economy and especially crop production of the rural household in the Sokoto metropolitan area, owing to a number of reasons. Firstly, the youth who form the majority of the and energetic group of rural residents often migrate to urban centres because of push factors like lack of job opportunities, social amenities and infrastructures in rural areas. Secondly, the preference to relocate to urban centres to search for jobs and make remittance back home in rural areas to support family members financially in order to meet the cost of food and other necessities for the welfare and to improve the socio-economic status of the household. As result of the findings it was concluded that migration of the youth (wards) to the cities is more appreciable to their parents than remaining home idle. Based on these realities, these types of researches are crucial for the development of the Sokoto Metropolitan area and its rural environs.

Based on the above the following recommendations deemed necessary:

1. Provision of social amenities and establishment of cottage industries in the rural areas is necessary to enable the rural residents live a very descent and comfortable live.
2. Government should encourage private sectors to invest in the villages more especially on agricultural activities.
3. Provision of subsidized agricultural inputs such as fertilizers, pest and herbicides, animal drawn plough and irrigation facilities that enable the rural farmers to have appreciable yield is highly recommended.
4. Rural populace should be encourage to establish projects with the aid of community driven development projects in the rural areas.

Conflict of Interest

The author(s) have not declared any conflict of interests.

REFERENCES

Adamu M (2009). The impact of rural-urban migration on the economy of the rural areas in Wudil Local Government Area of Kano State, Nigeria. Techno. Sci. Afr. J. 3(1):1-9.

Adamu SS (1997). Socio-economic factors associated with rural-urban migration in Nigeria. A case of Maiduguri. Sixth Urban Symposium: P. 74.

Adepoju A (2003). Migration in West Africa. Development. 46(3):37-41.http://dx.doi.org/10.1177/10116370030463006

Adesiji GB, Omoniwa V, Adebayo SA, and Matanmi BM (2009). Factors associated with the youths rural-urban migration drift in Kwara State, Nigeria. Interdisciplinary J. Contem. Res. Bus. 1(8):69-77.

Adewale JG (2005). Socio-economic factors associated with urban-rural migration in Nigeria: A Case Study of Oyo State, Nigeria."J. Human Ecol. 17(1):13–16.

Agbonlahor MU, Enilolobo OS (2013). Rural immigrants agricultural labour supply and rural nonfarm economy in the South western, Nigeria. J. Appl. Agric. Res. 5(1):1-17.

Akangbe JA, Adesiji G, Akinpelu OI (2006). Effects of youth migration on farmers agricultural production in Egbedore Local Government Area of Osun State, Nigeria. Proceedings of 8th National Research Conference and Network Meeting of CYAP in Nigeria, held in University of Ilorin, Kwara State, November, 2006.

Akinyele O (2005). Poverty, malnutrition and the public health dilemma of disease. University of Post Graduate School Interdisciplinary Research Discourse 2005, Ibadan. University of Ibadan.

Ango AK, Abdullahi AN, Abubakar BB (2011). Role of socio-economic Parameters in Determining the Efficacy of Urban Agriculture in Providing Food Security in Birnin Kebbi Metropolitan Area, Kebbi State, North Western Nigeria. Int. Res. J. Agric. Sci. Soil Sci. 1(6):185-192.

Aworemi JR, Abdula-azeez IA, Opoola NA (2011). An appraisal of the factors influencing rural-urban migration in some selected Local Government Areas of Lagos State, Nig. J. Sustain. Develop. 4(3):6.

Chaudna RC, Bala R (1994). Urban growth and housing problems in India. Avadh J. Soc. Sci. 1:25-34.

Ehirim NC, Onyeneke RU, Chidiebere-Mark NM and Nnabuihe VC (2012). Effects and prospect of rural to urban migration on the poverty status of migrants in Abia State, Nigeria. Agric. Sci. Res. J. 2(4):147–149.

Fadayomi TO (1998). Rural Development and migration in Nigeria: Impact of the Eastern Zone of Bauchi State Agricultural Development Project. Nigeria Institute of Socio-economic Research, Ibadan, Nigeria.

Gimba Z, Kumshe GM (2000). Causes and effects of rural-urban migration in Borno State: A Case Study of Maiduguri Metropolis. Asian J. Bus. Manage. Sci. 1 (1):170-171.

Hunnes D (2012). Understanding rural to urban migration in Ethopia: Driving factors, analytical framework and recommendations. J. Global. Health Perspect. 2(1):12-31.

Kamaldeen O (2003). Impacts of migration of youth on agricultural production. A case of B/Kebbi Local Government Area. BSc. Project, Department of Geography, Usmanu Danfodiyo University, Sokoto, Nigeria.

Martin P, Taylor JE (2003). Farm employment, immigrants productivity and earnings costing analysis. J. Agric. Resour. Econ. 28(2):349-363.

National Population Council-NPC (1991). Handbook of 1991 Population Census in Nigeria, NPC Lagos.

Okpara JO (1983). The Impact of migration on the quality of Nigeria Rural Life. Nigerian Agricultural Research Management and Training Institute. Seminar Series 3:116.

Ray DP (2001). Effects of labour migration on agriculture production in Mohattari District. MSc Thesis, Dept of Agricultural Extension and Rural Sociology, IAAS, Rampur, Chitwan, Nepal.

Singh BR (1995). Soil management strategies for the semi arid ecosystem in Nigeria: The Case of Sokoto and Kebbi State. Africa Soil 28:317-320.

Singh BR, Babaji GA (1989). Characteristics of soil in Dundaye District II, the Fadama Soils of the University Farm.

United Nations (2001). Millennium Development Goals, United Nations Department of Public Information.

Waliu U (2003). Factors associated with youth rural-urban drift in Zamfara State. Interdisciplinary J. Contemp Res. in Bus. 1(8):57.

World Bank (1986). Report on Poverty and Hunger: Issue and Option for Food Security in Developing Countries Washington D.C.

Zimmerer KS (2004). Cultural ecology: Placing household in human-environment studies in the case of tropical forest transitions and agro biodiversity change. Prog. Human Ecology, 28(6):795-806.

Production and productivity of maize subjected to modern and traditional methods of weed control options in Federal Capital Territory, Nigeria

Mutambara, J.[1], Ayoola, G. B.[2], Ejembi, E. P.[3] Avav, T.[4], Masvongo J.[1] and Moyo S.[5]

[1]Department of Agricultural Economics and Extension, Faculty of Agriculture, University of Zimbabwe, Box MP 167 Mt Pleasant Harare, Zimbabwe.
[2]Department of Agricultural Economics, College of Agricultural Economics, Extension and Management Technology, University of Agriculture, Makurdi, Benue State, Nigeria.
[3]Department of Agricultural Extension and Communication, College of Agricultural Economics, Extension and Management Technology, University of Agriculture, Makurdi, Benue State, Nigeria.
[4]Department of Crop and Environmental Protection, College of Agronomy, University of Agriculture, Makurdi, Benue State, Nigeria.
[5]Department of Crop Science Faculty of Agriculture, Faculty of Agriculture, University of Zimbabwe, Box MP 167 Mt Pleasant Harare, Zimbabwe.

This study was designed to evaluate the implications of using alternative weed control options on viability, production and productivity of maize crop in the Federal Capital Territory (FCT), Nigeria. Data were collected using a structured questionnaire from 451 randomly selected smallholder farmers. Gross margins, regression and correlation analysis were used to analyze the data. About 72% of farmers who grew maize practiced manual weed control while 40% used integrated modern and manual weed control. Results revealed that viability was higher in maize subjected to integrated (119,664.38 naira/ha) than in manual (4211.63 naira/ha) weed control. Seed rate (β = 0.155), variety (β = 0.283), herbicides (β = 0.593), fertilizer (β = 0.100) and labour (β = 0.535) were positive and significant determinants of maize output with a coefficient of determination of 0.823. Productivity of seed, labour, and fertilizer were higher under integrated weed control option than manual weed control. There was a positive association between use of herbicides and resource productivity (seed quantity (r = 0.607), herbicides (r = 0.526), fertilizer (r = 0.347) and labour (r = 0.770). Farmers were therefore encouraged adopt herbicides to improve viability, production and resource productivity. There was need for further research to identify intervention areas for inducing adoption of herbicides by farmers.

Key words: Integrated weed control, manual weed control, gross margins, regression.

INTRODUCTION

Food and Agriculture Organization (FAO) projections in food demand suggest that cereal demand will increase by almost 50% towards 2050 (FAO, 2003). This is in line with the expected increase in population over time, the past 100 years have seen the world's human population increasing by nearly fourfold (UN population Division,

2007); and it is projected to increase from 6.7 billion (2006) to 9.2 billion by 2050.

To increase crop production in line with increasing demand for food, three primary factors should be considered and these are; increased cropland and rangeland area (15% contribution in 1961 to 1999); increased yield per unit area (78% contribution); and greater cropping intensity (7% percent contribution) (FAO, 2006). Thus for food production to keep pace with population demand, there is a need to invest in more efforts to increase yields, continued expansion of cropland by conversion of natural habitats, or by optimizing food or feed energy efficiency from production to consumption.

One of the serious challenges to productivity increases, cropland expansion and intensification is the threat from invasive alien species such as pests and diseases that have been estimated to cause an annual loss of US$12.8 billion in yield of eight of Africa's principal crops and may reduce yields in developing countries overall by around 50% (Nellemann et al., 2009). Rossman (2009) estimated that alien invasive weeds and pathogens are estimated to be responsible for about 8.5 and 7.5% in yield reduction, respectively, equivalent to US$24 billion and US$21 billion of acrop value of US$267 billion. Across Africa and worldwide, a weed species of the genus *Striga* has a direct impact on local livelihoods; it affects more than 100 million people and as much as 40% of arable land in the savannas.

Weed management is thus an important operation in crop production that should be effectively and efficiently carried out to ensure desired production and productivity increases for improved livelihoods and welfare. One of the main objectives in agricultural societies is attainment of an optimally high level of living with a given amount of effort, thus, an increase in the productivity of resources employed in farm production amounts to progress and according to Gianessi (2009), unless weeding is improved among other things, farmers in Africa will not obtain the optimum from their crops. Thus efficiency and efficacy in weed management, is one strategy that can ensure achievement of food production goals.

There are five common methods used in weed management namely manual, mechanical, cultural, biological and chemical weed control (Anyanwu et al., 2003). It is recommended that farmers should chose methods that enhance crop production and profitability. The choice among weed control option by rational individuals is however based on the economic implications of different options of weed control among other factors. Given the various alternatives to weed control, rational farmers are expected to use cost effective methods that are more productive to ensure greater returns from their farming.

Outcomes from experimental trials on weed control have been favouring use of chemical weed control options given the high returns exhibited from experimental trails (Ayoola and Adedzwa, 2005). However, given the differences that exist between experimental situations and real farmer situations there is a need to evaluate and verify experimental outcomes in real (farmer) circumstances. This study was carried out to evaluate implications on viability, production and productivity of alternative weed control options practiced by smallholder maize farmers. The study is instrumental in guiding resource adjustments and policy strategies in weed control for successful crop production.

MATERIALS AND METHODS

Study area

The study was carried out in Federal Capital Territory (FCT) of Nigeria with a total human population of 1,405,201. Out of this population size, the farming population is 446,506 and the total farming households is about 93,092 distributed in 6 council areas (FCTFDP, 2007).

Sampling

Random sampling technique was used to draw a proportionate sample from each council area to build up a total sample of 451 households. A structured questionnaire was used to collect data on crop production, socio-economic environment and weed control practices from farmers with the assistance of Fadama project enumerators.

Data analysis

Descriptive statistics, gross margins, regression, production function and correlation analysis were used to evaluate viability, production and productivity of alternative weed control options in maize crop.

Gross margin analysis

Gross margin budgeting is one technique that can be used to evaluate economic viability of alternative activities and provide a guide in deciding among different options (Vere et al., 1997). In this study, total income (TI) was compared with total variable cost (TVC) to provide a gross margin. A positive margin will imply that the activity is viable and a negative margin implies a non viable activity. Comparing the gross margins under alternative weed control options will provide a guide on the most desirable options.

Regression analysis

Regression analysis was used to estimate the production function of maize including the influence of weed management option on maize output, different forms of production functions were fitted to the maize production data and the best fitting model selected, the model is specified as:

$$Y = F(X_1, X_2, X_3, X_4, X_5, X_6, U) \quad (1)$$

Where output (Y) of maize was expressed as a function of hectares of land cultivated (X_1), amount of seed (X_2), variety of seed (X_3),

Table 1. Crop production in the FCT, Nigeria (N = 451).

Major crops	Percentage of farmers producing
Maize	87.4
Yam	81.8
Sorghum	54.3
Groundnuts	43
Rice	39.1
Cassava	36.9
Okra	18
Millet	16
Pepper	14.2
0thers (sesame, beans, cowpeas, ginger, melon, tomatoes and pumpkins)	20.1

Source: Field survey.

total labour hours used for all operations (X_4), amount of fertilizers used (X_5), and amount of herbicide used (X_6).

The different forms of production function are as specified in Equations (2) to (5) as a linear, Cobb-Douglas, exponential and semi log functions of the inputs respectively.

$$Y = a_0 + a_1 X_1 + a_2 X_2 + a_3X_3 + a_4X_4 + a_5X_5 + a_6X_6 + U \quad (2)$$

$$\text{Log } Y = a_0 + a_1 \log X_1 + a_2 \log X_2 + a_3 \log X_3 + a_4 \log X_4 + a_5 \log X_5 + a_6 \log X_6 + \log U \quad (3)$$

$$\text{Log } Y = a_0 + a_1 X_1 + a_2 X_2 + a_3X_3 + a_4X_4 + a_5X_5 + a_6X_6 + U \quad (4)$$

$$Y = a_0 + a_1 \log X_1 + a_2 \log X_2 + a_3 \log X_3 + a_4 \log X_4 + a_5 \log X_5 + a_6 \log X_6 + \log U \quad (5)$$

Where a_0 to a_6 = unknown parameters. U= Error term. The best fitting equation was chosen on the basis of the magnitude of the coefficient of multiple determination (R^2), significance of coefficients, significance of overall production function as given by F value, and the appropriateness of signs of the regression coefficients based on a priori expectations. The t-test was used to determine the significance of variables in the model.

Productivity analysis

The average physical product for each input was calculated by Y/X_i, the marginal physical product as dy/dx_i and the marginal value product as $(dy/dx_i)P_{maize}$). The ratio of MVP/MFC was used as an indicator of optimality of resource use was determined by dividing the marginal value product by the marginal cost of input (cost per unit input). To determine the significance of any differences in resources productivity between farmers using manual and chemical weed control options, a Z-test was used.

Correlation analysis

In order to ascertain the relevance of herbicides in maize productivity, correlation analysis was used to measure the degree of association between use of herbicides and resource productivity. The correlation coefficient (r) was calculated using the formula:

$$r = \frac{n\sum X_iY - \sum X_i\sum Y}{(n\sum Y^2 - (\sum Y)^2(n\sum X_i^2 - (\sum X_i)^2)} \quad (6)$$

Where Y was amount of herbicides used as a measure of use of chemical weed control option. X_i variable representing resource productivity of input i (i=seed, labour, fertilizer and herbicides).

RESULTS

Crop production and maize cropping practices by farmers in the FCT

Farmers were producing a variety of crops. The majority of farmers were producing maize (87.3%), yam (81.8%) and sorghum (54.3%). Other crops produced were groundnuts, rice, cassava, okra, millet, pepper, sesame, beans, cowpeas, ginger, melon, tomatoes and pumpkins in order of decreasing importance (Table 1). Maize was the dominant crop produced, hence it was chosen in the gross margin analysis as a case study. On average 1.2 ha of maize were produced per household and 61.4% of farmers who grew maize used a local variety of maize while 38.6% used improved varieties. The seed rate was on average 16.8 kg/ha.

About 72% of farmers who grew maize practiced manual weed control where weeds were removed by hoes, cutlasses, and hand pulling and 40% used integrated modern weed control methods that involved applying herbicides in addition to manual weeding. Common herbicides used in maize were gramoxozone, 2-4-D and atrazine in order of decreasing importance.

Labour was the major input in maize production. On average labour usage per hectare in all farming operations was 58 labour days. About 71% of farmers did not apply any fertilizer in maize production while 29.4% applied fertilizer at an average rate of 41.2 kg/ha

Gross margins analysis

Average output was higher for users of herbicides than in manual weed control and the margins were very large

Table 2. Maize output, yield and gross margins under manual and chemical weed control.

Variable	Manual (N=282)	Chemical (N=112)	Z-value
Average output (Kilograms)	892.65	3223.57	-7.559**
Average land (Hectares)	1.2	1.5	-1.80
Average labour days	74.84	51.89	3.661**
Average fertilizer use (Kilograms)	32.98	129.2	-3.829**
Average seed (Kilograms)	17.88	19.99	0.994
Average yield (Kilograms/hectare)	743.88	2149.1	-11.882**
Gross margins/ha (1- (2+3+4)	4211.63	119664.38	-10.015**
Value of output	44632.45	161178.75	-7.559**
Cost of labour	37418.35	25947	-3.661**
Cost of fertilizer	1433.40	5616.52	-3.983**
Cost of seed	1569	2578.80	-2.909**
Cost of chemicals	0	7371.88	-8.221**

Source: Field Survey, 2008. **Z value was significant at 1 percent, *Z value was significant at 5%.

Table 3. Impact of herbicide use on maize production.

Coefficients	Linear	Cobb Douglas	Exponential	Semi log
Constant	-111.838 (-1.180)	3.551 (12.649)**	5.811 (103.926)**	1320.835(1.517)
Land	88.009 (1.060)	0.245 (1.192)	-0.315 (-6.435)**	4387.325 (6.886)**
Seed	14.503 (2.165)*	0.155 (2.104)*	-0.00014 (-0.036)	149.671(0.655)
Variety	393.396 (3.586)**	0.283 (4.072)**	0.433 (6.685)**	191.397 (1.282)
Herbicides	238.949 (13.150)**	0.593 (13.794)**	0.173 (16.176)**	467.832 (3.506)**
Fertilizer	5.911 (14.705)**	0.100 (9.078)**	0.0013 (5.270)**	287.046 (8.378)**
Labour	4.988 (2.324)*	0.535 (5.282)**	0.013 (9.956)**	-1038.922 (-3.306)**
R^2	0.783	0.823	0.663	0.623
F	237.089**	299.387**	126.813**	106.709**

Source: Field Survey. **t value was significant at 1%, *t value was significant at 5%.

(2331 kg). Usage of variable inputs (land, fertilizer and seed) was also high for users of chemical weed control option than in manual weed control option. However, use of labour was higher for users of manual methods (74.84 days) than it was for users of chemical control methods (51.89 days).

Output and average yields per hectare were higher for users of chemical weed control than for manual weed control. Overally, maize crop subjected to chemical weed control had higher gross margins (119,664.38 naira) than in manual weed control (4211.63 naira). This can be attributed to high cost and usage of labour among other things. In manual weed control about 44% of farmers had negative gross margins while in chemical weed control, maize production was viable for all the farmers. Table 2 is the summary of the gross margin analysis.

Impact of use of herbicides on maize production

The Cobb-Douglas production function was the best fitting equation and thus formed the basis of this analysis. Land (β = 0.245) was positively associated with output but the relationship was not significant at 5%. Seed (β = 0.155), variety (β = 0.283), herbicides (β = 0.593), fertilizer (β = 0.100) and labour (β = 0.535) were all positively associated with output and the relationships were significant at 5%. All the explanatory variables together explained about 82.3% of the total variation in value of output of maize (Y) as indicated by value of R^2 of 0.823. Table 3 is a summary of results from the regression analysis.

Productivity of maize under manual and chemical weed control

Productivity of seed, labour, and fertilizer were higher under chemical weed control option than manual weed control with margins, APP_{seed} (111 kg), APP_{labour} (59 kg) and $APP_{fertilizer}$ (6 kg). The same applied to MPPs and MVPs as they are derived from APPs according to formula in the methodology. At the prevailing prize of maize output of 50 naira per kilogram, the MVPs were

Table 4. Resources productivity in maize crop under traditional and modern weed control options.

Resource	Traditional	Modern	Difference (margin)	Z-value	R
Seed					
N	282	112			
APP_{seed} (kg)	53.15	163.77	110.62		
MPP_{seed} (kg)	8.24	25.38	17.14	-13.380**	0.607**
MVP_{seed} (naira)	411.90	1269.22	857.32		
MVP/MFC	6.11	11.42			
Labour					
N	282	112			
APP_{labour} (kg)	2.58	61.55	58.97		
MPP_{labour} (kg)	6.73	32.93	26.2	-23.288**	0.770**
MVP_{labour} (naira)	336.51	646.37	309.86		
MVP/MFC	0.42	2.06			
Fertilizer					
N	57	59			
$APP_{fertilizer}$ (kg)	13.85	19.90	6.05		
$MPP_{fertilizer}$ (kg)	1.39	1.99	0.6	-6.445**	0.347**
$MVP_{fertilizer}$ (naira)	69.24	99.48	30.24		
MVP/MFC	1.39	1.99			
Herbicides					
N		112			
$APP_{herbicides}$ (kg)		556.64			
$MPP_{herbicides}$ (kg)		330.09			0.770**
$MVP_{herbicides}$ (naira)		16504.44			
MVP/MFC		13.73			

Source: Field Survey, 2008; **-value was significant at 1%, *-value was significant at 5%. r = Correlation coefficient between resource productivity and use of herbicides.

also higher for maize under chemical weed control than manual weed control. The differences were significant at 5 percent level across all inputs (Table 4).

The MVP_{seed}/MFC_{seed} ratio was 6.11 under manual weed control and 11.42 under chemical weed control and the differences were significant at 5%. The $MVP_{labour}/MFC_{labour}$ ratio was 0.42 under manual weed control and 2.06 under chemical weed control and the differences were significant at 5%. The $MVP_{fertilizer}/MFC_{fertilizer}$ ratio was 1.39 under manual weed control and 1.99 under chemical weed control and the differences were significant at 5%. The ratio of MVP/MFC of herbicides was 13.73 (Table 4). This indicated inefficiency in the use of all variable resources by all farmers.There was a positive association between use of herbicides and resource productivity (seed quantity (r = 0.607), herbicides (r = 0.526), fertilizer (r = 0.347) and labour (r = 0.770). The relationships were significant at 5% level (Table 4). This indicated the significant role played by herbicides in improving resource productivity.

DISCUSSION

The results indicated that there were higher yields, output and gross margins in using chemical weed control as compared to manual weed control in smallholder maize production. This implied that chemical weed control improved viability of smallholder maize production. This outcome was similar to research findings by Ayoola and Adedzwa (2005) which proved that use of herbicides was most effective in soybean and cassava in terms of high yields, least cost of production and high margins than hoe weeding and intercropping. Similar research findings in Kenya by Kibata et al. (2002) also proved that herbicides increased net benefits by 61% in a maize/bean intercrop and 46% in a maize monocrop. The higher gross margin under chemical weed control confirms the role of modern technology in improving viability of farming. Johnson (1995) discovered that 80% of smallholder farmers would increase the size of their cultivated land if weeds were less of a problem. In addition, when using manual weeding, weed infestation in crops may be so severe that weeding is not always worthwhile; therefore, fields are effectively abandoned. In Malawi, nationwide survey data reports one third of the area planted to maize by smallholders was either left unweeded or incompletely weeded (Orr et al., 2002). Similarly, in Zimbabwe, 21% of cotton farmers abandon more than 20% of their crops each year as a result of weed infestation (Mavudzi et al., 2001).

The higher margins from users of chemical weed control can thus be associated with higher yields obtained as weeds are controlled effectively and within the critical growth stage. According to Indian Council of Agriculture Research (ICAR) (2006) chemical weed control is easier, faster and in many cases less costly. Manual weed control is labor intensive, time consuming, causes chronic pain, spinal deformation and is usually avoided or delayed such that by the time it is done, considerable yield loss would have occurred. Chikoye et al. (2004) expressed that although a lot of energy in consumed in removing weeds manually, crop yields are generally very low due to weed competition caused by untimely and ineffective weed control by this method. Use of time is of prime importance to the smallholder farmer as there is competition for labor between weeding and other farm and non-farm operations, if the farmer can use time saved from manual weeding to carry out other profitable operations, then there is a good case for using herbicides (Benson, 1982). Thus it is evident from this analysis that the benefits from herbicides use are more than the direct yield benefits as there are other productivity benefits accruing from labor saved for other farm and off farm operations.

Another reason why it was less viable in manual weed control was that smallholder farmers spend 50 to 70% of their total labor time weeding (Chikoye et al., 2007) given the demand nature of weeding manually, with the high cost of labor in Nigeria's FCT of about 800 naira per day, the cost of production becomes too high. Furthermore, in Africa, women contribute more than 90% of the hand weeding labor (Ukekje, 2004). In addition seven out of every 10 farm children between the ages of 5 and 14 are forced to leave school and work in the agricultural sector at the peak period of weeding (Ishaya et al., 2008). Thus besides the potential benefits of herbicide use in the current study of increased incomes and reduced drudgery, there are also social benefits salient to women, children and the poor for example when children are able to attend school because they are not needed to weed fields.

The production function indicated that the efficacy for herbicides, variety, seed, fertilizer and labour were positive and significant at 5% hence these inputs are important determinants of maize productivity. Summing up the coefficients of all explanatory variables ($\sum B_i = (0.283 + 0.593 + 0.1 + 0.535 + 0.155 + 0.245) = 1.911$) revealed increasing returns to scale implying that additional inputs of productive resources resulted in larger increases in product than the preceding unit. This outcome indicates that there is potential for farmers to increase production and efficiency by inputting more factors of production, however smallholder farmers are constrained in most cases cannot attain the optimal levels of production (Gianessi, 2009). There is therefore, a need to investigate other constraints related to input usage that is preventing attainment of optimal utilization of inputs among smallholder farmers.

The benefits associated with use of herbicides of higher yields associated with its effectiveness and efficiency in weed control as earlier alluded to in the discussion is a major explanation to the positive association between output and use of herbicides. The outcomes from this research implies that improved use of herbicides have potential to increase crop output. The crop will be subjected to less competition from weeds as weeding can be done on time within the critical stages of weeding as prescribed by ICAR (2006) and this will yield higher output. Another possible explanation why herbicides use in maize production is positively associated with higher output is the complementarities of the technology with conservation agriculture that is being promoted of late. As a way of mitigating the effects of soil degradation and climate change, conservation agriculture based on integrated weed control, minimum tillage, residue retention and rotations has been proposed (Thierfelder and Wall, 2009) and if tillage is abandoned, weed control becomes a major challenge for smallholder farmers which calls for an integrated approach to control the weeds (Steiner and Twomlow, 2003). Mazvimavi et al. (2012), highlighted that small holder yields under conservation agriculture are higher and gross margins are subsequently higher under conservation agriculture which is based on minimal mechanical working of the soil than conventional methods with output elasticities indicating positive responses for labor and seed in CA, and negative responses in conventional farming (Mazvimavi et al., 2012).

The comparative analysis of resources productivity between maize crops subjected to chemical weed control and manual weed control indicated that all resources productivity (APP, MPP, and MVP) were significantly higher for maize subjected to chemical weed control option than the one subjected to manual weed control options. The differences in performance of farmers may be directly and entirely related to differences in productivity of inputs caused by differences in technology and managerial efficiency. The role of technology and technology transfer in inducing sustainable agriculture development is supported through productivity increases attained by adopters of modern technology to weed management.

The ratio of MVP/MFC compared for all inputs between the two groups was used as an indicator of optimal utilization of resources. In both groups, the ratios for all inputs (seed, labour and fertilizer) were not equal to one indicating non optimal use of resources. Herbicides that were only limited to chemical weed control were also inefficiently utilized. This could be attributed to poor application methods and techniques, herbicide resistance by some weed species among others. This outcome is consistent with developmental theories that states that agricultural development can only be induced through multiple faceted interventions. There is a collaborative and interactive efforts of institutional innovation, improvements in human capital as well as changes in the availability of biological and physical capital (Kanyenze et al., 2011) that collectively work together to ensure economic utilization of resources. The productivity differences in agriculture are increasingly a function of scientific and industrial capacity and in the education of people rather than natural resource endowments. Thus advancement of modern technologies such as seeds, fertilizers and herbicides alone is not enough but will require complementary institutional and human capital investments to ensure effective utilization.

The significant positive association between use of chemical control and resource productivity of all the inputs (seed, labour, fertilizer, chemicals) signifies the importance of herbicides in improving returns to production. The association implied that an increase in use of herbicides will result in an increase in productivity of other resources. This outcome indicates complementarities of resources used in production systems as explained earlier by Kanyenze et al. (2011). Rukuni (2006) further emphasized the need to raise rural incomes through increased agricultural productivity requires the expanded use of productivity enhancing inputs and technologies. The use of purchased inputs like herbicides in this study was found to be associated with improved productivity of complementary inputs implying higher output per unit input and more income to farmers.

The paper concludes that it is economically rational for farmers to move from traditional to chemical weed control in maize production to improve effectiveness and efficiency. Extension practitioners from government, private and non-governmental organizations who have been promoting use of herbicides should continue to encourage farmers to use herbicides to obtain higher output, yields and profits from production. However, there is a need to step up extension efforts as adoption of appropriate production technology is low in smallholder agriculture, herbicide use is less than 5% and smallholder farmers in Africa generally do not use herbicides (Magyembe, 1997). Furthermore, research to explore causes of non optimal resource allocation in maize production and appropriate socio-economic environment for adoption of advanced technologies such as herbicides will be instrumental in improving efficiency and providing policy direction for improve weed control through adoption of herbicides.

ACKNOWLEDGEMENT

The authors of this paper are grateful to the following: i. FCT Agricultural Development Projects and Fadama staff for technical assistance provided during data collection. ii. University of Agriculture, Makurdi, Benue state, Nigeria especially the College of Agricultural Economics, Extension and Management Technology, Department of Agricultural Economics for the guidance and assistance during the three years of PhD studies by the first author.

REFERENCES

Anyanwu AC, Anyanwu BO, Anyanwu VA (2003). A Textbook of Agricultural Science For Schools and Colleges. 6th Edition, Ibadan, AFP Publishers. pp. 357-361.

Ayoola JB, Adedzwa DK (2005). A partial budget analysis for evaluating economic viability of integrated management options for imperata cylindrical in small holder farms of Benue state, Nigeria. PAT. Int. J. Agric. Res. 2(1):1-8.

Benson JM (1982). Weeds in Tropical Crops: Review of Abstracts on Constraints in Production Caused by Weeds in Maize, Rice, Sorghum-millet, Groundnuts and Cassava. FAO Plant Prod. Protect. 32(1):33-38.

Chikoye DF, Ekeleme F, Akobundu IO (1997). Weed composition and population dynamics in intensified smallholder farms in West Africa, The 1997 Brighton Crop Protection Conference, Weeds.

Chikoye D, Schultz S, Ekeleme F (2004). Evaluation of integrated weed management practices for maize in the Northern Guinea Savanna of Nigeria. Crop Protect. 23:895-900.

FAO (2006). World Agriculture, towards 2030/2050. FAO, Rome. Available online at: http://www.fao.org/es/ESD/AT2050web.pdf [Accessed on the 20 January 2009].

FAO (2003). World agriculture: towards 2015/2030. FAO, Rome. Available online at: ftp://ftp.fao.org/docrep/fao/004/y3557e/y3557e.pdf [Accessed on the 20 January 2009].

Federal Capital Territory Fadama Development Project (FCTFDP), (2007). FCT Fadama II Project of Agriculture and Rural Development (ARD) project: Our Legacy/ Achievements. P. 54.

Gianessi L (2009). Solving Africa's weed problem: Increasing crop production and improving the lives of women. Crop Protection Research Institute 202-872-3865. www.croplifefoundation.org.

Indian Council of Agriculture Research (ICAR), (2006). Handbook of Agriculture. 5th Edition, pp. 481-508. New Delhi 110 012. India. pp. 466-480.

Ishaya DB, Dadari AS, Shebayan JAY (2008). Evaluation of herbicides for weed control in sorghum (Sorghum bicolour) in Nigeria, Crop Protection, 26, pp 1697–1701.

Johnson DE (1995). Weed management strategies for smallholder rice production, The 1995 Brighton Crop Protection Conference, Weeds, pp 1171-80.

Kanyenze G, Kondo T, Chitambara P, Martens J (2011). Beyond the Enclave. Towards a Pro-Poor AND Inclusive Development Strategy for Zimbabwe. Weaver Press, ANSA, LEDRIZ and ZCTU. pp. 30-72.

Kibata GN, Maina JM, Thuranira EG, Musembi FJ, Nyanya G, Muthamia JGN, Okuro JO, Mutura I, Amboga S, Micheni AN, Mureithi F, Overfield D, Terry PJ. (2002). Participatory development of weed management strategies in maize based cropping systems in Kenya, Thirteenth Australian Weeds Conference. pp. 343-344.

Nellemann C, MacDevette M, Manders T, Eickhout B, Svihus B, Prins AG, Kaltenborn BP (Eds). (2009). The environmental food crisis – The environment's role in averting future food crises. A UNEP rapid response assessment. United Nations Environment Programme,

GRID-Arendal, www.grida.no.
Mavudzi Z, Mashingaidze AB, Chivinge OA, Ellis-Jones J, Riches C (2001). Improving weed management in a cotton-maize system in the Zambezi Valley, Zimbabwe, The 2001 Brighton Crop Protection Conference, Weeds. pp. 169-74.
Mazvimavi K, Ndlovu PV, An H, Murendo C (2012). Productivity and Efficiency Analysis of Maize under Conservation Agriculture in Zimbabwe Selected Paper prepared for presentation at the International Association of Agricultural Economists (IAAE) Triennial Conference, Foz do Iguaçu, Brazil.
Magyembe JM (1997). Herbicide use in Uganda, 16th East African Biennial Weed Science Conference Proceedings, pp. 165–169.
Orr A, Mwale B, Saiti D (2002). Modeling agricultural 'performance': smallholder weed management in southern Malawi. Int. J. Pest Manage. 48(4):265-278.
Rossman A (2009). The impact of invasive fungi on agricultural ecosystems in the United States. Biological Invasions 11 (1):97-107.
Rukuni M (2006). The evolution of agricultural policy: 1890-1990. In: Rukuni M, Tawonezvi P, Eicher CK, Munyuki-Hungwe M, Matondi P (eds), Zimbabwe's agricultural revolution revisited. University of Zimbabwe Publications, Harare.
Steiner KG, Twomlow S (2003). Weed management in conservation tillage systems. African conservation tillage network Information Series P. 8.
Thierfelder C, Wall PC (2009). Effects of conservation agriculture techniques on infiltration and soil water content in Zambia and Zimbabwe. Soil. Tillage Res. 105:217-227.
Ukekje E (2004). Modernizing Small-holder Agriculture to Ensure Food Security and Gender Empowerment: Issues and Policy, Intergovernmental Group of Twenty Four, http://www.g24.org/researchpapers.html.
UN Population Division (2007). UN 2006 population revision. UN, New York. Available online at: http://esa.un.org/unpp/ [Accessed on the 20 January 2009].
Vere DT, Jones RE, Griffith GR (1997). An integrated economic methodology for evaluation of impacts of weeds in agricultural production systems and the farm and industry benefits of improved weed management. C.R.C for weed management systems, Waite campus, University of Adelaide, Australia. 2:33.

Seasonality in cocoa spot and forward markets: Empirical evidence

Helyette Geman* and Seth Sarfo

Commodity Finance Centre, Birkbeck University of London and ESCP Europe, United Kingdom.

This paper first describes the main features of supply and demand in cocoa spot markets. A state-variable model is proposed to describe the random evolution of cocoa forward curves over time, which essentially adapts to agricultural commodities, introduced by Borovkova and Geman (2006) for energy. In contrast to most of the literature on the subject, the first state variable is not the spot price, as it combines seasonal and stochastic features and may not be observable, instead, the average value of all liquid futures contracts is a quantity devoid of seasonality and conveys a robust representation of the forward curve level. The second state variable is a quantity analogous to the stochastic convenience yield, which accounts for the random changes in the shape of the forward curve. We conduct estimation procedures for the cocoa market over the period of 1980 to 2009 and exhibit an interesting result on cocoa seasonality as well as an extension of the Samuelson effect.

Key words: Cocoa Spot markets, forward curve stochastic modeling, future curve average value.

INTRODUCTION

Cocoa originated in South America millennia ago. Archeological evidence in Costa Rica indicates that cacao, a chocolate drink named after cocoa, was drunk by Maya traders as early as 400 BC. After a Spanish expedition in 1519 brought back to Spain the recipe for cacao, cocoa was planted in Spanish territories like the Dominican Republic or Trinidad. Around 1660, France introduced cocoa to Martinique, Brazil and Grenada. England had cocoa growing in Jamaica by 1670; prior to this, the Dutch had taken over plantations in Curacao when they seized the island in 1620.

Lately, the explosion in demand brought by chocolate's affordability required more cocoa to be cultivated. Cocoa from Brazil was planted in Nigeria in 1874 and Ghana in 1879. In Cameroon, cocoa was introduced during the period 1925 to 1939. The natural habitat of cocoa trees is in the lower level of the evergreen rainforest as climatic factors, particularly temperature and rainfall, are important for optimal growth; variations in the yield of cocoa trees across years are affected more by rainfall than any other climatic factor. Today cocoa is produced in countries located in a band around the Equator, the largest producers being Cote d'Ivoire, Ghana and Indonesia as developed below.

Moreover, 95% of the world's cocoa is grown almost entirely in small farms with the whole family working together, in more than 50 countries located in Africa, Asia, the Caribbean and Latin America. In these countries, cocoa serves as the main source of cash income for millions of people whose rural livelihoods depend on it. Despite some investments made possible by the high spot prices of the marketing year 2009 to 2010, the formulation of agricultural policies in cocoa-dependent countries would benefit from a sound understanding of the price evolution in the cocoa futures market, as this one accounts for more than 70% of cocoa beans sales.

Countries recognized as cocoa producers include Ivory Coast[1] - which is the world's biggest supplier and accounts for more than 35% of global output; Ghana,

*Corresponding author. E-mail: hgeman@hotmail.com.

[1]A number of Ivorian cocoa trees, planted more than 25 years ago, have already passed their peak of productivity with many farmers switching to the more lucrative rubber.

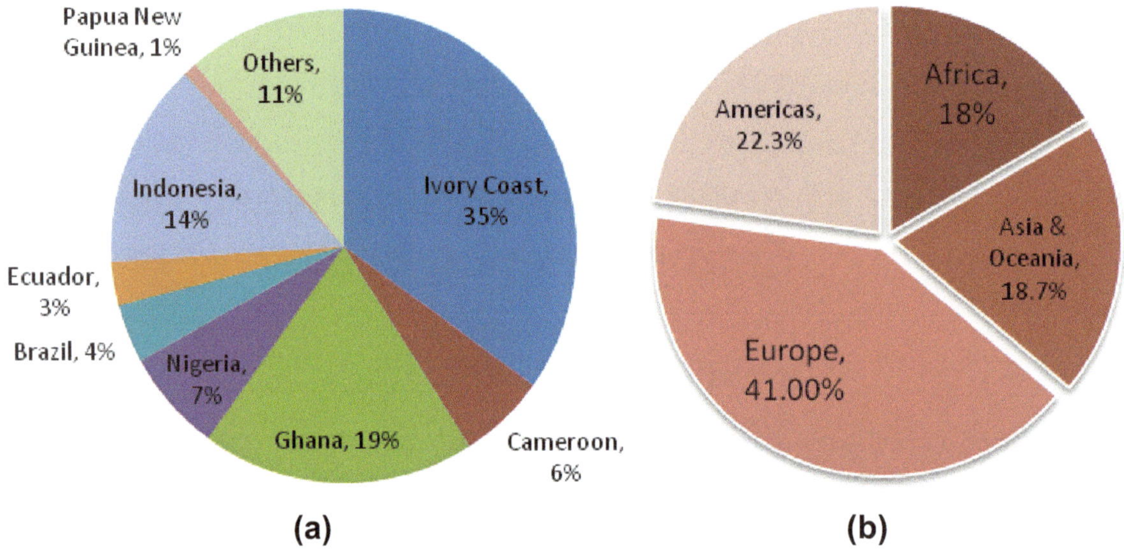

Figure 1. (a) 2008/2009 share in world production; (b) Share in world cocoa consumption (source: ICCO Annual Report, 2008/2009).

which is the second largest one; Indonesia, Cameroon, Nigeria, Brazil, Ecuador, Malaysia and others. Ivory Coast planted millions of cocoa trees during the period 1980 to 2005, thus becoming the top world producer. Using data from the International Cocoa and Coffee Association (ICCO), we display in Figure 1(a) the shares of the major cocoa producing countries. We can already observe, as it will matter for the discussion of seasonality in the next sections of the paper, that the main producers are immediately close to the Equator, in the Northern hemisphere like Ivory Coast or the Southern hemisphere like Brazil, which implies that harvests are spread over the calendar year.

In all these countries (except to a small extent for Malaysia and Indonesia), production comes from small entities with the average farm size not exceeding four hectares. Production from countries like Togo, Papua New Guinea, Dominican Republic, Peru, Mexico is individually insignificant but constitute together 10% of the world supply.

Apart from Ghana and to some extent Cameroon, where production levels have bounced back from the 1990s slowdown, the output from the other countries has been on the decline due to aging cocoa trees, lack of necessary investments in the cocoa sector and competing opportunities in the rubber and oil palm plantations (for example, in Ivory Coast). This explains why, after six years of stagnant world production during the period 2003- 2009, cocoa prices hit in July 2010 their highest level in three decades of $3500/tonne (Figure 2a).

Crinipellis perniciosa, commonly referred to as witches' broom, is a fungal cocoa disease that has for the past decade caused a decline in the production and yields of cocoa beans in Brazil. However, in Ecuador, the Latin America's second largest cocoa bean producer after Brazil, cocoa production has been on the increase due to the successful introduction of a new variety of plant resistant to the witches' broom disease. In West Africa, it is the so-called black pod which ruins the bean's quality; and a bad weather event is an intense episode of *harmattan,* the dry and dusty wind that blows from the Sahara and may jeopardize the flowering of the cocoa trees.

Production in both Indonesia and Malaysia takes place mostly on large plantations either privately or state owned. A downward trend in Malaysian cocoa production has been observed since the early 1990s, when the outbreak of disease coincided with the deterioration of the country's macro-economic conditions. In addition, farmers switched production from cocoa to more lucrative crops, such as palm trees, in response to the fall in world cocoa prices during the 1990s. The downward trend in the Malaysian cocoa sector is expected to continue due to the expansion of urban settlements. Therefore production levels observed a couple of decades ago are unlikely to be observed soon.

Government policies in Indonesia in the 1980 to 1990s have encouraged the expansion of production. Most of the increase during the last two decades was bulk cocoa coming from hybrid trees. While the expansion in Indonesia has slowed down since the late 1990s, yields in the country are still the highest among major cocoa producing countries in Asia. Like in Ghana, where producers' selling prices are linked to world market prices, cocoa farmers in Indonesia receive a high proportion of the market price and that encourages reinvestment in cocoa inputs, and in turn contributes to

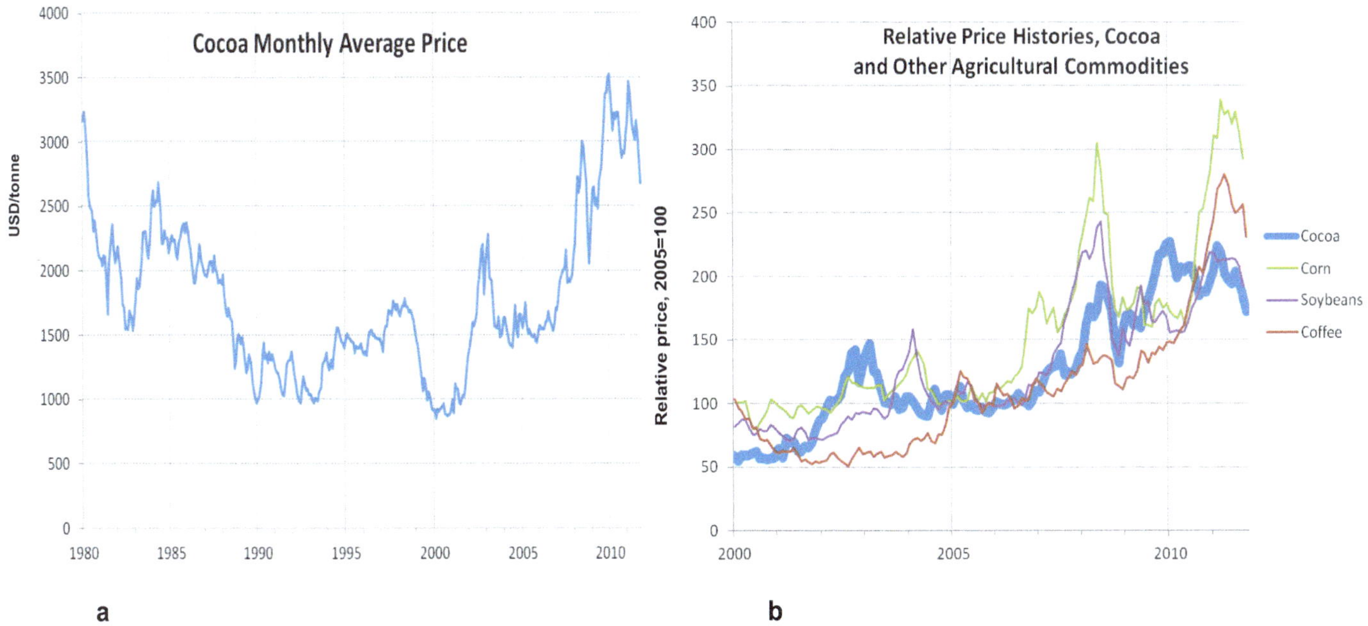

Figure 2. (a) Monthly cocoa price history, US$/tonne, 1980 to 2011(Source: World Bank); (b), Comparison of cocoa price history with other agricultural commodities, 2000 to 2011 (Source: World Bank).

improvement in yields.

There is currently an organic cocoa market but this represents a very small share of the total cocoa market. The ICCO estimates organic cocoa supplied to the world market to be less than 0.5% of total cocoa production. Production of certified organic cocoa is also estimated to be about 15,500 tonnes and sourced from countries including Madagascar, Tanzania, Uganda, Belize, Bolivia, Brazil, Costa Rica, Dominican Republic, El Salvador, Mexico, Nicaragua, Panama, Peru, Venezuela, India and Sri Lanka. Demand for organic cocoa products is growing at a very strong pace, as consumers are increasingly concerned about the safety of their food supply along with other environmental issues.

Like all agricultural commodities', cocoa prices are volatile. International prices fluctuate according to macroeconomic conditions in OECD countries (the main consumers of cocoa and cocoa products), whose annual demand has increased over the past two or three decades with an average annual growth over 3%. Cocoa is also susceptible to the vagaries of disease and weather, which can provoke wide fluctuations in world production from one year to the other. And there is a time lag between increase in demand and increase in production, as it takes at least three to five years on average for new trees to get into the production line. This time lag further contributes to the price volatility. Price volatility is magnified by trading of cocoa on futures markets where speculation is large - the volume of cocoa traded in futures contracts is ten times higher than the actual world production. The attention paid by speculators to the world's commodity markets, including agriculture, may also help to explain the co-incident spikes of many agricultural commodities including cocoa in the period 2007 to 2009 (Figure 2b) as well as the recent return of high prices.

Price volatility makes cocoa farmers vulnerable as they never know in advance what the international price will be when the harvest comes and cannot plan their production accordingly. Production prices are generally low due to lack of efficiency and fairness along the cocoa supply chain. World production of cocoa increased rapidly over the course of the 20th century, prompted by rising demand. From an annual production of less than 125,000 tonnes in the early 20th century, the annual global output has risen to 3.5 million tonnes at present, an annual average growth rate of 3.4%. Interestingly, this number is fairly similar to the growth in demand mentioned earlier, resulting in a fairly stable picture of the supply- demand balance

The remainder of the paper is as follows; A review of the literature related to agricultural commodities; description of the two state- variable model that we wish to apply to cocoa futures; the display of calibration results for the period of 1980 to 2011; conclusion.

LITERATURE REVIEW

The cocoa spot and futures markets

The International Cocoa Standards require cocoa of

merchantable quality to be fermented, thoroughly dry, free from smoky beans, free from abnormal odors and any evidence of alteration. It must be reasonably free from living insects, broken beans, fragments and pieces of shell and foreign matter and reasonably uniform in size. Throughout the world, the standards against which all cocoa is measured are those of Ghana cocoa, Ghana producing the best type of cocoa. Cocoa is graded on the basis of the count of defective beans in the cut test (source: UNCTAD- information on cocoa). Nearly all cocoa sold in the spot market either from farmers' associations to merchants or from individual farmers to a chain of middle men who then sell to exporters.

Forward and futures markets

While cocoa in the spot market is immediately delivered, there is a lag between the transaction and the physical delivery in the case of a forward trade. As a brief reminder, we recall that a forward contract signed between a buyer and a seller at date 0 involves no cash flow nor any exchange of goods, but just an agreement about the 'forward price', number of dollars to be paid at a fixed maturity T by the buyer to get delivery of the commodity by the seller of the forward contract. A future contract is a similar transaction, except that now both buyer and seller interact with the exchange, whose clearing house is demanding margin deposits from both parties before any trade is placed, and *margin calls* to be paid on a daily basis by the position whose value went down from one business day to the other (namely, the buyer of the future in the case of a decline in future prices). This 'marking-to- market' of the positions through margin calls ensures the integrity of the clearing house; the existence of the exchange platform when trading futures contracts takes away the counterparty risk present in forward contracts. We can observe that, in agricultural commodity markets, not only are futures contracts quite liquid, but forward contracts as well since the old market players like Cargill or Louis Dreyfus trust each other to enter a forward transaction and gain the benefit of the secrecy of their trade that would otherwise be reflected in the change of the *open interest* (namely the number of existing contracts with a buyer and a seller at each end) posted by the Exchange on a daily basis.

Like all other futures, cocoa futures are standardized commitments in which the contract buyer agrees to take delivery from the seller a specific quantity of cocoa, namely 10 tonnes, the *contract size* for cocoa at the future price prevailing on the purchase day, for delivery date in March, May, July, September or December. These contracts allow on the one hand producers to secure ex-ante their crop revenues and on the other hand investors to get exposure to changes in cocoa prices while avoiding the hurdles of storage. There are only two places where cocoa futures contracts are traded: the London International Financial Futures Exchange (LIFFE) in the UK, and the ICE[2] futures in the United States. These organized exchanges provide the facility and trading platform that bring buyers and sellers together. The last trading day is eleven business days prior to the last business day of delivery month.

The price of the future contract at any time is the value reached by the bids of buyers and sellers and reflects the anticipation of the spot price that will prevail at the expiration of the future contract. Factors entering these bids include information about the crop outlook, whether the maturity of the contract belongs to the harvest period, as well as world cocoa inventory levels and global demand. Hence, the terminology of *'price discovery'* brought about spot prices in the future by the daily values of Futures contracts posted by the exchange

Inventory and consumption

Once harvested, fermented, dried and transported, the cocoa bean is then stored for only a very short time before it is processed in readiness for commercial consumption. The storage takes place in jute bags for brief periods; whole beans can be stored for 5 to 6 months safely. Storage extended over more than 6 months may result in loss of quality due to problems of insect infestation, mold contamination and moisture exchange between atmosphere and the beans. Other problems of prolonged storage are beans size variation. If the water content is less than 6% of the bean weight, cocoa beans become brittle, while a water content of greater than 8% may cause the risk of vapor and mold damage. Quality differences and weight, obviously, have a bearing on the trading value.

In general, the amount of world grinding of cocoa beans is used as a proxy for world cocoa consumption. Although largely produced in developing countries, cocoa consumption is concentrated in industrialized countries (Figure 1b). World demand for cocoa and cocoa products has been in a steady rise since the late 1970s (Figure 3a). Worldwide consumption has increased in tandem. Worldwide stocks, measured at the end of the crop year in late September, just before the major harvests of West Africa, have remained relatively steady. A useful statistic is the so-called 'stocks to grindings ratio' (called 'year-end carryover' in most agricultural commodities and constituting an important indicator for the strategies of commodity trading advisors), representing the percentage of one year's consumption remaining in stock just prior to the new harvest. This peaked in 1990 at 70% and has declined gradually to 45% since then (Figure 3b), as consumption has increased with no commensurate increase in stocks. However, 45% should still be considered ample carryover, compared with the 2011

2. The Inter Continental Exchange (ICE) was first established in Atlanta and has acquired other exchanges like the International Petroleum Exchange in London.

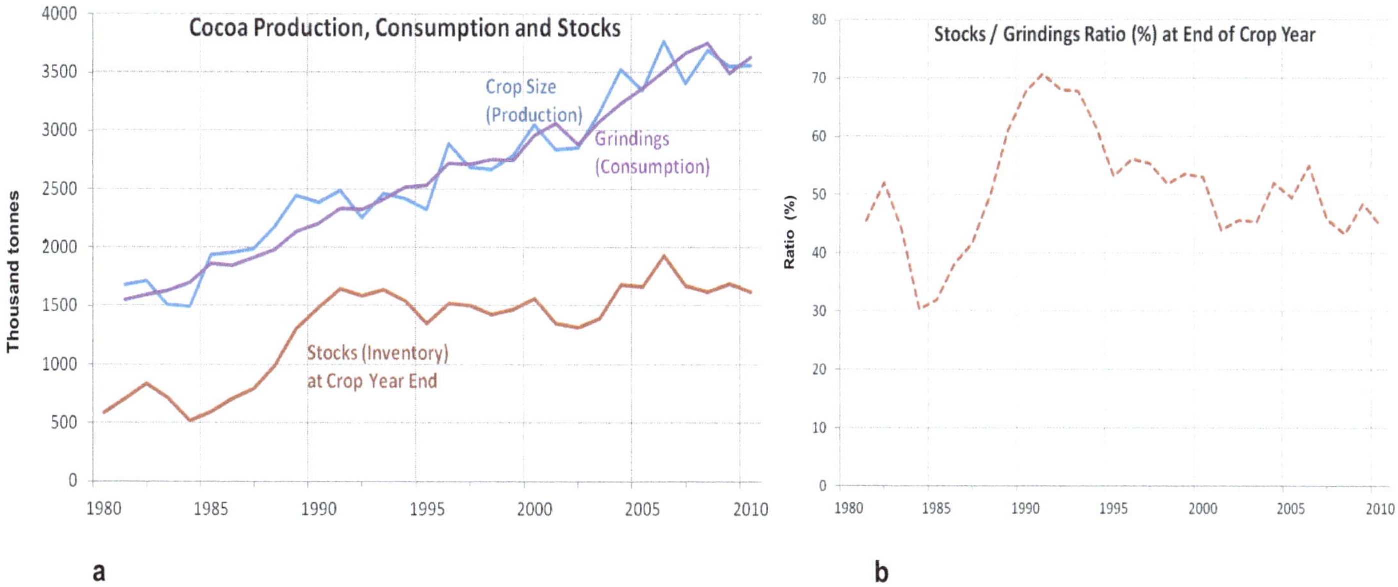

Figure 3. (a) Cocoa production, consumption ('grindings') and year-end stocks (Source: International Cocoa Organization) (b), stocks to grindings ratio, a measure of sufficient inventory.

worldwide estimates for wheat (30%), corn (16%) and rice (22%) (International Grain Council, 2011).

Besides the well- known cocoa bean, there are four other intermediate cocoa products - cocoa liquor, cocoa butter, cocoa cake and cocoa powder - and chocolate. Although the market for chocolate is the largest user of cocoa in terms of beans equivalent, intermediate products such as cocoa powder and cocoa butter are used in several areas such as the manufacturing of tobacco, soap and cosmetics.

It is worthwhile to note that the ranking of cocoa importing countries depends on the composition of the goods imported: trade is not only tracked by cocoa beans but also by semi-finished products of cocoa. The information provider global trade atlas tracks cocoa import data such as cocoa shells and cocoa paste. The Netherlands, as one of the main ports into Europe, leads in imports of beans; the United States, with significant production of cocoa complementary food products, leads in imports of powder; and France, one of the biggest chocolate markets in terms of consumption, leads in retail chocolate (Source: World Cocoa Foundation).

Prices of primary agricultural commodities used to be characterized by bursts in short- term prices and low prices on the long term. Cocoa prices, like many other commodity prices, have on the average been on the rise in the last few years as a result of factors such as excess world demand over supply following poor weather; under investment and political unrest are also persistent in some of the major producing countries.

The production of agricultural commodities continues to play a major economic role in many developing countries. Over the decades 1970 to 2000, prices of agricultural commodities on the whole have shown a pattern of long term price falls and short term price instability (International Monetary Fund, 2000). Further evidence of lack of any positive upward trend in commodity prices in the 1980s and 1990s can be found in Grilli and Yang (1988) and Cuddington (1992). Deaton (1999) shows that cocoa and coffee prices in his period of analysis were lower relative to the United States consumer price index than they were a century ago. Since October 2000 when prices reached a 27-year low, cocoa has seen steady price increases as displayed in Figure 2. These increases in prices have been attributed to poor harvests, speculative activities and a general strong demand growth.

Seasonality plays an essential role in managing risk in agricultural commodities and it must be carefully analyzed when modeling cocoa spot and futures prices. The seasonal behavior of many commodity prices has been documented in numerous studies. Sorensen (2002) considers the evolution of agricultural commodity futures (corn, soybean, and wheat) by adding a deterministic seasonal price component to the two-factor model of Schwartz and Smith (2000). Geman and Nguyen (2005) model soybean prices through a deterministic seasonal component and two state variables representing the spot price and spot volatility, the latter being a proxy for inventory.

Choi and Longstaff (1985) note that there exists seasonality in most agricultural markets, the supply of goods being determined by harvesting cycles. As the degree of price uncertainty changes through the year, the volatility of commodity future prices shows a strong seasonal pattern. In the period prior to the harvest when the amount of crops is unknown, the price uncertainty is higher than after the harvest when crop yields are known to the market participants. Hence, the price volatility increases from the previous harvest to the information arrival on the new one, resulting in a seasonal pattern in volatility in addition to the price level seasonality. For instance, the severe drought in the former Soviet Union combined with floods in Canada during the year 2010 created a severe rise in wheat prices together with a high volatility as of the beginning of June, a long time before the harvest.

Many other factors influence the price of agricultural commodities, including natural disasters, political intervention like subsidies, and diseases. These factors ensure that agricultural prices rarely trade in a similar manner to one another or to other asset classes. Hence they provide to investors unique diversification benefits. In recent years, while supply has been constrained by a number of factors including land and water scarcity, demand has been driven further up by rapidly developing emerging markets as well as an increased consumption in developed countries.

Earlier studies have modeled seasonal trends in agricultural commodity prices by decomposing economic time series into their temporary and permanent components or detrending by filtering. Whereas the former ones are unable to capture the underlying data generation process, the latter are either one-sided filters or based on *ad hoc* procedures.

Historically, the majority of the work related to the modeling of seasonality in prices has concentrated on the seasonal pattern in the spot price; but this spot price is not observable for many commodities. Moreover, conclusions from such studies are mixed. Kaldor (1940) concludes that "supply curves of agricultural commodities are much less elastic and weather effects cause frequent and unpredictable shifts in the supply curve". In other words, producers are not able to rapidly adjust the supply level in response to price changes because of the time involved in planting new areas , while exogenous factors (for example, weather) do have random effects on supply levels. Another reason given by Fama and French (1987) is the high cost of storage due to their perishable nature. Other authors have pointed out that the dynamics of supply and demand result in a mean-reverting behavior of commodity prices (Brennan (1991), Gibson and Schwartz (1990).

In recent years, there has been a growing importance in obtaining reliable forward curves for commodities because of their price discovery virtues. Agricultural commodities are too perishable to be stored for long time periods or there may simply not be a cash-and-carry market at the time of trading. In such markets, commodity forward curves have become increasingly popular and important, in particular because of the key role of Futures contracts in hedging activities and the information conveyed in their prices. We can note that cocoa crop forecasters are known as 'pod counters' as they literally travel around the plantations in West Africa counting the number of cocoa pods in each tree to estimate output. Consequently, at any given date, there is a component of 'true expectation' of future spot prices present in the Future price since this one results from the confrontation of buy and sell orders for a given horizon date.

Hedging commodity price risk has been a standard activity for farmers and producers eager to protect their revenues by selling futures. In the early days of the Chicago Board of Trade (CBOT, 1848), the buyers of these futures used to be 'speculators' willing to take risk in order to make profits, and so providing liquidity to the hedgers. Today, they include hedge funds, pension funds and all actors wishing to be exposed to commodity prices, both for their diversification benefits and long-term returns due to the explosive growth of demand for food in a growing world population, as well as the biofuels sector. In such markets, the analysis of commodity forward curves has become increasingly important as the market players, hedgers (farmers and agrifood companies) and financial players consider the whole spectrum of liquid maturities available to them.

Seasonality is obviously one of the features which distinguish commodities from equity and fixed income markets. Its presence in the spot prices of agricultural and some energy commodities, naturally arising from seasonal patterns in supply (for example, harvest) and demand (for example, cold weather) is relatively well understood. It has been studied by a number of authors, for example, Milonas (1991) and is usually represented by a deterministic component. For the remaining part, and its possible representation by a factor model, the first state-variable has always been in the literature on the subject the commodity price. However, this spot price combines seasonal and stochastic features and may be unobservable. We will propose in this paper to use instead the average value of liquid contracts forward price, which is devoid of seasonality and conveys a more robust representation of the forward curve level. The second factor will be a quantity analogous to the stochastic convenience yield and account for the random changes in the shape of the forward curve while seasonality will be classically accounted for by a deterministic seasonal premium.

In the case of cocoa like for other commodities, the shape of the forward curve depicts anticipated prices and provides insight for hedging, storage and production decisions (Geman, 2005). Figure 4 shows a small sample of cocoa forward curves from the LIFFE.

On a number of observation dates, cocoa futures prices as observed on the LIFFE tend to be higher for deliveries in May, July and September (highest in July) than in March and December. This price trend is traced to the harvesting cycle of cocoa from the major producing countries as supply drives seasonality, like in most agricultural commodity markets, and the risk premium is highest in July when there is release of the information

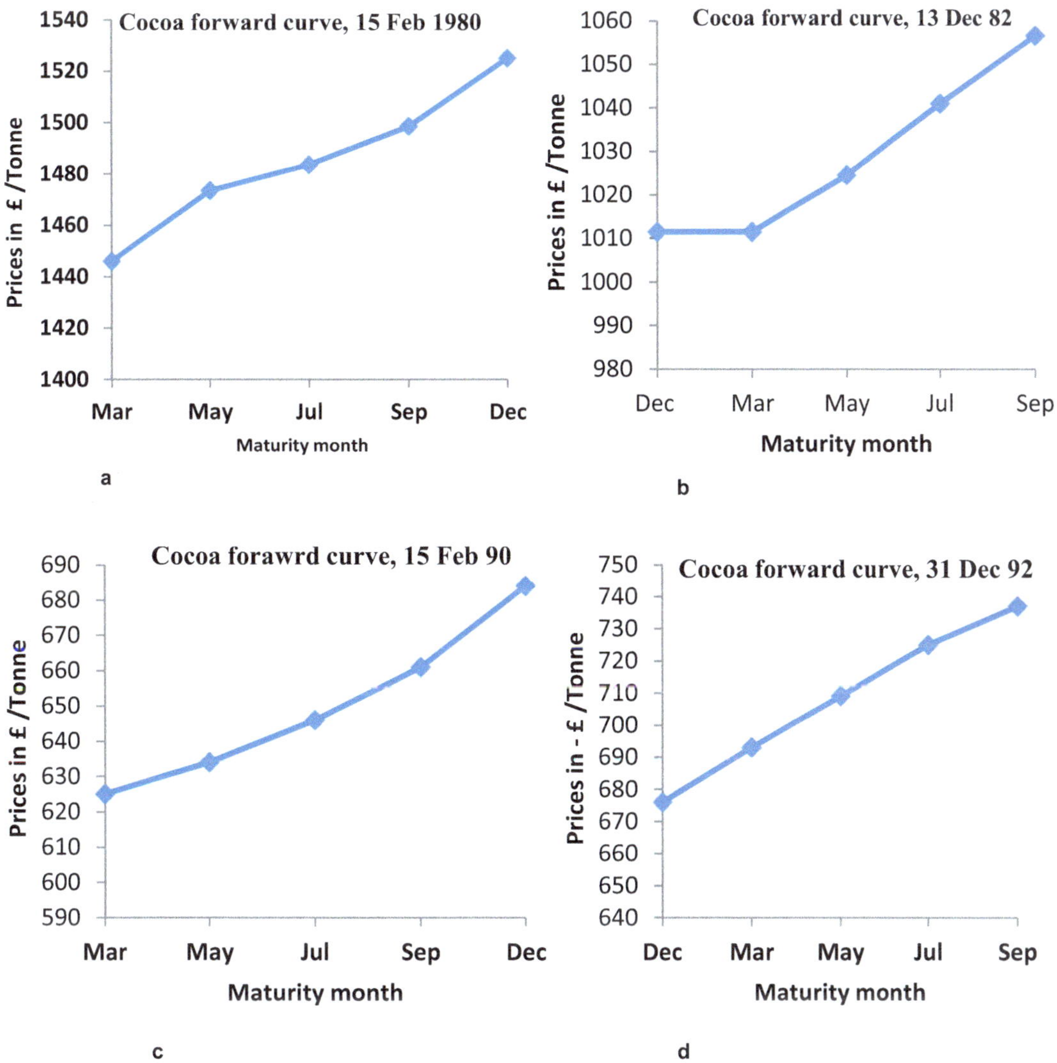

Figure 4. A sample of cocoa forward curves.

about the next harvest.

In general, we see strong seasonal price premia in those commodities which have strong seasonality in supply and/or demand. The price seasonality is more pronounced when inventories are low, since in these cases extreme shortage may occur just prior to a new harvest. However, in the case of cocoa there is only moderate seasonality in supply, because of the long harvest period of around 5 months. Compounding this, cocoa trees normally produce a smaller 'mid-crop' in between the main crops, accounting for around 20% of total production. In Table 1 we tabulate the main-crop months ('X') and mid-crop months ('o') for the main producing countries (UNCTAD, 2011). There are only 5 months of the year when main-crop harvesting is not occurring, and even then the mid-crop is being harvested. The combination of relatively high stocks and only weak seasonality explain why we find only mild seasonal premia in our later analysis.

As said before, the modeling framework aims at separating seasonal from stochastic features in cocoa forward curves. We follow Borovkova and Geman (2006) - hereafter BG - in not using the spot price as a state variable and introduce instead the average value of liquid forward contracts. Our framework is based on their approach and applied to cocoa futures data.

The price of any forward contract under the BG model is represented as the sum of an average forward price – meant to convey a global representation of the current forward curve, a deterministic seasonal component and a random component analogous to a stochastic convenience yield accounting for the random changes in the shape of the forward curve. Modeling the seasonal component as a deterministic variable is possible since

Table 1. Main-crop harvest months ('X') and smaller mid-crop harvest months ('o') for the main producing countries (UNCTAD, 2011).

Country	Proportion of worldwide production (%)	Calendar Month of Harvest											
		J	F	M	A	M	J	J	A	S	O	N	D
Ivory Coast	35	X	X	X		o	o	o	o		X	X	X
Ghana	19	X	X	X		o	o	o	o	X	X	X	X
Indonesia	14			o	o	o	o	o		X	X	X	X

the stochastic elements in seasonality will be embedded in the stochastic term. As we shall see, this representation significantly improves the well known cost-of-carry relationship.

Practical application exists in policy strategies meant to manage effects like price declines and volatilities in agriculture - dependent countries, many of which are among the least developed in the world.

MODEL DESCRIPTION

A factor model of the forward curve term structure allows one to capture stylized facts that are observed in Futures contracts of different maturities and at different moments in time in the most parsimonious and tractable form, namely through the dynamics of a small number of *state variables*. Term structure models were first developed in the fundamental case of interest rates, in order to account for the 'consistencies' between maturities implied by the no arbitrage assumption on the changes in spot rates. The same difficult problem holds for the term structure of commodity forward prices, with the further hurdle that such characteristics as seasonality and physical properties related to storability must be taken into account. For decades and today more than ever, has there been an acute interest in understanding and interpreting patterns of contango, backwardation, or kinks (existence of an inflection point in the curve) in commodity forward curves, from crude oil to gold or agricultural commodities.

The traditional theory of storage as originally proposed by Working (1932) provides a link between the term structure of futures prices and the level of inventories. This link, also known as "cost of carry relationship", establishes that in order to induce storage, futures prices have to be high enough over time to compensate inventory holders for the costs associated with storage. In a situation of contango, the futures price corresponds to the spot price increased by a positive storage cost. This explanation does not apply to explain the backwardation shape since the cost of storage is in nonnegative; hence, the necessity of introducing a *convenience yield* related to inventories and translating the benefit of owning the physical commodity readily available in the case of disruption in the production as opposed to a paper contract. The term convenience yield was introduced by the famous economist Kaldor (1939). It appeared in a finance setting in the spot- forward relationship proposed by Brennan and Schwartz (1985):

$$F(t,T) = S(t)e^{(r+c-y)(T-t)} \qquad (1)$$

whereby a futures contract at time t, for delivery at time T, is denoted by $F(t,T)$ and is a function of the current spot price S, foregone interest rate r and convenience yield y (which are both assumed constant[3] over time in this paper without any loss of generality), and cost of storage c. Note that c and r are positive; y is positive or negative depending on the period. The case $r+c-y<0$ gives rise to backwardation and corresponds to a positive net convenience yield[4]. The contango shape occurs when $r+c-y$ is positive. Both shapes obviously reflect the situation at the date of analysis of supply, demand and inventory, as well as the anticipations of production and geopolitical factors in the future period, reflected in forward prices.

A look at historical data on daily quote for cocoa futures prices for the five nearby maturities in LIFFE shows the forward curve since 1980 to the late 2010 (except in a few instances) has generally been in backwardation, which means futures contracts further in maturity are priced higher (Figure 4 and A5 to A8 in the appendix).

In a general situation as considered in Gibson

[3] Note that all the results of the paper can easily be extended to stochastic interest rates following the methodology introduced in Geman (1989).

[4] As introduced by Kaldor (1939) and Working (1948), y accounts for the benefit of owning the physical commodity rather than a Futures contract written on it. The crucial value of this relationship is fully appreciated by the market participants. Defining the convenience yield y as net of cost of storage shows that the forward curve at date t is an increasing or decreasing function of the maturity T, depending on the sign of $(r-y)$, respectively called contango or backwardation.

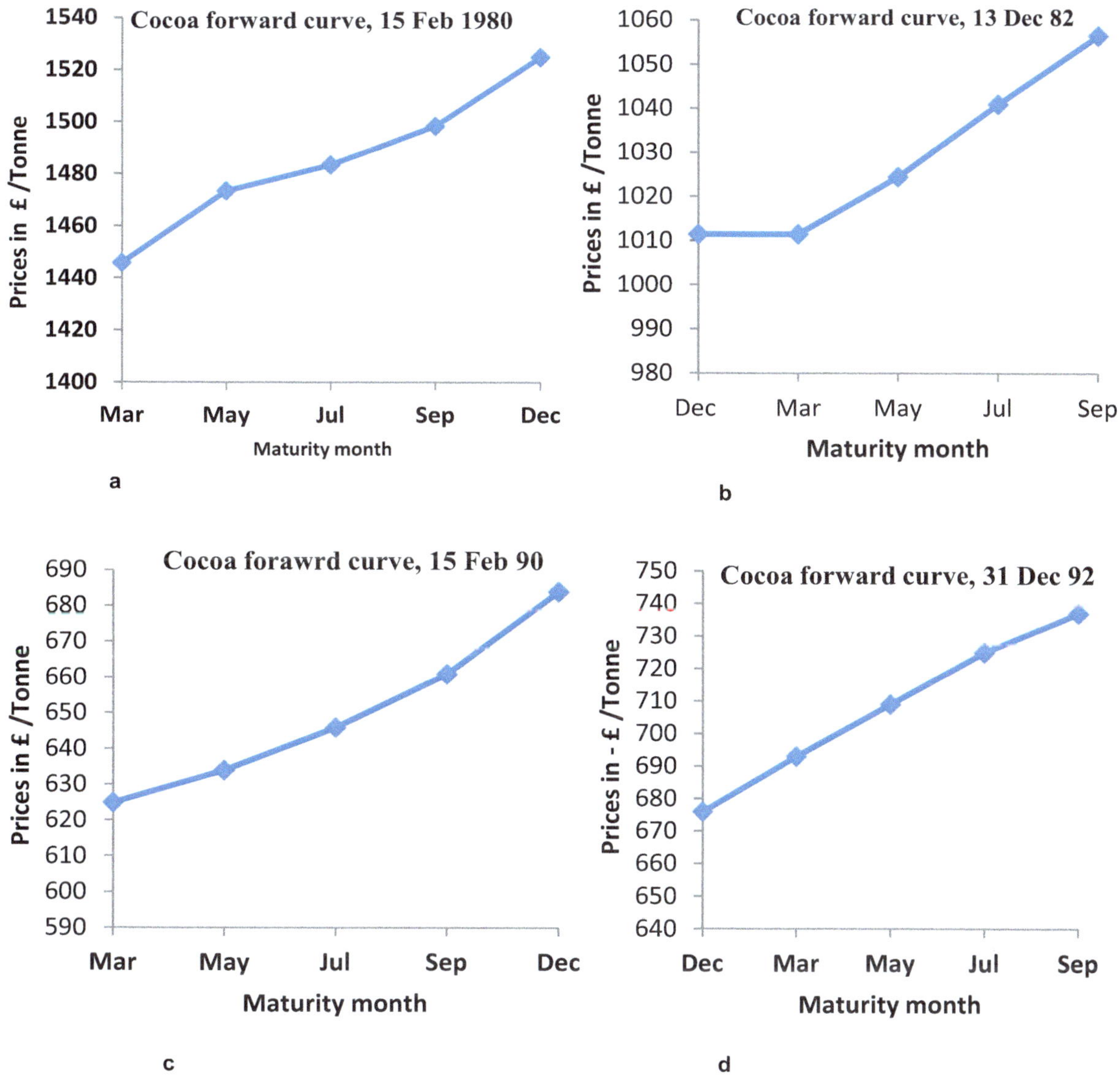

Figure 4. A sample of cocoa forward curves.

and Schwartz (1990), Schwartz (1997) and Litzenberger and Rabinowitz (1995), the convenience yield is defined as a stochastic process, *y(t)*. This conveys the fact that the "reward" received by the holder of the physical commodity changes with the world inventories and, in turn, economic agents' preferences for the physical good rather than a paper contract.

$$F(t,T) = S(t)e^{[r(t)-y(t)](T-t)} \quad (2)$$

Seasonality in the Futures price $F(t,T)$ lies in the cost of carry relationship and is embedded in the dependence on T of the convenience yield y related to the period (t,T). This is in sharp contrast with the case of a stock making a continuous dividend payment at the rate $g(t)$ as in Merton (1973), where the spot-forward relationship is given by;

$$F(t,T) = S(t)e^{[r(t)-g(t)](T-t)} \quad (3)$$

and where the rate $g(t)$ may vary with time but does not exhibit any dependence on the maturity T. We can rewrite equation (2) as:

$$F(t,T) = S(t)e^{[r(t)-y(t,T)](T-t)} \quad (4)$$

This expression reflects the character of time-spread

option embedded in convenience yield, as discussed for instance in Heinkel et al. (1990).

Since defined by the intersection of futures buyers and sellers' quotes, a futures price conveys the expectation, conditionally to the available information at a date t, of the spot price at a future date. Futures prices of seasonal commodities are also driven by stochastic factors other than such extreme weather circumstances (outside the average seasonal pattern), political crises within producing countries, market risk aversion among the actors. We choose to express these external influences by a stochastic convenience yield, while directly incorporating a deterministic seasonal premium within the convenience yield.

Seasonal cost-of-carry model[5]

The first state variable is the average level of the forward curve, which we denote by $\overline{F}(t)$, defined as the geometric average of the current forward prices since we want to be in line with the geometric Brownian motion representation of prices in the classical finance literature- the assumption of geometric Brownian motion itself for spot or forward price being totally unnecessary in this paper.

$$\overline{F}(t)=\left(\prod_{T=1}^{N}F(t,T)\right)^{\frac{1}{N}} \tag{5}$$

where *N* is the most distant liquid maturity.

For the seasonal cost-of-carry model for any maturity T, we write

$$F(t,T)=\overline{F}(t)e^{s(T)-\gamma(t,T-t)(T-t)} \tag{6}$$

or equivalently,

$$\ln F(t,T)=\ln\overline{F}(t)+s(T)-\tau\gamma(t,\tau) \tag{7}$$

where $s(T)$ is the deterministic seasonal premium, and $\gamma(t,\tau)$, $\tau=T-t$, defined by the relationship above, is the second state variable, the *stochastic convenience yield net of seasonal premium*, observed at date t for time to maturity T.

Note that, in the seasonal cost-of-carry model above, we have separated the dependence on the maturity date *T* from the dependence on time to maturity $\tau=T-t$.

[5] Readers are referred to Borovkova-Geman (2006) for the use of this model in energy markets

The *maturity date* (calendar month in fact) influences the futures price via the seasonal premium s(T), while the time-to-maturity τ affects the futures price via the stochastic convenience yield $\gamma(t,\tau)$, aggregated up to maturity in the product $\tau\gamma(t,T)$. The average forward price $\overline{F}(t)$ only depends on the current date *t* of observation. The choice of this first state variable is a key feature of our approach.

Next, we specify the dynamics of the state variables. Defining $X(t)=\ln\overline{F}(t)$, we describe the dynamics of $X(t)$ and $\gamma(t)$ under the real probability measure P (no arguments of "risk-neutrality" are necessary in the developments below) by the stochastic differential equations

$$dX(t)=\alpha(\mu-X(t))dt+\sigma dW_1(t) \tag{8}$$

$$d\gamma^{\tau}(t)=-a^{\tau}\gamma^{\tau}(t)dt+\eta^{\tau}dW_2 \tag{9}$$

where σ, the volatility of $X(t)$ is assumed to be constant (but, brought together with volatilities of γ^{τ}, still leads to a term structure of Futures prices volatilities, as shown below in equation (8)). The convenience yields of all maturities $\gamma^{\tau}(t)$ are subject to a single source of uncertainty, represented by the Brownian motion W_2, uncorrelated with the Brownian motion W_1 driving the average forward price. We view γ^{τ} as fluctuating around zero over time, since shocks to inventories get eventually absorbed by adjustment of the production; $X(t)$ reverts to a long-term value *μ*, here assumed to be constant. The speeds of mean reversion are a^{τ} and α, respectively. The set (η^{τ}) $\tau=1,2,...,N$ represents the stochastic convenience yield volatilities for different maturities. Note that the assumption of independence of W_1 and W_2 is totally reasonable given the way the state variables were defined.

Substituting the stochastic differential equations into equation (6), we can derive the dynamics of the Futures log-prices $Y(t,T)=\ln F(t,T)$ under the real probability measure:

$$dY(t,T)=\left[\alpha(\mu-X(t))+\gamma^{\tau}(t)(a^{\tau}\tau+1)\right]dt+\sigma dW_1(t)-\eta^{\tau}\tau dW_2(t) \tag{10}$$

So $Y(t,T)$ is obtained by integrating the above differential

Table 2. Parameters of the mean-reverting process $X(t) = \ln \overline{F}(t)$ for the period February 1980 to March 2009.

Parameter	Calibration result	Annualized value
α	0.001143	0.2857
μ	6.9367	6.9367
σ	0.014544	0.2300

equation with the initial condition $F(0,T) = \overline{F}(0)e^{s(T)-\gamma^{\tau}(0)(T-0)}$

Then $F(t,T)$ is log-normally distributed with variance

$$\delta^2(t,\tau) = \sigma^2 + (\eta^{\tau}\tau)^2, \qquad \tau = T - t \tag{11}$$

APPLICATION TO COCOA FUTURES AND EMPIRICAL RESULTS

We obtained daily futures price data from LIFFE, in British Sterling Pounds per tonne, for the period 1980 to 2009, and calibrate using the technique as described in Borovkova and Geman (2006). In total 7583 days' of observations were used. We restricted our analysis to the first 5 contracts, comprising a futures curve of one year, since only the months of March, May, July, September and December trade on LIFFE.

Results of the calibration for the mean-reverting log-price process $X(t) = \ln \overline{F}(t)$, Equation (8), are presented in Table 2. The mean reversion speed α of 0.2857 indicates that the log-price (and hence the price) reverts to the mean in 1/0.2857 years, or in 3.5 years. This is a typical result for mean reversion in commodity prices Bessembinder et al (1995). The long term reversion value μ of 6.9367 for the log-price process corresponds to 1029 for the long-term reversion value of the price process itself. Note that the units of price are £ sterling as traded on LIFFE, compare with Figure 2(a) with cocoa prices in US dollars. Finally, the annualized volatility of 0.23 (23%) confirms cocoa as having moderate price volatility, typical of agricultural commodities. Note that the averaging effect of computing $\overline{F}(t)$ leads to a lower volatility value than would normally be calculated as the volatility of the spot price, using the first-to-mature futures price as a proxy. This is for two reasons. Firstly, the 'Samuelson effect' leads the first-to-mature futures price to be more volatile than later contracts, which $\overline{F}(t)$ includes. Secondly, since the daily changes to the empirical prices $F(t,T)$ are not perfectly correlated across all the contracts, the averaging procedure results in a lower overall volatility than the volatility of any one contract. This stability of $\overline{F}(t)$, and its explicit lack of seasonality, are two of the benefits of the BG (2006) model.

Figure 5a shows the seasonal premia for cocoa futures prices ranges (-0.00365 to 0.002868). The seasonal premia in coffee futures are supplied for comparison purposes in Figure 5b.

In the case of cocoa, the seasonal premia $s(T)$ are very small; they represent the proportion by which a given month exceeds the average forward price $\overline{F}$. As expected, the seasonal premium is negative in December and March, which represents the major harvesting period for cocoa in West Africa. Futures expiring in May, July or September are at premium with respect to the average futures price level. This is also in line with the mid-crop or lean harvest when demand for cocoa is expected to outstrip supply and stocks are being depleted. Based on these small but varying premia, we can infer (as is the case) that the main harvest occurs between September and December when the premium goes from positive to negative. The premium is highest in July when inventories are at their lowest and uncertainty about the next main crop is maximal (the time called 'weather period' in the wheat market, indicating the importance of any weather news). In September, the harvest has started and more precise levels of crop outlook are available, hence a reduction in the uncertainty risk premium, although some uncertainty still remains since the remainder of the harvest period could still experience disruption.

In the case of coffee, we see higher seasonal premia. Coffee together with cocoa and tea form what are generally termed 'the tropical beverages'. Both cocoa and coffee are produced in the tropics. Coffee, however, is harvested in only specific months of the year, with a clear harvest and no-harvest period. This contributes to the greater seasonal premia. The values in Figure 5(b) implies a mean premium of 1.5% for September coffee over the mean, and a May discount of 1%. These are still low compared to the extreme seasonality noted in BG (2006) for the case of natural gas, in which December expiry gas futures exceeded the mean by almost 30%.

We obtain the estimates for the stochastic convenience yield series for all available times to maturity from the residuals of equation (10) as:

$$\gamma(t,\tau) = \frac{1}{\tau}[\ln \overline{F}(t) - lnF(t,T) + s(T)]$$

and display these in Table 3. The values of a^{τ} a, the mean reversion speeds for the various maturities, indicate rapid reversion in $\frac{1}{a^{\tau}}$ years, with the annualized values corresponding to reversion speeds between 0.174 and 0.237 years, that is, the stochastic convenience yield - net of seasonal premium - reverts rapidly to 0. Thus we

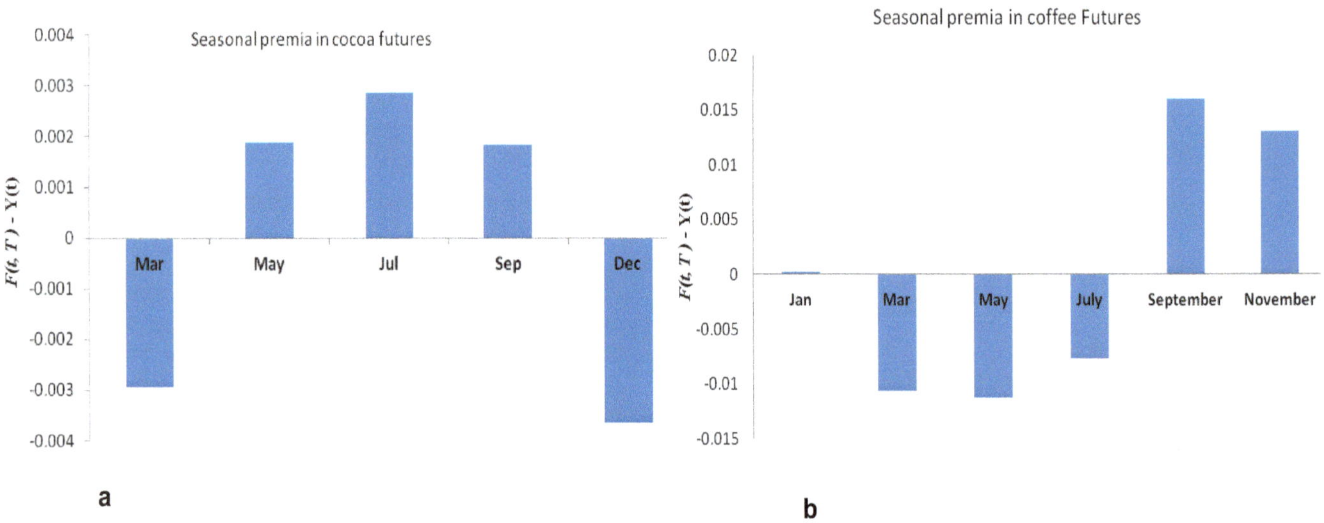

Figure 5. (a) The set of Seasonal Premia $s(T)$ obtained for cocoa; (b), the set of seasonal premia $s(T)$ obtained for coffee analyzed over an identical study period.

see that the overall premium of a given log-futures price $F(t,T)$ over the average value $\overline{F}(t)$, while stochastic, reverts rapidly to that calibrated as the fixed seasonal premium $s(T)$, lending credence to the utility of the BG (2006) model.

Figure 6a depicts the version of the "Samuelson effect" corresponding to our setting, namely the decreasing volatility of the stochastic convenience yield with the time to maturity of the future contract, as observed by BG (2006) for natural gas and electricity. For the sake of comparison and in order to exhibit another instance of this property in the class of 'softs', we plot in Figure 6b the term structure of the convenience yield volatilities in the case of coffee. These results tells us that convenience yield varies more rapidly as we approach maturity, just as price itself varies more rapidly in the 'original' Samuelson effect, which refers to the volatility of price.

Conclusion

Agricultural commodities represent a fascinating asset class because of their indispensable place in human life and from a theoretical standpoint. Having a better understanding of the evolution over time of the forward curve is crucial for farmers and agrifood companies who wish to hedge their revenues across the year and investors who wish to get exposure to the commodity markets.

In the case of cocoa, a food crop which is consumed in various forms around the world, we have shown that a two- state variable model which avoids the opaque spot price, allows one, together with seasonality, to calibrate in a very satisfactory manner the data for the period extending from 1980 to 2009. Prices of food crops typically follow a seasonal pattern, falling immediately after the harvest and rising thereafter until the next harvest, as farmers and merchants store some supplies to meet consumer demand throughout the year. Cocoa however has a very mild seasonal price trend because it is harvested over the entire year (varying only in harvest quantities between major and minor harvests).

The knowledge of the seasonal pattern of commodity prices is essential to effective potential price stabilization policies on the world market. For economies that depend on cocoa export for foreign revenue, knowledge of the seasonal pattern of cocoa price and its magnitude is essential to the formulation of effective policies for stabilizing foreign revenue and the income of farmers. The results of our analysis provide a meaningful start point for cocoa price policy. Although cocoa prices currently display only weak seasonality, we expect the seasonality would rise if the stocks to grindings ratio were to fall substantially. However, despite most production being concentrated in West Africa, the length of the harvest period, and the presence of the inter-harvest mid-crop secondary harvest, means that cocoa will never display the strong seasonality witnessed currently in the case of corn or wheat.

The cocoa market is subject to a moderate degree of volatility due to global cocoa production being limited to just eight countries (that is, eight countries serving the global demand). This presents attractive hedging and trading opportunities for cocoa traders around the world. The results of our analysis provide traders one less factor to worry about as seasonal trend in cocoa price is empirically shown to be insignificant.

Table 3. Gamma parameters for the period February 1980 to March 2009.

Contract maturity	Calibrated values			
	a , daily	*a*, annualized	η	η, annualized
March	0.018051	4.513	0.022871	0.361
May	0.022197	5.549	0.015536	0.245
July	0.022976	5.744	0.01136	0.179
September	0.017917	4.479	0.008268	0.131
December	0.016854	4.213	0.006503	0.103

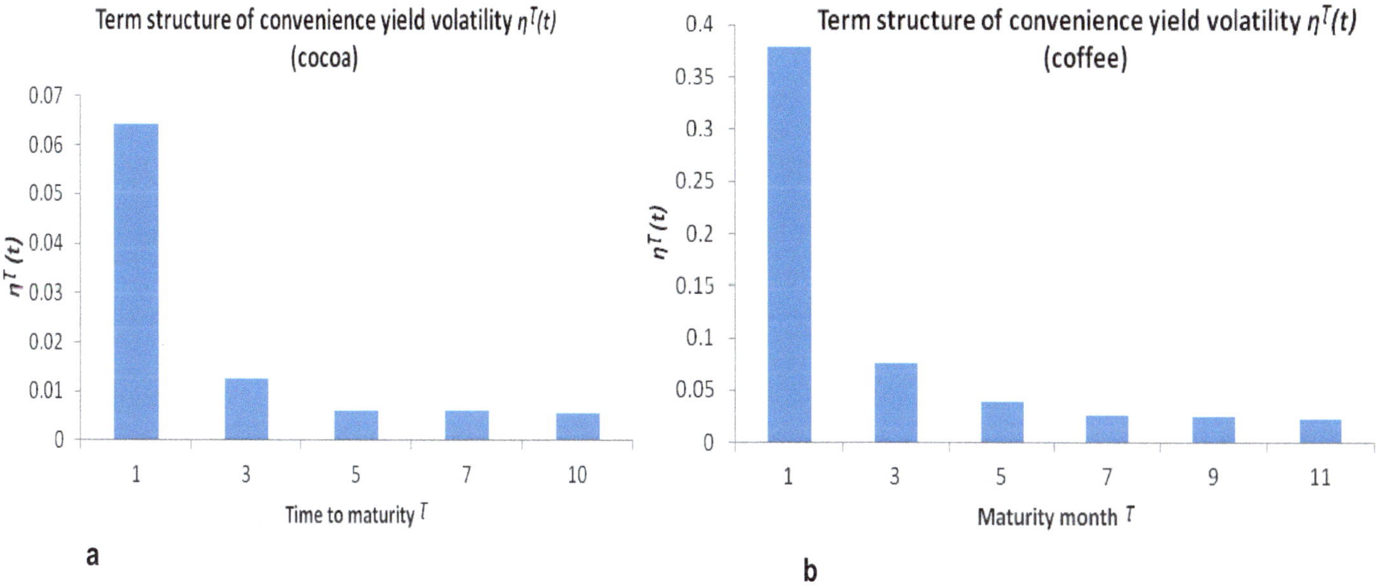

Figure 6. (a) Term structure of stochastic convenience yield volatility for cocoa; (b), term structure of stochastic convenience yield volatility for coffee.

REFERENCES

Bessembinder H, Coughenour JF, Seguin PJ, Smoller MM (1995). Mean Reversion in Equilibrium Asset Prices: Evidence from the Futures Term Structure. J. Finance, 3: 361-375.

Borovkova S, Geman H (2006). Seasonal and Stochastic Effects in Commodity Forward Curves. Review of Derivatives Research, 9: 167-186.

Deaton A (1999). Commodity Prices and Growth in Africa. J. Economic Perspectives, 13: 23-40.

Fama EF, French KR (1987). Commodity Futures Prices: Some Evidence of Forecast Power, Premiums and the Theory of Storage. J. Business, 60: 1.

Geman H (2005). Commodities and Commodity Derivatives. Wiley Finance, pp. 35-39.

Geman H, Nguyen V (2005). Soybean Inventory and Forward Curve Dynamics. Manage. Sci., 51(7): 1076-1091.

Gibson R, Schwartz ES (1990). Stochastic Convenience Yield and the Pricing of Oil Contingent Claims. J. Finance, 45(3): 959-976.

Grilli RE, Yang MC (1988). Commodity Prices, Manufactured Goods Prices, and the terms of trade of developing countries. World Bank Economic Rev., 2: 1-48.

Heinkel R, Howe ME, Huges JS (1990). Commodity Convenience yield as an option profit. J. Futures Markets, 10(5): 519-533.

International Grain Council (2011). Grain Market Report, 3rd November 2011, pp. 1-4.

Kaldor N (1939). Speculation and Economic Stability. Rev. Econ. Stud., 7: 1.

Kaldor N (1940). A Note on the Theory of the Forward Market. Rev. Econ. Stud., 7: 3.

Litzenberger RH, Rabinowitz N (1995). Backwardation in oil futures markets: Theory and empirical evidence. J. Finance. 50(5): 1517-1545

Milonas NT (1991). Measuring Seasonalities in Commodity Markets and the half-month effect. J. Futures Markets, 11(3): 331-346.

Schwartz ES, Smith JE (2000). Short-term variations and long-term dynamics in commodity prices. Manage. Sci., 46(7). 893-911.

Sorensen C (2002). Modeling Seasonality in Agricultural Commodity Futures. J. Futures Markets, 22(5): 393-426.

UNCTAD (2011). InfoComm – Market Information in the Commodities Area: Cocoa Crop, http://www.unctad.org/infocomm/anglais/cocoa/crop.htm.

Working H (1932). Price Relations between July and September Wheat Futures at Chicago since 1885. Wheat Studies of the Food Research Institute 9(6).

Working H (1948). The Theory of Price of Storage. J. Farm Econ., 30: 1-28.

Appendix

We display below a sample of cocoa forward curves, observed at different dates in the year and across the period of observation.

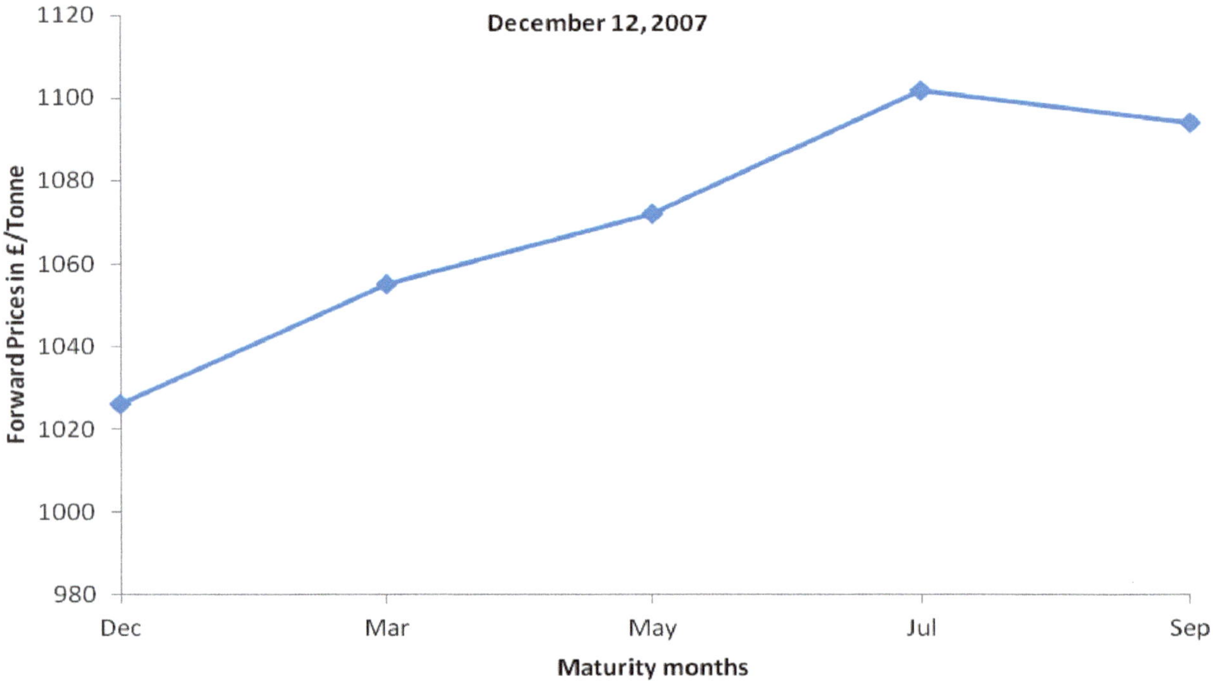

Figure A1. Cocoa forward curve; December 12, 2007.

Figure A2. cocoa forward curve; December 15, 2008.

Figure A3. Cocoa forward curve; December 15, 2009.

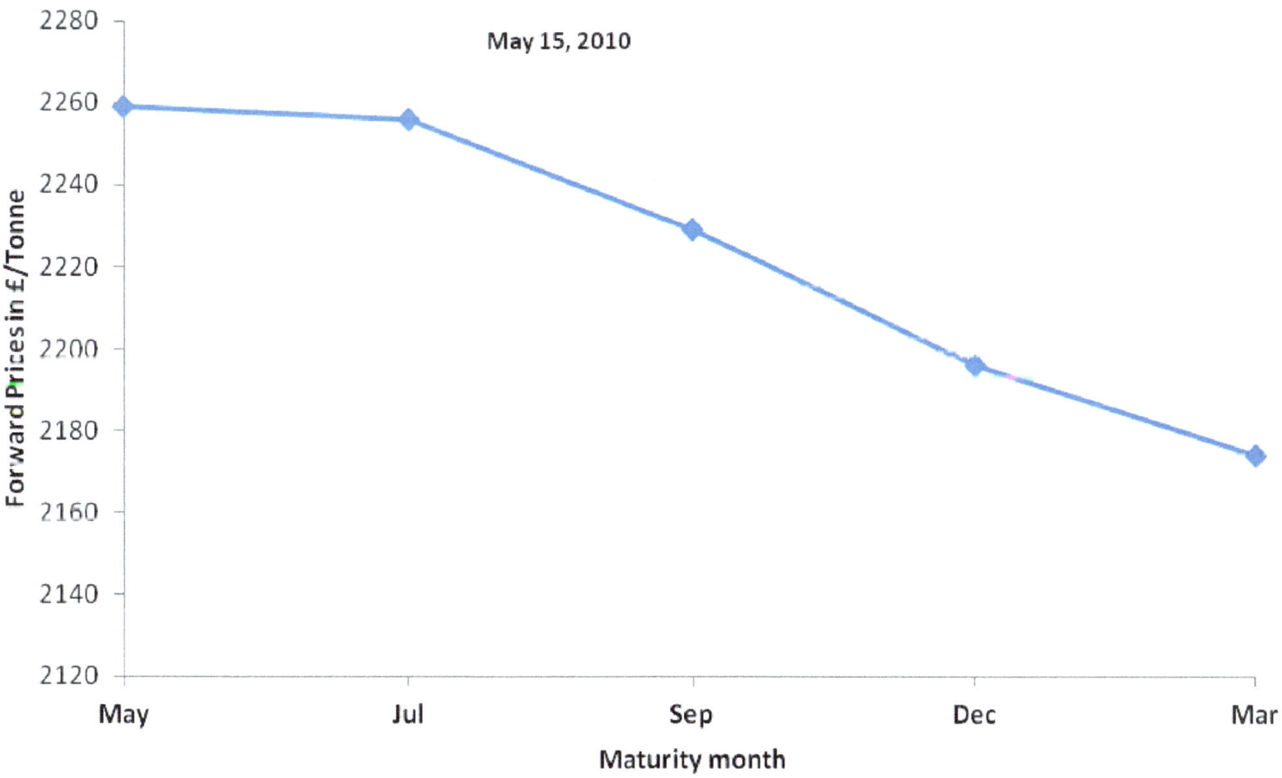

Figure A4. Cocoa forward curve; May 15, 2010.

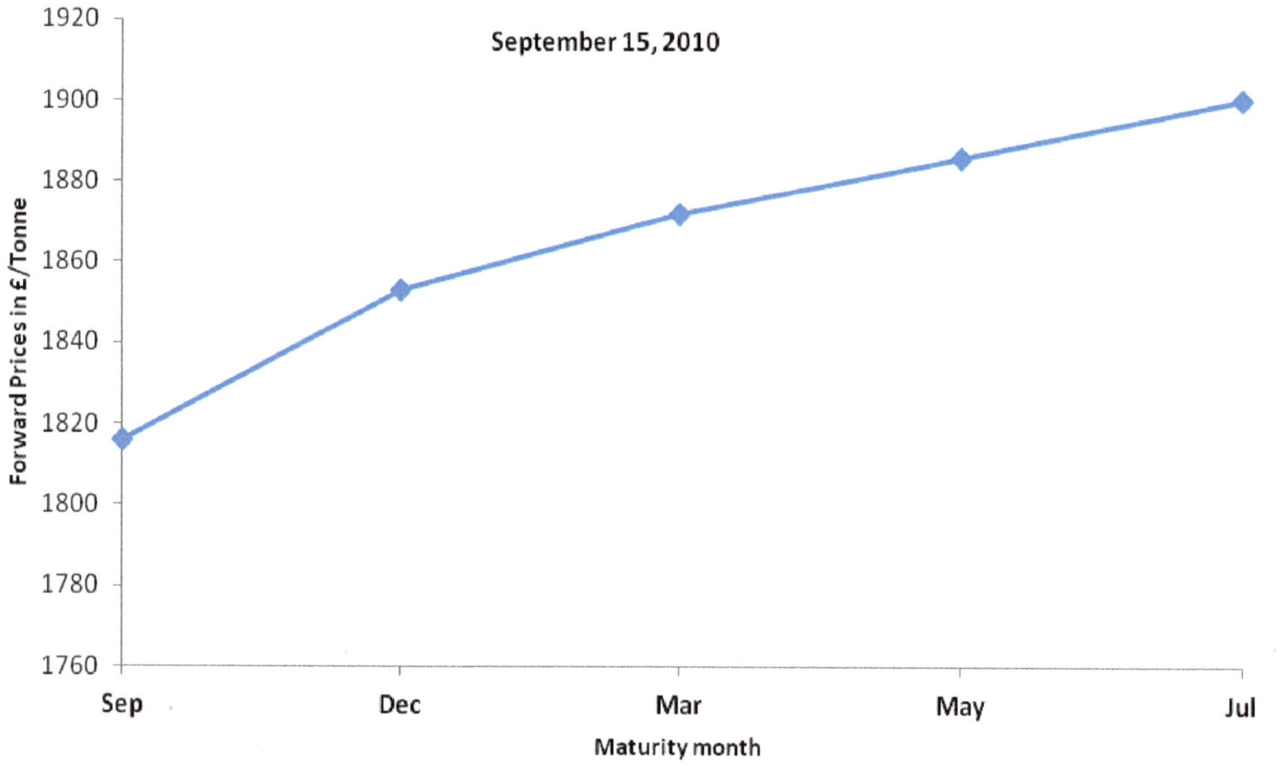

Figure A5. Cocoa forward curve; September 15, 2010.

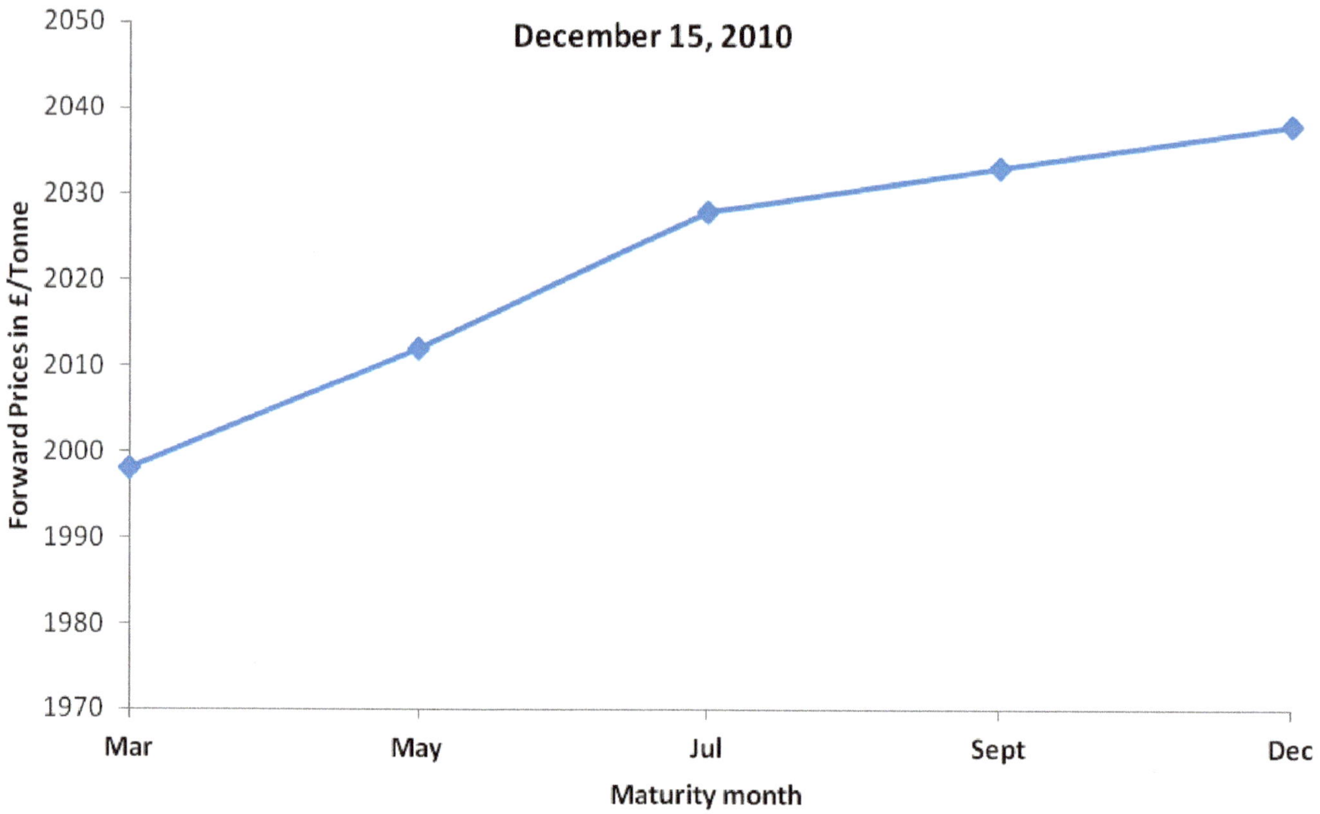

Figure A6. Cocoa forward curve; December 15, 2010.

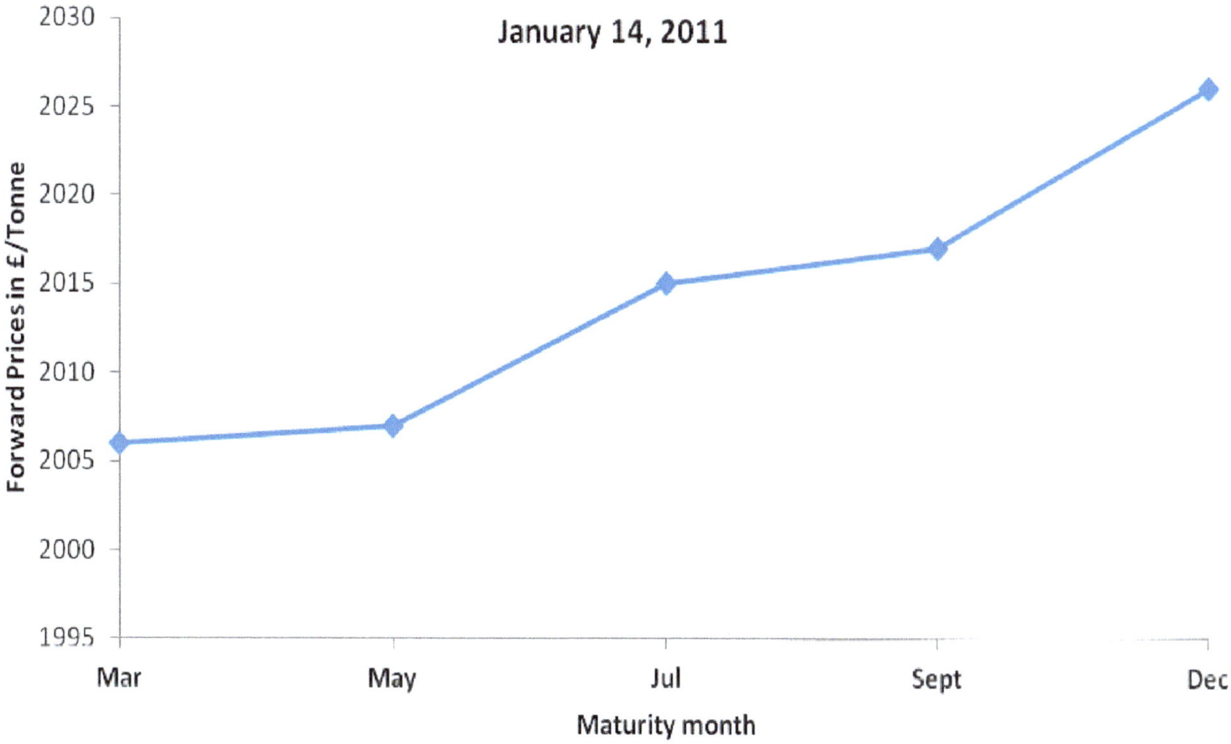

Figure A7. Cocoa forward curve; January 14, 2011.

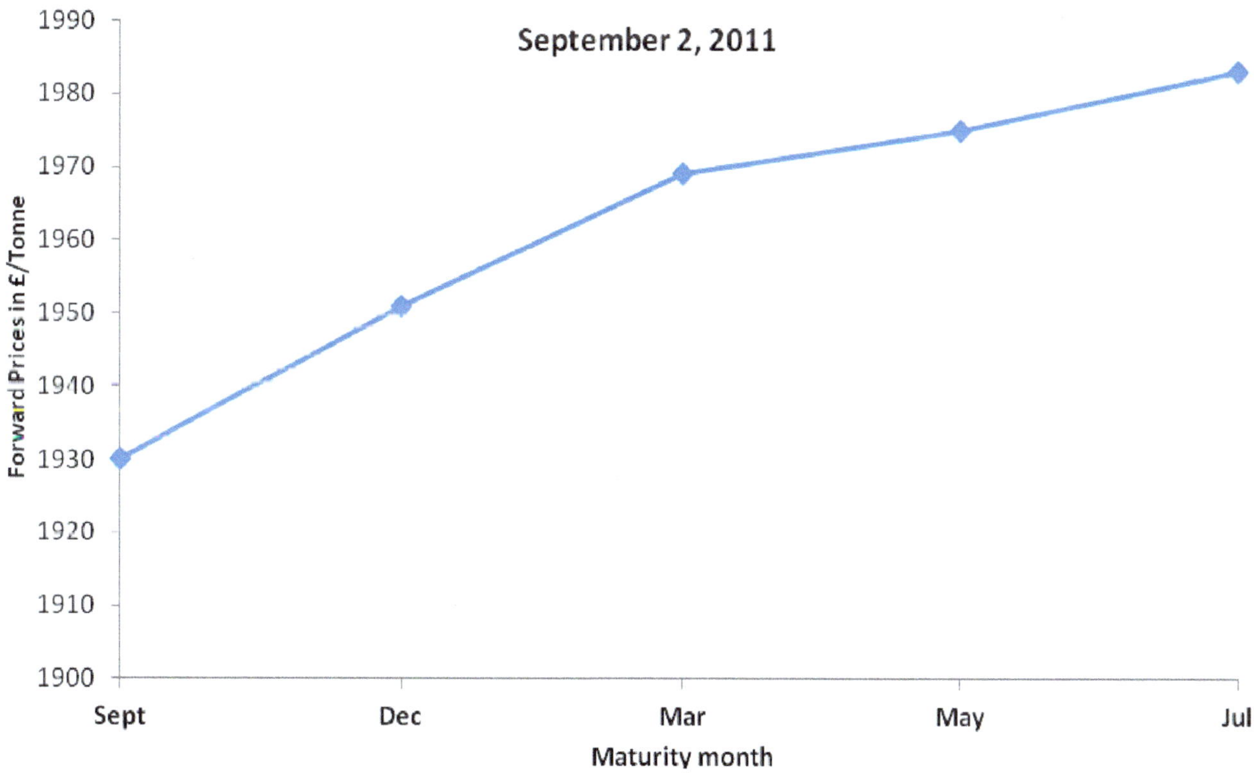

Figure A8. Cocoa forward curve; September 2, 2011.

12

Weight gain, carcass characteristics and economic feasibility of Horro rams supplemented with different substitution levels of maize grain and molasses

Chala Merera[1], Jemal Dekebo[2] and Ulfina Galmessa[1]

[1]Ambo University, P. O. Box 19, Oromia, Ethiopia.
[2]Bako Agricultural Research Center, P. O. Box 03, Oromia, Ethiopia.

A total of forty Horro rams with an average initial body weight 19.5 kg were randomly assigned to one of the following five treatments with eight replicates: 184.2 g ground maize grain (T1), 140 g ground maize grain + 60 g molasses (T2), 90 g ground maize grain + 120 g molasses (T3), 50 g ground maize grain + 180 g molasses (T4), 240 g molasses (T5). In addition to the treatment diets, all groups were supplemented with equal amount of noug cake (190 g/d) and 1% of salt to provide the protein and minerals requirements, respectively. Cyndon dactyl hay was used *ad libitum* as a basal diet. There was a significant ($p<0.05$) difference between Treatments 1 and 5 for total weight gain and average daily weight gain of Horro rams but there was no significant difference between and within the reaming treatments. Rams fed on Treatment 1 showed higher daily weight gain (73.4 g/d/h) followed by Treatments 2, 3, 4 and 5 (59.1, 53.8, 51.9 and 40.9 g/d/h, respectively). Rams fed on Treatment 1 had higher carcass weight (11.6 kg) and dressing percentage (45.2%) followed by Treatments 2, 3, 4 and 5, respectively. The cost benefit analysis indicated that substituting 44.2 g of maize grain by 60 g molasses had showed higher net benefit of Birr 20.69 next to the previously recommended amount of maize grain in Treatment 1 with a net benefit of 30.71 Birr/head of animal. This indicates that fattening using Treatments 1 and 2 were found to be economical and increased a net return at least by 50.0%. The results showed that Horro rams fed on Treatment 2 had showed higher live weight gain, carcass traits and net benefit next to the already recommended Treatment 1. Thus, recommended levels of molasses can replace maize grain and one can use molasses as energy sources in a concentrate supplements for fattening Horro rams.

Key words: Weight gain, carcass, economic, Horro sheep, molasses, maize grain.

INTRODUCTION

Natural pasture and crop residues are the major sources of livestock feed in the western part of Ethiopia. Because of the steadily conversion of grazing lands into cropping fields, crop residues are becomes more important in livestock feed. However, crop residues are known for its low protein and fermentable carbohydrate levels. They can supply only sub-maintenance requirement of animals when they are fed alone (Seyoum and Zinash, 1998). Supplementation of crop residues and natural pasture hay by feeds of high protein and energy sources was for the efficient utilization of the former to enhance animal production and productivity (Lemma et al., 1996). Among the high nutrient density diets suggested for supplementation, noug cake and molasses are abundantly found in the western part of the country and hence small-scale farmers can easily get access to finish their animals by supplementation with this agro-industrial by products.

High-energy concentrates are fed for energy. Grains and grain products commonly fed are shelled corn, sorghum, oats and wheat. Liquid feedstuffs, such as molasses, can also be used as a source of dietary energy. Alternative energy sources, such as fodder beet and sweet potato can be fed to growing and finishing sheep and goats. The use of these feeds depends on the price differential in utilizing them for fattening compared to other alternative uses. Fattening of sheep based on agro-industrial byproducts is also practiced in Ethiopia. Fattening using agro-industrial byproducts like sugar processing byproducts is feasible in places for instance in parts of Wellega where valuable feed resources such as molasses (from the Finchaa sugar factory) and corn (grain and residue) are widely available (Alemu, 2008).

Molasses had been used previously in animal feeds, but always at relatively low levels (usually less than 20% of the diet) (Preston and Leng, 1986). The most appropriate role for small amounts of molasses in ruminant diets is as a vehicle for other nutrients (e.g. urea and minerals). A drought feeding strategy based on the use of liquid molasses supplements containing from 8 to 10% urea is now an established practice in Australia (Nicol et al., 1984) and has been introduced successfully in Africa (Preston and Leng, 1986). The incorporation of urea and other nutrients in molasses-based (multi-nutritional) blocks promises to be an even more attractive technology, especially for smallholder-village farmers, for supplementation of locally available crop residues which are of low digestibility and also deficient in fermentable nitrogen (Sansoucy et al., 1986).

Molasses is an agro-industrial by-product used as energy supplement for livestock in different parts of the world where sugar cane processing factories are available. From different works done in Ethiopia (Amsalu et al., 1998) and elsewhere in the world (Sibanda et al., 1987), promising results were obtained from livestock finishing studies where molasses was used as one of the major constituent of the ration. Maize grain, which is the constituent of the already recommended concentrate in the region, has become hardly possible to use as livestock feed due to increased competition as human food. In such system it is apparent to find alternative energy sources in livestock ration.

Thus, the objectives of this study were to determine the levels of molasses that substitute maize grain and to evaluate economic feasibility of fattening Horro rams supplemented with different substation level of maize grain and molasses.

MATERIALS AND METHODS

Description of the study area

Bako Agricultural Research Center is located 250 km west Addis Ababa at an altitude of 1650 masl. The center received mean annual rainfall of 1200 mm in a bimodal distribution, 80% of which falls from May to September. The area had a mean relative humidity of 59% and mean minimum and maximum temperatures of 13.5 and 27°C, respectively.

Animal feeding and management

A total of forty Horro rams with an average initial body weight of 19.5 ± 0.73 kg and similar age group were purchased from the nearby local markets. Rams were randomly assigned to the following five treatments (T) with eight replicates: T1 = 184.2 g/h/d ground maize grain; T2 = 140 g/h/d ground maize grain + 60 g/h/d molasses; T3 = 90 g/h/d ground maize grain + 120 g/h/d molasses; T4 = 50 g/h/d ground maize grain + 180 g/h/d molasses; T5 = 240 g/h/d molasses.

The treatments were based on previously recommended amount of 400 g/h/d concentrate (49% maize grain, 49% noug cake, and 1% salt) for finishing Horro rams (Solomon et al., 1991). In addition to the treatment diets, all groups were supplemented with equal amount of noug cake (190 g/h/d) and 1% of salt to provide the protein and minerals requirements respectively. Cyndon dactyl hay was used *ad libitum* as a basal forage diet. Water was also offered *ad libitum*. All animals were drenched and dipped against endo and ecto-parasites. A group of rams were kept in individual pens and fed the treatment diet for 90 days after 15 days adaptation period.

Statistical analysis

Data on daily feed intake and body weight of the rams were recorded fortnightly during the experimental period but for the edible and non-edible carcass traits four rams from each treatment were slaughtered at the final date of the trial. The data were analyzed using the general linear model of statistical analysis system (SAS, 1996). During analysis, treatment was considered as independent variable where as final body weight, total weight gain, daily weight gain, carcass traits and feed intake were considered as dependent variables. Pearson's correlation coefficient was used to assess interrelation ship of live body weight and some carcass traits.

Economic analysis

Three local live animal dealers were made estimation of the final price of rams and the average price of rams was used for cost benefit analysis. The cost benefit analysis was made using partial budget, dominance and marginal rate of return. The prices of rams, maize, molasses and noug cake were used for analysis. The price of noug cake and maize was taken as an average of five years market price. Besides, the stability of the result was complemented by sensitivity analysis by varying the prices of inputs.

RESULTS AND DISCUSSION

Growth performance of Horro rams

Least-square means of total weight gain and average daily weight gain of Horro rams fed on different levels of molasses and maize grain are indicated in Table 1. There was significant ($P < 0.05$) difference between Treatments 1 and 5 in total weight gain and average daily gain of Horro rams but there was no significant difference between and within the other treatments. Horro rams fed on Treatment 1 showed higher daily weight gain

Table 1. Least squares means (± SDE = Standard Error) of initial body weight (IWt), final body weight (FWt), total weight gains (TGain) and average daily weight gain (ADG) of Horro Rams fed on different levels of molasses and maize grain.

Treatment	IN Wt	F Wt	T Gain (kg)	ADG (g)
1	19.7 ± 0.73	27.0 ± 0.94	7.3 ± 0.74[b]	73.4 ± 7.45[b]
2	20.6 ± 0.73	26.5 ± 0.94	5.9 ± 0.74[ab]	59.1 ± 7.45[ab]
3	19.6 ± 0.73	25.0 ± 0.94	5.4 ± 0.74[ab]	53.8 ± 7.45[ab]
4	19.3 ± 0.73	24.5 ± 0.94	5.2 ± 0.74[ab]	51.9 ± 7.45[ab]
5	19.4 ± 0.73	23.5 ± 0.94	4.1 ± 0.74[a]	40.9 ± 7.45[a]

Means with different letters of superscription differ significantly (P<0.05).

Table 2. Analysis of variance of final body weight (kg) and daily weight gain (g/d) of Horro Rams fed on different levels of molasses and maize grain.

Source of variation	Final body weight	Average daily weight gain
Treatment	25.29	55.81
Error mean square	2.66	21.07
Error DF	35.00	35
R-square (%)	21.67	22.55
CV (%)	10.5	17.74

(73.4 g/d/h) followed by those on diet 2, 3, 4 and 5 (59.1, 53.8, 51.9 and 40.9 g/d/h, respectively). In agreement with these results Nnanyelum and Bala (1981) reported that the results of average daily weight gain were improved with molasses replaced maize in fattening rations without detrimental effects (Table 2). Similar average daily weight gain was also reported by Solomon et al. (2000). Since, there was no significant (p > 0.05) difference observed from Treatment 1 through 4 in body weight gain it is possible to improve the use of fourth treatment as it has less cost to substitute more of the grain with molasses only looking at input costs but further economic result using sheep final price is more apparent as the final net benefit is not a mere result of body weight. These findings are in general agreement with those in the literature (Meyreles et al., 1982; Meyreles and Preston, 1982).

Feed intake

Almost all supplements offered to each group of treatments were completely consumed regardless of molasses level. This is probably due to the DM intake which was not materially affected by energy level, but it was rather significantly influenced by protein level. A better weight gains and body condition performances were observed as the level of molasses substitution decreased from 180 to 60 g of supplementation. Feed intake and conversion efficiency were higher for rams fed from diet 1 up to 4 in comparison to diet 5. Similarly, Nnanyelum and Bala (1981) revealed that the results of feed intake and feed conversion efficiency showed that molasses replaced 30% maize in fattening rations without detrimental effects. Preston and Leng (1986) also reported that molasses had been used previously in animal feeds, but always at relatively low levels (usually less than 20% of the diet). Molasses could seriously be considered as an alternative to cereal grains as a means of intensifying animal production in the tropics.

Carcass and non carcass components

Least square means of edible and non-edible carcass traits of Horro Rams are shown in Tables 3 and 4. There was significant (P < 0.05) difference in edible and non-edible carcass traits of Horro rams supplemented with different substitution level of molasses and maize grain between Treatments 1 and 5, but there was no significant difference observed between the remaining treatments. Rams fed on Treatment 1 had higher carcass weight (11.6 kg) and dressing percentage (45.2%) followed by Treatments 2, 3, 4 and 5 (10.5 kg and 43.2%, 10.5 kg and 42.2%, 10.9 kg and 43.7%, 8.8 kg and 41.6%, respectively). The carcass weight and dressing percentage of Horro rams fed on different level of molasses and maize grain was almost similar to the result reported by Ulfina et al. (2004) and Alemu (2008) who stated that there was improvement in carcass weight and dressing percentage with increased level of concentrates (Table 5). Nnanyelum and Bala (1981) also investigated that carcass data showed that rams which received the low and medium molasses diets had better

Table 3. Least square means (± SDE) of edible carcass traits of Horro rams fed on different levels of molasses and maize grain.

Traits	TRT				
	1	2	3	4	5
Carcass weight (kg)	11.6 ± 0.77[b]	10.5 ± 0.77[ab]	10.5 ± 0.77[ab]	10.9 ± 0.77[ab]	8.8 ± 0.77[a]
Dressing percentage (%)	45.2[b]	43.2[ab]	42.2[ab]	43.7[ab]	41.6[a]
Kidney fat free (g)	82.5 ± 4.25	73.8 ± 4.25	76.3 ± 4.25	88.8 ± 4.25	80.0 ± 4.25
Heart (g)	137.5 ± 10.54	131.3 ± 10.54	120.0 ± 10.54	130.0 ± 10.54	111.3 ± 10.54
Liver (g)	430.0 ± 24.43	396.3 ± 24.43	437.5 ± 24.43	500.0 ± 24.43	388.8 ± 24.43
Omental fat (g)	146.3 ± 36.99[ab]	167.5 ± 36.99[b]	105.0 ± 36.99[ab]	103.8 ± 36.99[ab]	50.0 ± 36.99[a]
Kidney fat (g)	60.0 ± 9.03[b]	50.0 ± 9.03[ab]	57.5 ± 9.03[ab]	51.3 ± 9.03[ab]	30.0 ± 9.03[a]
Tail weight (kg)	0.6 ± 0.10	0.6 ± 0.10	0.4 ± 0.10	0.6 ± 0.10	0.4 ± 0.10
Fat thickness at 12th rib area (mm)	0.4 ± 0.11	0.4 ± 0.11	0.3 ± 0.11	0.5 ± 0.11	0.3 ± 0.11
Rib eye area (cm $^{2)}$	9.9 ± 1.07	8.3 ± 1.07	8.2 ± 1.07	8.5 ± 1.07	7.8 ± 1.07
Visceral empty (kg)	1.7 ± 0.11[ab]	2.0 ± 0.11[b]	1.8 ± 0.11[ab]	1.8 ± 0.11[ab]	1.6 ± 0.11[a]

Means with different letters of superscripts differ significantly (P<0.01, 0.05).

Table 4. Least square means (± SDE) of some non-edible carcass traits of Horro rams fed on different levels of molasses and maize grain.

Traits	TRT				
	1	2	3	4	5
Head (kg)	1.5 ± 0.07[b]	1.4 ± 0.07[ab]	1.4 ± 0.07[ab]	1.5 ± 0.07[b]	1.3 ± 0.07[a]
Fore legs (g)	267.5 ± 12.62[ab]	261.3 ± 12.62[ab]	287.5 ± 12.62[b]	276.3 ± 12.62[ab]	246.3 ± 12.62[a]
Hind legs (g)	261.3 ± 13.46	257.5 ± 13.46	271.3 ± 13.46	246.5 ± 13.46	235.5 ± 13.46
Lungs +Trachea (g)	445.0 ± 42.16[ab]	456.3 ± 42.16[ab]	541.3 ± 42.16[b]	450.0 ± 42.16[ab]	398.8 ± 42.16[a]
GIT full (kg)	5.1 ± 0.41[ab]	5.8 ± 0.41[ab]	6.3 ± 0.41[b]	5.6 ± 0.41[ab]	4.8 ± 0.41[a]
Skin (kg)	2.4 ± 0.18	2.3 ± 0.18	2.4 ± 0.18	2.7 ± 0.18	2.1 ± 0.18
Spleen (g)	71.3 ± 7.99	76.3 ± 7.99	67.5 ± 7.99	60.0 ± 7.99	61.3 ± 7.99
Blood (kg)	1.0 ± 0.09	1.0 ± 0.09	1.0 ± 0.09	1.0 ± 0.09	0.9 ± 0.09
Testicles (g)	352.5 ± 34.09	343.8 ± 34.09	301.3 ± 34.09	292.0 ± 34.09	265.0 ± 34.09
Penis (g)	52.5 ± 6.19	48.8 ± 6.19	55.0 ± 6.19	61.3 ± 6.19	48.8 ± 6.19

*Means with different letters of superscripts differ significantly (P<0.01, 0.05).

Table 5. Analysis of variance of carcass weight (kg) and dressing percentage (%) of Horro rams fed on different levels of molasses and maize grain.

Source of variation	Carcass weight	Dressing percentage
Treatment	10.47	13.21
Error mean square	1.54	2.66
Error DF	15.0	15.0
R-square (%)	32.93	22.44
CV (%)	14.69	6.16

dressing percentage and wholesale cuts. Some non-edible carcass traits were not significantly differing except head, lung and trachea and visceral full. The result of correlations indicated that live body weight and some carcass traits of rams were positively correlated value except kidney fat and kidney fat free. There was a significant (P < 0.01) positive correlation of liver with carcass weight and final live weight, initial live weight with

Table 6. Correlations of live body weight and some carcass traits of Horro rams fed on different levels of molasses and maize grain.

Variable	CWT	DP	KFAT	OMFAT	FWT
LIV	0.56*	0.23	0.36	0.36	0.66**
HRT	0.69**	0.62**	0.71**	0.44*	0.60*
KFF	0.22	0.11	-0.04	0.02	0.25
IWT	0.52*	0.20	0.43	0.35	0.65**

** P< 0.01; * P< 0.05. CWT = Carcass weight; IWT = initial weight; DP = dressing percentage; FWT = final weight; KFAT = kidney fat; LIV = liver; OMFAT = omental fat; HRT = heart; KFF = kidney fat free.

Table 7. Partial budget, dominance and marginal analysis of Horro rams fed on different levels of molasses and maize grain.

Item	**Treatments**				
	1	**2**	**3**	**4**	**5**
Av. Inwt	19.66	20.63	19.63	19.28	19.38
Av. Fwt.	27.00	26.53	25.00	24.47	23.47
Adjusted wt.	26.73	26.26	24.75	24.23	23.24
Gross benefit/head	163.88	153.00	135.50	125.63	117.13
Costs that vary					
Maize	14.59	11.09	7.13	3.96	0
Molasses	0	0.39	0.59	0.89	1.19
Noug cake	6.58	6.58	6.58	6.58	6.58
Initial price	112.00	114.25	109.38	107.63	103.13
TCV	21.17	18.06	14.3	11.43	7.77
Net benefit/head	30.71	20.69	11.82	6.57	6.23
Dominance analysis using tabular method (NB are arranged in descending order)					
NB in descending order	30.71	20.69	11.82	6.57	6.23
TC for each treatment	133.17	132.31	123.68	119.06	110.9
Dominance analysis		ND	D	D	D
MRR (%)		50			

D, Dominated; ND, non dominated.

carcass weight and final body weight and heart with carcass weight, dressing percentage, kidney fat, omental fat and final weight as indicated in Table 6. Ulfina et al. (2004) reported similar results on finishing old Horro ewes on concentrate. The correlation between final body weight, carcass weight, dressing percentage and kidney fat were positive and high. There was no significant (P > 0.05) difference observed from Treatment 1 up to 4 in carcass traits.

Economic feasibility

The partial budget analysis indicated that substituting 44.2 g of maize grain by 60 g molasses had higher net benefit of Birr 20.69 next to the previously recommended amount of maize grain in Treatment 1 with a net benefit of 30.71 Birr/head of animal (Table 7). Similar net benefit was reported by Solomon et al. (1991) and Solomon et al. (2000). Nnanyelum and Bala (1981) also indicated that the economic advantage in the use of molasses in fattening rations is attractive, given the wide price differential between maize and molasses. The net benefits of Treatments 1 and 2 were at least twice higher than the net benefit of the other treatments. This indicates that fattening using Treatments 1 and 2 were found to be economical and increased a net benefit at least at minimum acceptable return (50.0%). Since Treatment 1 was already recommended, the marginal rate of return (MRR) in treatment 2 be accepted and gave the highest net benefit as compared to Treatments 3, 4, and 5. The recommended rate of feeding is economically feasible if each of the prices of maize and molasses per kilogram is not exceeding Birr 0.99 and 0.55, respectively (Table 8). Besides, information of market price in the area is also important to decide when to start the

Table 8. Sensitivity analysis for changing prices of maize and molasses in fattening Horro rams on different levels of molasses and maize grain.

Item	Treatments				
	1	2	3	4	5
Average initial price of young rams (Birr/kg)**	5.79	5.54	5.57	5.58	5.32
Average final price of young rams (Birr/kg)*	6.07	5.77	5.42	5.13	4.99
Average price of maize (Birr/kg)**		0.99			
Average price of molasses (Birr/kg)**		0.55			

*Changes greater than the indicated figure are acceptable; **Changes less than the indicated figure are acceptable; Data of feed prices during experimental period: Maize grain price = 0.90 Birr/kg; Noug cake price = 0.36Birr/kg; Molasses price = 0.05 Birr/kg; weight down adjusted to: 1%

fattening activities.

CONCULUSION AND RECOMMENDATIONS

The results showed that Horro rams fed on Treatment 2 had showed higher live weight gain, carcass traits and net benefit next to the already recommended Treatment 1. Fattening with the amount of molasses in Treatment 2 can be economical and recommended. It is also advisable and economical to plan the start of fattening program immediately after maize harvest. The stability of this recommendation was supplemented by sensitivity analysis by varying the prices of inputs. It could be advisable and economical to conduct the fattening of rams with the recommended level of molasses with maize grain as compared to using only maize grain since there was no significant difference observed between Treatments 1 and 2 in live weight gain, carcass traits and net benefit also use of molasses can reduce the increased competition of human being with livestock for maize grain. Thus, recommended level of molasses can replace maize grain and one can use molasses as energy sources in a concentrate supplements for fattening Horro rams.

REFERENCES

Alemu Y (2008). Short term intensive fattening of sheep and goats before slaughter for rapid improvement in weight and condition and also producer incomes. Ethiopia Sheep and Goat Productivity Improvement Program (ESGPIP). Techn. Bull. 11:1-11.

Amsalu S, Tesfaye A (1998). The effect of molasses, urea block and concentrate supplementation on growth rate of Arsi bulls. Proceeding of fifth National Conference of Ethiopian Society of Animal Production (ESAP) pp. 228-232.

Meyreles L, Pound B, Preston TR (1982). The use of *Leucaena leucocephala* or sugarcane tops as sources of forage in cattle diets based on molasses/urea, supplemented with chicken litter and/or wheat bran. Trop. Anim. Prod. 7:92-97.

Meyreles L, Preston TR (1982). The role of poultry litter in molasses/urea diets for the fattening of cattle. Trop. Anim. Prod. 7:138-141.

Nicol DC, Venamore PC, Beasley RC (1984). Fortified molasses systems for beef properties. Anim. Prod. Austr. 15:216-219.

Nnanyelum NU, Bala A (1981) The Replacement Value of Sugarcane Molasses for Maize in Sheep-fattening Rations. J. Sci. Food. Agric. 32(5):489-492.

Preston TR, Leng RA (1986). Matching Livestock Systems with Available Feed Resources. International Livestock Centre for Africa, Addis Ababa. P. 331.

SAS (1996). Statistical Analysis system SAS Institute Inc. Cary, North America.

Seyoum B, Zinash S (1998). Utilization of teff straw as livestock feed: Research review. In: Proceedings of the Fifth National Conference of Ethiop. Soc. Anim. Prod. pp. 173-185.

Sibanda S, Chakanyaka C, Mlilo T (1987). Effect of level of cane molasses in fattening diets on performance of beef steers and heifers. Utilization of agricultural by products as livestock feeds in Africa. Proceeding of a workshop held at Ryalls Hotel, Blantyre, Malawi. pp. 228-232.

Solomon A, Fikru T, Ulfina G, Gemeda D (2000). Growth performance and carcass characterstics of Horro lambs castrated at different ages. In: proceedings of the 8rd National conference of the Ethiop. Soc. Anim. Prod. pp. 108-120.

Solomon G, Solomon A, Asfaw N (1991). Growth Responses of Horro sheep to different levels maize and noug cake supplements. In proceedings of the 4th National Livestock Improvement Conference. pp. 113-118.

Ulfina G, Gammada D, Solomon A, Girma A, Solomon G, Shiv P, Fiqru T (2004). Pre-Market Supplementary Feeding of Aged Horro Sheep in Relation to Weight Gain, Carcass Yield, and Economic Response in Western Oromia. Indian. J. An. Sci. 74(3):327-329.

13

Characterization of a barley-based farming system in the Arsi Highlands, Ethiopia

Bedada Begna, Mesay Yami and Eshetu Lemma

Ethiopian Institute of Agricultural Research, Kulumsa ARC, Ethiopia.

In order to minimize the gap between the demand for barley, particularly malt barley, and its supply, this study was initiated to assess the existing farming systems of the selected agro ecologies in order to develop appropriate malt barley technologies and major production constraints. In line to this, two districts with a potential in barley production were purposively selected from the Arsi highlands of Ethiopia and totally 120 farmers were selected and used to conduct the study. During 2009 to 2010 production season, both food and malt barley varieties were grown by about 95% of surveyed farmers. Among the surveyed farmers, about 82% have knowledge or knowhow to improved barley varieties while 19% did not. Yielding potential and market value are the major reasons of farmers for preferring to produce some malt barley varieties over the others. Holker, miscal-21, and local food barley varieties are the relatively high-yielding as well frequently cultivated varieties in the study areas. The present and foreseen increase in number of breweries and their expansion currently and in the future indicates the demand for malt barley product in the future. Unpredictable rainfall, shortages of labor, improved seed, working capital, low quality of seed, untimely supply of inputs, and problem of good market facilities were the major constraints of barley production in the study area.

Key words: Characterization, barley, barley based farming system and Arsi highlands.

INTRODUCTION

Barley is one of the major cereal crops that are largely produced in the central and south east mid and high altitude areas of Ethiopia. It is the fifth important cereal crop after Tef, Maize, Wheat and Sorghum with annual production of about 1.2 million tons cultivated on an area of about 1.0 million hectares (CSA, 2005) and it is the most desirable crop in the highlands where there is limited alternative crop (Yosef et al., 2011). Barley is grown in about 70 million hectares in the world. Global production is 160 million tons.

Barley (*Hordeum vulgarre* L) which includes both food and malt barley species is cultivated in Ethiopia. Barley grain is mostly used as feed for animals, malt, and food for human consumption. Traditionally barley is used for making local recipes and drinks such as 'dabo', 'kolo', 'ganfo', 'kinche', 'baso,' tela', 'borde' and other types of food. Its straw is a good source of animal feed (Yosef et al., 2011) and it is also used for thatching of roofs. Malt is the second largest use of barley.

Malt barley, at the present time, is considered as one of the cash crops and its demand by malt factory has increased due to its increased capacity of malt barley processing and the expansion of breweries and beer consumption levels in the country (AMF, 2012). Beer and bread are some of the industrial products produced from malt and food barley, respectively. As reported by Mohammed and Getachew (2003) and presented by Ethiopian-Barley-Business (2012) malt barley is among crops demanded in good quantity that lacks supply in which its impact is directly connected with national economy, as Beer Factory import it from abroad with high hard currency. The report of Malt factory indicates that as the share of inland production of malt covers only 35 to 40% and the left 60 to 65% is imported on the behalf of beer factories which exhausting the country's foreign currency reserves. Similarly in 2011, breweries in Ethiopia imported 60% of the malt primarily from international producers (Ethiopia-Barley-Business-Case presentation, 2012).

Agricultural lead industrialization policy also demands agricultural production as an input. In spite of the demand created from agro-industry side, supply of products in quality and quantity hinders the activity of these industries and their potential growth. Since few years due to the scaling up activities by various stake holders including the Asella malt factory, barley production has been increasing in major malt barley producing areas of Arsi and West Arsi zones. However, the availability of improved seed is not satisfactory in all suitable barley producing agro ecologies.

Research on barley improvement in Ethiopia started in 1955. Since then, research system has produced substantial amount of technologies, information and knowledge. Due to the weak linkage between research and extension, however; these technologies delayed to reach to growers to bring the required impact. Coordination and cooperation among the different stakeholders in agricultural development is at its low level. Therefore, this study was initiated to assess the current problems and technology needs, identify the weaknesses and strengths of barley varieties in use as well as varieties of interest and major production constraints. The result can help the selection and dissemination of barley varieties for suitable barley producing areas through participatory approach with farmers.

Objectives of the surveys

The objectives of the farm surveys were to overview the existing farming systems of the selected agro-ecologies in order to develop appropriate malt barley technologies and to identify production constraints and opportunities for technological interventions in the study and the surrounding areas.

METHODOLOGY

Selection of the study area and sample respondents

Characterization of barley based farming system was designed to be carried out in potential barley producing areas of Arsi zone in Ethiopia. To execute the survey both purposive sampling and two stage random sampling procedures were followed, and two representative districts (Honkolowabe and Chole) were purposively selected from the potential barley producing districts of the zone. In two stage random sampling procedures; at first stage, the lists of peasant associations (PA), those potentially producing barley in their respective districts, were made by respective agricultural development officeers in consultation with the research team and two PA were randomly selected from each district. Accordingly, we selected Tajiwalkite and Lemu Kara from Honkolowabe. Similarly from Chole, Gadoyabeta and Tulu ego were selected for the study. In the second stage, totally 120 farmers were selected by equal distribution method which means 60 from each districts or 30 household heads were randomly selected from a list of farmers in each selected PAs. Finally, among the farmers selected randomly in each PA, household heads who cultivated barley on their lands and who did not cultivate were identified and used for conducting the study.

Data collection

The data used for this study were collected from both primary and secondary sources. Both informal and formal surveys were used to collect the primary data. The informal survey was conducted through group discussions, interviewing key informants, district agricultural officers, PA level development agents, group of farmers, and barley seed multiplying farmers in all districts and selected PA. The informal survey was mainly focused on the assessment and identification of barley production problems and technology needs, weaknesses and strengths of barley variety in use, variety of interest and criteria for the selection of varieties, and crops used and needed in the community.

The discussions were entirely participatory. The assessment results of the informal survey were used for forming questionnaire to collect quantitative data. The structured and semi-structured questionnaire were pre-tested and specifically designed and used to perform the formal survey. The informal survey was conducted in the months of March and April in 2010, and the formal survey was carried out in April and May, 2010.

Data analysis

The quantitative data collected through the formal survey was analyzed using SPSS computer program and various descriptive statistics and mathematical methods like mean, percentages and various analytical methods were used based on the objectives of the survey. Tables and figures were used to present summary statistics such as mean, standard errors and percentages.

Agro-ecological and climatic features of the study area

The selected districts are mainly characterized as highland areas with prolonged rainy months and various soil types. Some of agro-ecological and climatic conditions of the areas are given below according to the information gathered from districts' agricultural development offices. The altitude of Honkolowabe is in the range of 1600-3843 masl and the Chole's altitude varies from 1140—3574 masl. Based on the information obtained from districts' agricultural offices, the three major agro-ecologies namely high altitude (Dega),

Table 1. Total population of the districts' with their sex composition.

Variables	Honkolowabe		Chole	
	No.	Percent	No.	Percent
Male	35,521	49.7	44,466	50.4
Female	35,811	50.3	43,720	49.6
Total population	71,367	100	88,190	100

Source: District agricultural development offices (2011).

mid-altitude (Weinadega), and lowland (Kola) are found in the two districts. Accordingly; 30, 65 and 5% of the total areas of Honkolowabe fall in highland, mid-altitude and lowland areas respectively. Similarly, 60, 18 and 22% of the land area of Chole district fall in highland, mid-altitude and lowland, respectively. The temperature of Honkolowabe and Chole varies from 15.5 to 20°C and 15 to 25°C respectively. Average temperature of Honkolowabe is about 16°C while Chole is 20°C. The two districts are characterized by bimodal rainfall which usually starts in February and prolongs until the end of April (short rainfall period). During this period, farmers grow "belg" season crops except in the low land areas where rainfall is erratic and farmers mainly grow barley, Faba bean, field pea, maize and wheat. In the main season ("Meher"), the rainfall starts in the month of June and lasts up to October. The districts' main/ longer rainy months is from June to October and from June to September in Honkolowabe and Chole respectively. Yearly rainfall in Honkolowabe and Chole districts was 900-1400 mm and 800-1200 mm respectively. In the main season, most agricultural crops are produced by the farmers.

Soils of Honkolowabe are composed of 65% vertisol, 5% sandyloam and 30% redish brown, and Chole also has vertisol, reddish brown and sandy loam soil types with unknown proportions. The Chole district has rivers that have water throughout a year. The major rivers include Legbuna, Qoro, Giersa, Guaya, Sinkise and Gobesa. Out of these rivers some are used by a few PAs for irrigation purpose. These rivers are used for growing various vegetables, coffee, sugarcane and "chat". The total irrigated land coverage is about 1,253 ha (District Agricultural Development office, 2011).

RESULTS AND DISCUSSION

Socioeconomic characteristics

The most common socioeconomic characteristics of the households including sex, age, family size, education, occupation, land use, land holding, cropping pattern, and other socioeconomic conditions and services were studied in the area. Table 1 depicts the total population of each district and the number of household heads that have private agricultural lands and their sex composition.

Total population and their sex composition

The total population of an area can determine the agricultural activity and utilization pattern of the land. The total population of Honkolowabe was 71,367 out of which 49.7 and 50.3% were male and female, respectively. Similarly, Chole the total population of Chole was 88,190 where 50.4 and 49.6% were male and female, respectively (Table 1). From the total interviewed household heads 92.5% were male and 7.5% were female.

Age and family structure of households

It is clear that age of household head, age composition of a family and family size can determines agricultural activities and production of a family as well the community. The life of rural community mainly relies on agriculture which embraces animal rearing and crop cultivation which requires more labor.

The mean age the households considered in the study was 41.87 years (Table 2). Age composition of a family is also important factor in determining families' agri-business. Out of the families considered in the study; the average of total number of family members (TFMS) less than 15, between 15 and 25, between 25 and 55, and above 55 years was 3.74, 2.21, 1.97 and 1.21 respectively. This implies that in each family a number of children was slightly higher (3.74) than other family member groups. Family size also plays a significant role in rural agriculture dependent parents likewise male to female composition of the family. Although there was family with 18 members, the average total family size was 7.29 and there was similarity in mean of male to female composition in the study area (Table 2).

Household occupation

The major occupation of the majority of household heads (86.7%) in the study area was agricultural activity that includes both crop production and animal husbandry. There were also some household members who were involved in non-farm work such as petty trade and daily labor work in addition to agricultural activity. Among the household heads interviewed 5.8% depends only on crop production to lead their livelihoods and livelihoods of about 1.7% household heads depends only on livestock husbandry.

There is division of labor by sex and age. Women and children (above 12 years) contribute to weeding, harvesting, threshing and transporting grain. And ploughing is mainly done by men. The survey result

Table 2. Household characteristics of sample households heads.

Family structure	Minimum	Maximum	Mean	Std. deviation
Age of the respondent	24	69	41.87	11.625
Total family size	1	18	7.29	3.003
Male family size	1	9	3.83	1.908
Female family size	1	9	3.62	1.734
TFMs below 15 years	1	12	3.74	1.897
TFMs between 15 and 25 years	1	9	2.21	1.493
TFMs between 25 and 55 years	1	7	1.97	0.968
TFMs above 55	1	2	1.21	0.419

Table 3. Family labor activities.

Family members labor division	Mean	Std. deviation
Fully involved in agricultural activity	3	1.48
Partly involved in agricultural activity	2	1.72
Fully involved in studying	2	1.72
Not involved in any work	2	1.31

indicated the average labor divisions among family members that is, about 3 individuals are fully involved on farm activities when about 2 family members are partly involved. On the other hand two of the family members are not involved in any activities while two members are fully involved in studying (Table 3).

Educational status of the household heads

Education is one of the important variables which increase an individual's ability to acquire, process, and use agricultural information. Low level of education and high illiteracy rate is typical in developing countries like Ethiopia. In fact, education level of individuals assumed to increase the ability to use the obtained information in a better way.

In contrary to the conditions of developing countries more of the respondents in the study area about 33.3% can read and write whereas the illiterate respondents were only about 22.5%. Furthermore, the rest largest proportion of interviewed farmers attended primary, junior and high school educations formally (Table 4).

Farm resource utilization

Land holdings and land use pattern of the study districts

According to the information obtained from the respective agriculture development bureaus of Honkolowabe and Chole district land holdings were 34,884 and 44,979.3 ha respectively (Table 5). In both districts the land used for crop production takes more proportions. About 74.8% of land in Honkolowabe and about 64.4% of land in Chole was cropped land. Out of the cropped land more was cultivated through rainfed in both districts. However, Chole district holds more total land (Table 4) when compared with Honkolowabe's the average land holdings per household remain lower at Chole (Table 6). This could be due to the conserved/ protected land (18.6%) in Chole is more and high population holdings (Table 1).

Land use type

Rural agricultural families allocate their land for variety of purposes based on the total land they have. Accordingly, the land use patterns of farmers in the study area were cultivated land, grazing land, fallow land, homestead area and wood lands in which the average land holding size was summarized in Table 5. Farm households in both districts allocate more part of the lands for cropping that is, 46.55 and 61.72% of the household's land was cropped land in Honqolowabe and Chole respectively. The average cultivated land holding occupied by individual farmer in both locations was about 1.81 ha. The farmers of Honkolowabe have more grazing land (0.86 ha) and fallow land (0.67 ha) when compared with (0.49 ha) and (0.19 ha) that of Chole. Generally, the Honkolowabe farmers own more land and allocate more land for different purposes than farmers in Chole district (Table 6) that is, the average land holding size per household was 3.91 and 2.90 ha respectively in Honkolowabe and Chole district.

Table 4. Educational status of the household heads.

Educational status	Frequency	Percent
Illiterate	27	22.5
Read and write	40	33.3
Primary school (1-6)	21	17.5
Junior school (7-8)	23	19.2
High school (9-10)	9	7.5
Total	120	100

Table 5. Land use pattern and total land holdings of the districts.

Land use pattern	Total land holdings in ha			
	Honkolowabe		Chole	
	ha	Percent	ha	Percent
Cropped land	26,120	74.8	28,987.3	64.4
(i) Rain-fed	25,840	74	26,636.3	59.2
(ii) Irrigated	280	0.8	2,351	5.2
Forest	1,764	5	1,208	2.7
Grazing	2,600	7.5	3,506	7.8
Settlements	3,500	10	2,909	6.5
Conserved/protected	900	2.6	8,369	18.6
Total	34,884	100	44,979.30	100

Source: Respective agriculture office (2011).

Table 6. Land use type and average size per household at Honkolowabe and Chole.

Land use type	Average land size (ha)			
	Honkolowabe		Chole	
Cropped land	1.82	46.55%	1.79	61.72%
Grazing land	0.86	22.00%	0.49	17.00%
Fallow land	0.67	17.14%	0.19	6.60%
Homestead	0.29	7.41%	0.20	7.90%
Wood lot or forest land	0.27	6.90%	0.23	7.93%
Total	3.91	100	2.90	100

Land leased in or out pattern in 2010

In the study areas, farmers practice both land leased in and land leased out for performing the agricultural activities. From 120 households involved in the study 29.17, 20.83 and 21.17% were practicing land leased in for barley production, for other crop production and for grazing respectively. On the other hand, 5 and 0.83% of the farmers were engaged in land leased out for agriculture and for grazing respectively (Table 7). The survey result also showed that more farmers in Chole practiced land leased in for barley production (74.3%) and for other crops (64%) than farmers in Honkolowabe district (Table 7). In connection to land leased in or out, considering the lands' rental value is crucial. Land rental value can vary based on a variety of factors like agro ecology, soil condition, and land cultivation history and so on. In land leased in and out system of the study area the average rental value of agricultural land and grazing land was 2438.85 and 1596.05 Ethiopian birr/ha respectively during the survey period.

Livestock production

Livestock are one of the important sources of household's income for many rural communities. Livestock have multiple uses aside from income

Table 7. Land leased in or out pattern in 2010.

Land leased in or out	Total No. of farmers		Honkolowabe		Chole		Min. (ha.)	Max. (Ha.)	Mean (ha.)
	No.	percent	No.	Percent	No.	percent			
Leased in for barley crop production	35	29.17	9	25.7	26	74.3	0.25	1.75	0.69
Leased in for other crops production	25	20.83	9	36.0	16	64	0.25	2.50	0.70
Leased in for grazing	26	21.17	14	53.8	12	46.2	0.25	3.00	0.91
Leased out for agriculture	6	5.00	1	16.7	5	83.3	0.25	2.75	0.83
Leased out for grazing	1	0.83	0	0	1	100	1.00	1.00	1.00

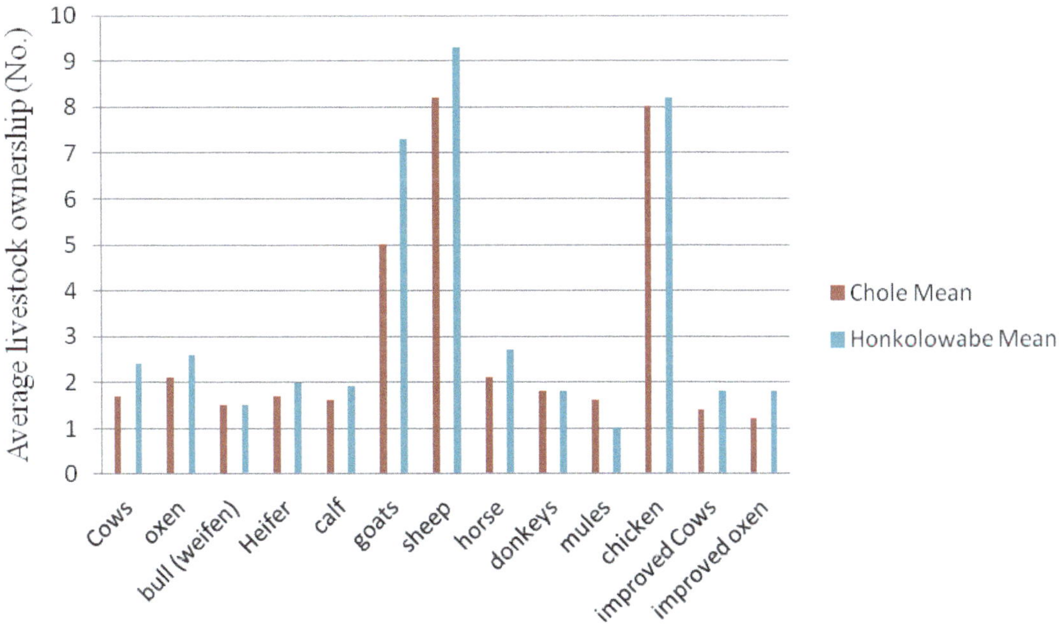

Figure 1. Average number of owned livestock per household in the survey area.

generation, including cash storage for those beyond the reach of the banking system, draught and pack services, milk and meat for household consumption, and manure for fuel and fertilizer (Sintayehu et al., 2010). Accordingly, Oxen have been used as drought power for land preparation, for transporting and threshing crops. Cows give milk to the family's consumption and the sale of milk products serve the family as the major sources of income. Donkeys are used for transporting crops from the field to home area and transporting the sell crops to marketable areas, while horses and mules are used as means of transportation. Average livestock that were owned by sample farmers are shown in Figure 1.

The overall mean of cows, oxen, bull, heifer and calf owned per the household of Honkolowabe district farmers were higher when compared to that of Chole's district (Figure 1). Generally, the result indicates at Honkolowabe the livestock holding per household were higher. Goat and sheep numbers were also relatively higher at this location. According to this survey the ownership of improved oxen and cows in Chole was higher than in Honkolowabe as shown in the figure. The main livestock diseases reported in the case study areas were: internal parasites, blackleg and anthrax. Shortage of grazing lands, unavailability of livestock medicine and long distance to implement artificial insemination are the most cattle production constraints of the study areas.

Crop production

It is clear that crop production pattern of an area depends mainly on agro ecology factors namely climate, soil types, crops type, community crop production habit and also marketing factors. Combining all these elements together Arsi zone is characterized by a cereal dominated farming system in which bread wheat and barley are the major crops (CIMMYT, 2000). Even though wheat and barley

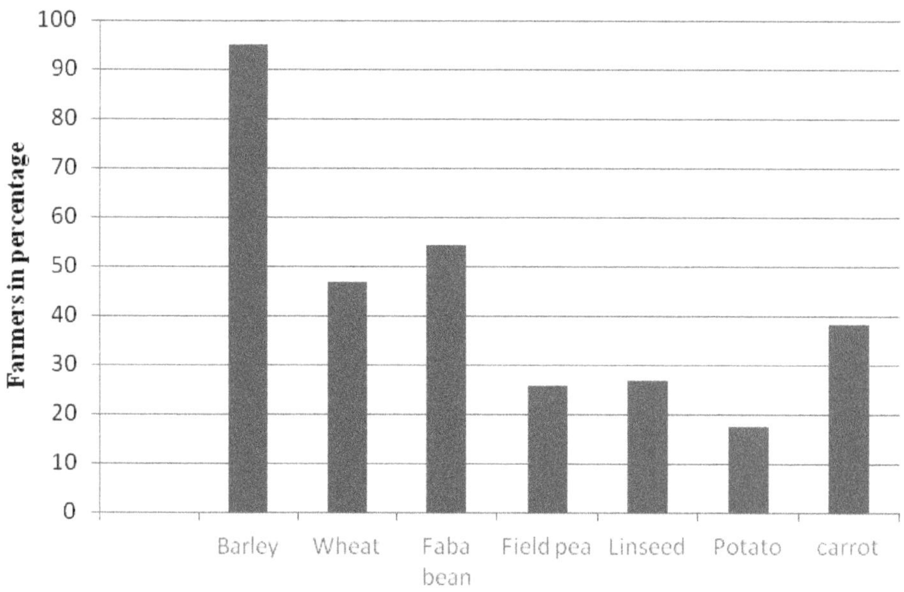

Figure 2. Major Crops under production in the study area.

are more popular in Arsi, crops like faba bean, field pea, linseed, potato, are mainly grown in the study areas. Some vegetables such as potato and carrot are also under cultivation by more farmers in the study area (Figure 2). Some crops like maize, tef, sorghum, rapeseed, onion, garlic and cabbage are also in cultivation in smaller amounts by some farmers. The study result showed that some variation between the districts; however, all the crops shown in Figure 1 are grown at both locations except carrot which is limited only to Honkolowabe district.

Malt barley is the crop grown mostly in the survey area. Yields of the main crops grown in the survey area in the year 2009 were indicted in Figure 3 based on the analysis of survey data. The best mean yield 1.779 t/ha was obtained from variety Holker (malt barley). The second best mean yield (1.608 t/ha) was from local food barley varieties. In spite of the importance of barley as a food and malting crop, its productivity in production fields has remained relatively low (about 1.3 t/ha compared with the world average of 2.4 t /ha) (Bayeh and Berhane, 2011) and they also indicated different biotic and abiotic stresses which are leading to low barley productivity. But, in recent years a concerted effort has been made by researchers, development agents and farmers to increase the outputs and incomes of small-scale farmers by improving the productivity of barley through participatory research and development efforts. As a result the average yield obtained from holker, miscal-21 and food barley varieties increased to 3.2, 3.5, and 4.3 t/ha according to the informations collected from the districts' agriculture offices. Mean yields from wheat and faba bean crops were equal. Almost about the same yield (2.080 and 2.020 t/ha) was also obtained from carrot and potato accordingly. There was also about equal yields of field pea and linseed (0.668 and 0.602 t/ha) respectively. According to the survey results obtained malt barley, faba bean, wheat and food barley were grown by many respondents.

Barley production

Barley production is common in the study areas for variety of purposes like marketing, home consumption in the form of injera, local drinks (tela, areqe), qolo, genfo, kinche, and bread. Barley is also produced for its straw for animal feed and it is used for thatching of roofs. Most of the interviewed farmers had grown local food barley varieties and about 57% of them were growing malt barley variety before five years of the survey time. The farmers accessed to improved barley varieties by AMF, agricultural offices, and research centers. Variety Holker was the most popular to the farmers' of both locations. In 2009/2010 production season, both food and malt barley varieties were grown by about 95% of the total respondent farmers. In similar season, Holker and Miscal-21 (malt barley varieties) were grown by about 62.5 and 12% of respondent farmers respectively. Even though the farmers' land allocation trend for barley was increased, there was farmers (14% of the interviewed) not cultivating barley at survey time.

Knowledge and introduction of improved barley variety

Knowledge is important for farmers to use the improved

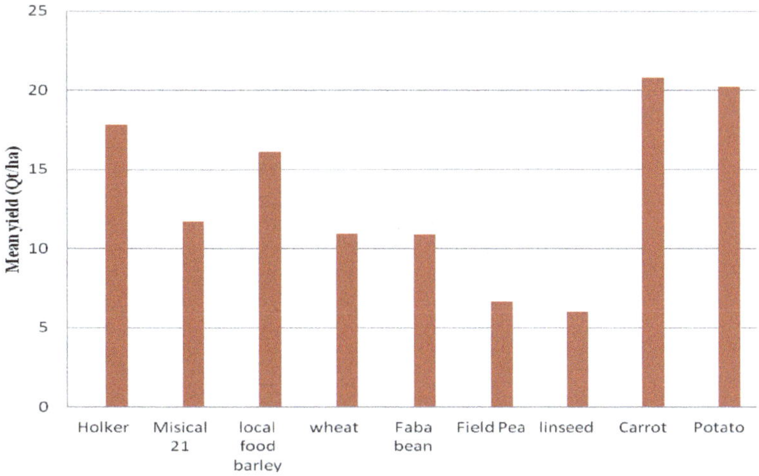

Figure 3. Mean yield (q/ha) of crops commonly grown in the study area.

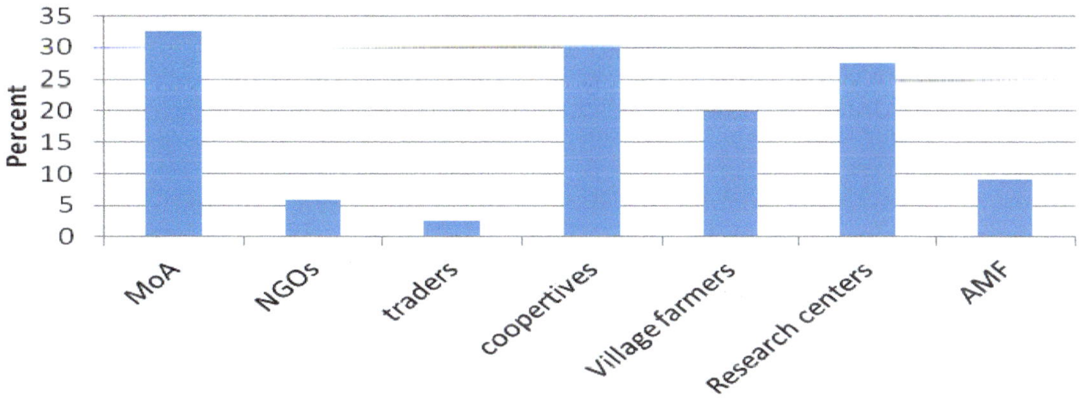

Figure 4. Stakeholders contribution in introducing improved barley varieties.

varieties introduced and knowledge level determines the utilization of the improved technology (Brij, 2006). Farmers of an area obtain information or knowledge from other farmers, bureau of agriculture, NGOs, research centers and the like.

Among the interviewed farmers about 82% have knowledge or knowhow of improved barley varieties while 19% did not. As indicated in Figure 4, bureau of agriculture (MoA) (32.5%), cooperatives (30%), Agricultural research centers (27.5%) and village farmers (20%) were played a significant role in the introduction and provision of improved barley varieties in order to enhance barley production. At the survey time (2009/2010) farmer cooperatives / unions, government (BoA), village farmers, and research centers were used as source of improved barley seeds while Assela malt factory, NGOs, and private seed company (traders) contributed less. The FAO - Crop Diversification and marketing Development Project (FAO –CDMDP), Assela (2010) reported that seed production of the most grown crops (Wheat, Barley, Pulses, and Oilseeds) in Arsi zone have been increased in recent years. According to the report, the newly established Oromiya Seed Enterprise (OSE), Oromiya Bureau of agriculture and Rural Development, FAO, Assela malt factory and other NGOs were facilitating the seed supply where the EIAR and ESE were the producer and supplier of the basic seed. Barley seed produced and the number of farmers engaged in production in 2006 to 2008 was indicated in Table 8.

Malt barley varieties

As to the information obtained from the farmers, the yielding potential and market value were the major reasons for producing some malt barley varieties or preferring its production over the others. The survey result indicated Holker and Misical-21 (malt varieties), HB 1307, Shege, Dimtu, HB-42 and local food barleys were

Table 8. Amount of seed produced by the farmers in 2006-2008.

Years	No of farmers	Production (tons)
2006	46	33
2007	461	375
2008	777	342
2009	411	47

Source: FAO – CDMDP (2010).

grown around Honkolowabe districts in 2009/10. At Honkolowabe malt barley was the most important cereal crop grown by all of the respondent farmers. Crops like linseed, Faba bean, field pea and potato were also grown. In the survey area next to malt barley carrot was the most preferred horticultural cash crop which was grown by about 77% of the respondent farmers.

Farmers in Chole district also grew three malt varieties namely Beka, Holker and Misical -21 and Food barley varieties include: HB-1307, Tesfaye, Nechgebse, Nazo, Zebale, Mwange, Warkiye and local food barley. At Chole food barley was widely grown more than malt barley due to late introduction of malt barley to the area. Other crops (like linseed, pulse crops (Faba bean and field pea), wheat, and potato) were dominantly grown in Chole district.

Generally three malt barley varieties Holker, Misical-21 and Beka were in production. Holker malt barley variety was more popular around Honkolowabe when compared to Chole. As to the discussion made with farmers group, Holker was more preferred for the yield potential and market price value. Farmers are more interested in production of Holker and Misical 21 for the sake of their yield potential, but AMF refused to use Misical- 21 due to lack of limited qualities required for the preparation of malt. Even though the straw of barley was preferable for cattle feed, the semi – dwarf of barley type which gives less straw was highly preferred than the long type which was susceptible to lodging and wind damage. In both locations malt barley produced, was mainly used as cash crop when food barley was used for family consumption.

Pest, disease, weed control and fertilization

In the past years most farmers were not using yield improving inputs through improving the crops' performance and growth. But, in recent years the input utilization level is increasing due to the efforts exerted through the extension activities to increase of level of awareness to inputs even if it is not at adequate level. In order to improve productivity of barley using inputs at right time, right type, and at right rate are important points to be taken into account (Getachew et al., 2011). According to the farmers' explanation, even though they were utilizing inputs, there were problems in application time, in type of inputs and in rate of inputs used.

In 2009/2010 production year, among the farmers involved in the study only 7% used pesticide on barley field to control aphids. Absence of serious pest problem, high price of chemicals and lack of chemicals were the reasons for the reduction of pesticide usage. Leaf rusts and leaf blights were the major barley diseases in the study area while smut and scold were also appeared as barley diseases as indicated by respondent farmers. As a factor of cereal based mono-cropping system grass weeds like *Avena fatua*, *Lolium temulentum,* and *Setaria pumila* were the so important weeds in the study areas. Due to the morphological similarity identification and controlling through hand weeding was difficult except after flowering. In the discussion made with the farmers, they explained grass and other weeds were highly contributing to the lower productivity of barley. In order to control broad leaf weeds 80% of farmers were used herbicides to control weeds in barley while the rest farmers used hand weeding. Similarly, to supply crops with sufficient nutrients important for plant growth and development adding fertilizers to the soil is crucial. From the total farmers interviewed 80% were used fertilizers for barley production. The type of fertilizers used was only DAP and manure as supplementary. Urea which is rich in nitrogen was not totally used. As the result obtained from interviewed farmers, the average rate of DAP fertilizer under usage was 73kg/ha which was much far below from the recommended rate. The survey indicated that there were farmers who used 200 kg/ha of DAP and there were also farmers that used 15 kg/ha DAP for barley production. The amount of fertilizer used by the farmers depends on different factors. As replied by farmers to determine fertilizer rate; some (22.5%) rely on recommended rate, some (41%) rely on crop rotation practiced, some (22.5%) rely on the amount of manure applied, and few (13%) depend on fertilizer price.

Household income and income sources

It is clear that the rural household livelihoods mainly depend on agriculture which includes livestock production and crop production. The life of farmers in the study area relies mainly on crop production which is cereal based farming in which livestock husbandry plays a significant

Table 9. Farmers' income sources and the farmers' percentage on district bases.

Income sources	Honkolowabe (n=60)		Chole (n=60)		Combined (N=120)	
	No.	Percent	No.	Percent	Total No.	Percent
Food Barley	27	45.0	23	38.3	50	41.7
Malt Barley	53	88.3	24	40.0	77	64.2
Other cereal crops	25	41.7	12	20.0	37	30.8
Pulse crops	14	23.3	38	63.3	52	43.3
Oil crops	13	21.7	11	18.3	24	20.0
Vegetable and fruits	42	70.0	13	21.7	55	45.8
Spices production	3	5.0	-	-	3	2.5
Tree wood and wood products	4	6.7	11	18.3	15	12.5
Livestock and Livestock product sales	27	45.0	37	61.7	64	53.3
Non-farm works	7	11.7	10	16.7	17	14.2
Remittance and gifts	-	-	3	5.0	3	2.5

Table 10. Annual income from different commodities.

Annual income	Combined (N=120)		Minimum (in ETB)	Maximum (in ETB)	Mean (in ETB)
	Total No.	Percent			
Food Barley	50	41.7	100.00	12000.00	2643.60
Malt Barley	77	64.2	200.00	26550.00	7156.70
Other cereal crops	37	30.8	160.00	11400.00	3019.46
Pulse crops	52	43.3	200.00	13600.00	2612.92
Oil crops	24	20.0	190.00	15000.00	2377.46
Vegetable and fruits	55	45.8	100.00	63000.00	4403.00
Spices production	3	2.5	3000.00	3000.00	3000.00
Tree wood and wood products	15	12.5	210.00	10000.00	2654.00
Livestock and Livestock product sales	64	53.3	250.00	28800.00	4145.94
Non-farm works	17	14.2	210.00	45000.00	4669.41
Remittance and gifts	3	2.5	2600.00	8000.00	4533.33

ETB= Ethiopian Birr, 1$=18ETB at the survey period.

contribution. The study result indicated that the farm households in the study area generate their income mainly from different cereal crops, pulse and oil crops, vegetable and fruits. Livestock and their products are among the important sources of income for many households. Among 120 interviewed farmers for more than 64% of them malt barley was source of income. Malt barley was income source for more than 88% farmers interviewed in Honkolowabe and for 77% farmers in Chole (Table 9). Next to malt barley, livestock and livestock product sales, vegetable and fruits, pulse crops, food barley, other cereal crops, oil crops and tree wood and wood products were used as income sources for 53.3, 45.8%, 43.3, 41.7, 30.8 and 12.5% farmers respectively (Table 9). Non-farm works was also sources of income for some (14.2%) farmers among the interviewed farmers while income from remittance and gifts was insignificant.

The amount of income earned from different Commodities varied with the suitability of an area for production and might be the farmers' dedication on management to the commodity production. Similarly, the farmers in the study sites generated income from crops production and livestock keeping and also few farmers from non farm works.

In the study area, more farmers generated more income from malt barley production i.e. the mean income earned from malt barley sales by more than 64% respondent farmers was 7,156.70 ETB which was the highest income in comparison with the mean income obtained from the sales of other commodities. The study showed that farmers could get minimum income of 200 ETB and on contrary farmers could earn income up to 26,550 ETB from malt barley crop sales (Table 10). Out of the farmers interviewed about 53.3 and 45.8% of them got mean income of 4145.94 and 4403.00 ETB per year from livestock and livestock product sales, and vegetable and fruits sales respectively.

Barley production opportunities and constraints

Nowadays many factories are engaged in beer production and a large number of beer factories are under construction and expansion, and this shows the presence of higher demand for malt barley products now and in the future. Due to inadequacy of malt barley products with in the country many beer factories are importing malt barley from abroad. So as to enhance its malt production Assela malt factory is also under expansion to maximize the malt supply to the beer factories. As to the information gathered from some experts, Arsi and Bale highlands are suitable and ample for malt barley production to fulfill high malt barley demands with in the country. Different research activities are also undergoing to provide farmers with productive and quality malt barley varieties, and extension and introduction of new improved malt barley to farmers is undergoing by different stakeholders. Combining all these facts together, the opportunity for barley production improvement in the future is expected to be high to fill the maximum demand gap.

Unpredictable rainfall, pests, diseases like scald and net blotch, lack of pesticide, shortages of labor (during weeding, harvesting and threshing), shortage of improved seed, shortage of working capital, low quality of seed, untimely supply of inputs, problem of good market facilities were the major constraints raised by the respondent farmers. Water logging problem in general unpredictable high rainfall in particular and frost problem, climate change from season to season were among the reasons resulting to low grain yield and low quality seed produce of the survey area. Fertilizer and other chemicals were sold to farmers mostly on a cash basis through Ministry of Agriculture and Service Cooperatives. Although farmers were aware of the positive effects of fertilizers on crops, due to financial shortage the resource poor farmers were not used the inputs totally or used under the recommendation rates. This had hindered the effect of inputs on their crop yields.

Insects such as shoot fly, aphids and cutworms caused considerable damage on barley crop. Pesticides against the pest problem were not easily available to farmers in the area. Net blotch, scald, leaf rust and head smut were the main barley diseases reported by the respondent farmers. Farmers mainly complained that they could not get good market facilities especially to malt barley and carrot produces. The outsider traders and brokers were the beneficiary of their product on behalf of the producers. Occasionally there were no malt barley product collectors in Chole district, and farmers were forced to sell with low price to local small traders which were occasionally used to sell to Asella Malt Factory.

Conclusion

Arsi zone is characterized by both crop production and livestock husbandry over which crop production dominates. Crops like wheat, barley, faba bean, field pea, linseed, potato and carrot are the major crops produced in the study locations and crops like tef, maize, sorghum are grown on small plots by some farmers. Crop production in Arsi is highly dominated by wheat cultivation in general whereas barley is commonly cultivated in highlands of the zone. Barley grain is produced for sale, home consumption and for its straw for animal feed. In the past five years household's land allocation for barley production was increased due to the wide works on improved varieties introduction and provision by different stakeholders (agricultural development office, research centers, Assela malt factory, agricultural cooperatives, etc). MoA, research centers, cooperatives/ unions, village farmers were played significant roles in the supply of improved barley seeds where research centers are engaged in improved variety or technology generation and MoA together with cooperatives/ unions works on the technologies extension and dissemination.

Holker, miscal-21, and local food barley varieties were the relatively high yielding varieties and the highly cultivated varieties in the study areas. In Honkolowabe, malt barley was the most commonly grown cereal crop while food barley was more common in Chole. Due to the variation of awareness creation activity in the areas, there was variation of crop cultivation pattern. For instance, farmers in Chole district were aware of the improved malt barley varieties than farmers of Honkolowabe. For enhancing productivity of barley using only the productive and quality improved varieties cannot lead us to the intended target unless we use full production packages at proper time and amount with proper type. In the study areas, the level of pest disease control was low because of shortage and unavailability of pesticides. Furthermore, DAP was only chemical fertilizer which was utilized by the smallholder producers and urea was not used totally.

The increase in number of breweries and their expansion currently and in the future indicates the demand for malt barley product in the future. Research on barley improvement has produced substantial amount of technologies, information and knowledge, and also on producing. More efforts are also underway for the extension of improved technologies to the farmers. These all together shows the opportunities for barley production improvement in the future.

Unpredictable rainfall, barley diseases, lack of pesticide, shortages of labor (during weeding, harvesting and threshing), shortage of improved seed, shortage of working capital, unavailability of quality seed, untimely supply of inputs, problem of good market facilities were the major constraints of barley production in the study area. Significant yield reduction was also appeared sometimes due to the occurrence of unpredictable rain during harvesting time. Hence, taking relevant actions by considering those indicated constraints, can improve barley production and productivity in the study areas.

Conflict of Interest

The authors have not declared any conflict of interest.

REFERENCES

AMF (2012). Assela Malt Factory (2011/2012 (2004EFY). Disclosure Journal. December, 2012. Assela, Ethiopia.

Bayeh M, Berhane L (2011). Barley research and development in Ethiopia – an overview. Ethiopian Institute of Agricultural Research (EIAR), Holetta AgriculturalResearch Center (HARC), P.O. Box 2003, Addis Ababa, Ethiopia.

Brij B, Sharma SD, Sharma RK (2006). Knowledge and Adoption Level of Improved Technology among Rural Women owing to Extension Programmes. Agric. Econ. Res. Rev. 19:301-310.

CIMMYT (2000). The eleventh regional Wheat Workshop for Eastern, Centeral and Southern Africa. Addis Ababa, Ethiopia: CIMMYT.

CSA (2005). Report on Agricultural sample survey: Area and production of crops 2003/04 Meher season.Central Statistical Agency, Addis Ababa, Ethiopia.

Ethiopia-Barley-Business-Case Investor Presentation (2012). The Business Case for Investing in a Malting Plant in Ethiopia, Investor Presentation June 2012. Addis Ababa, Ethiopia.

District Agricultural Development office (DADO), (2011). Chole and Honkolowabe District Agricultural Development office annual report.

FAO-Crop Diversification and Marketing Development Project (FCDMDP) 2010. Seed Value Chain Analysis as a means for Sustainable Seed System: A case of farmers based seed production and marketing in Arsi Zone, Oromia Region, Asella Ethiopia.

Getachew A, Minale L, Adamu M, Abraham F, Agdew B, Fite G (2011). Research achievements in soil fertility management in relation to barley in Ethiopia. Barley Research and Development in Ethiopia, 28–30 November 2006. Holetta Agricultural Research Centre, Ethiopia.

Mohammed H, Getachew L (2003). An Overview of malt barley production and marketing in Arsi. Proceedings of the Workshop on Constraints and Prospects of Malt Barley, Production, Supply, and Marketing Organized by Asella Malt Factory and Industrial Projects Service. March 15, 2003. pp. 1-25.

Sintayehu G, Samuel A, Derek B, Ayele S (2010). Diagnostic study of live cattle and beef production and marketing: Constraints and opportunities for enhancing the system. Accessed at: www.ata.gov.et/wp-content/uploads/Ethiopia on April, 20/2013.

Yosef GH, Kebede T, Senayt W (2011). Achievements of food barley breeding research for low moisture stressed environments of north east Ethiopia. Barley Research and Development in Ethiopia, 28–30 November 2006. Holetta Agricultural Research Centre, Ethiopia.

Consumers' attitude towards local rice production and consumption in Ondo State, Nigeria

Alfred S. D. Y. and Adekayode A. B.

Department of Agricultural Economics and Extension, Federal University of Technology, Akure, Ondo State, Nigeria.

The Nigerian soils are well adapted to the production of rice. In all the ecological zones of the country, rice production is well entrenched. Despite their favourable soil and ecological conditions for rice production, Nigerians still spend substantial proportions of their earnings on imported rice. Why? It is against this backdrop that this study was carried out in Ondo State, Nigeria to investigate the consumers' attitude towards local rice. Two Local Government Areas (LGAs) were selected for the study and from each, two communities were randomly selected. Five wards from each community were selected, from where five respondents per ward were interviewed. The results showed that 93% of the respondents consumed local rice but only 26% consumed more of local rice than the imported. Also, 42% of the respondents consumed local rice for its relative cheaper price in addition, 80% of the respondents purchased their rice from the market while 82% spent less than 24% of their income on rice consumption. The attitude of the respondents from the study showed that, they were indifferent in their preference of local rice to imported rice, but they showed unfavourable attitudes towards importation of rice at the expense of the locally produced one given the favourable natural resources at the nation's disposal. It was recommended that, since the people desired an improvement in the production of local rice, the improvement would likely be in the area of processing which has much to be desired compared with the imported in terms of neatness and attractiveness, efforts must therefore, be made to make the rice more attractive. This can be accomplished by adopting the use of much improved technology.

Key words: Consumers, attitude, rice, consumption, production.

INTRODUCTION

Rice has become a predominantly staple food in Nigeria and other developing countries, having emerged from being a "festivity food" as in few previous decades. Rice provides 20% of the world's dietary energy supplies and it is a good source of thiamine, riboflavin and niacin (Odusina, 2008). An average Nigerian now consumes 24.8 kg of rice per year representing nine percent of total calories in take, giving the fact that the status of rice in the average Nigerian diet has been transformed from being a luxury food item to that of staple (Odularu, 2010). As food production grows at the rate of 2.3% annually, Nigerian population increases at 3.2% (FAO, 2006).

Nigeria has therefore, not been able to produce enough rice for domestic consumption consequent to the mass rice importation with its unattractive consequences on the nation's economy (Daramola, 2005). Globally, there is the need for increase in rice production from 586 million metric tons in 2001 to 756 million metric tons by 2030 (WARDA/USAID, 2003)

Efforts of the various governments and donor agencies have been recorded in policy making, such as in the River Basin Authorities, which had the mandate of cultivating and irrigating the soil among other things for rice production, increase in awareness of improved inputs and credit facility (Bamidele et al., 2010). Despite that these efforts, coupled with those of researches and extension, have not correspondingly translated to sufficient rice production, much more so, there is a higher taste by the people for imported rice (Lancon, 2003). Some of the peoples' complaint for their distaste for local rice has been its poor quality comparative with imported rice. The general observation is that "local rice does not taste well; it is stony and such things". However, these observations are mainly attributed to poor processing methods which allow a mixture of rice products with pebbles (Kassali et al., 2010).

Evidence from the past studies has shown that, the Nigerian local rice has failed to meet the demand of her teeming population not only in quantity but also in quality, largely, due to neglect of the agricultural sector by successive governments (Ogazi, 2009). The crux of this study is therefore hinged on the attitude of the consumers towards local rice production and consumption. If rice must be produced to meet the needs of the populace, if there must also be drastic reduction in the rate of rice importation, local rice must no longer be treated with so much disdain. Specifically therefore, this study sought to:

1. Identify the socio-economic characteristics of the respondents.
2. Ascertain the degree of local rice consumption.
3. Determine the effect of respondents' socio-economic characteristics on local rice production and consumption.
4. Determine the consumers' attitude towards local rice production and consumption.

METHODOLOGY

The study was carried out in Ondo State. The ecological zone, under which the state falls, favors the growth of forest vegetation thus, making the area characterized with a vast of forest land with high relative humidity. The local food crop grown in the state includes rice, maize, beans, yam, cassava, plantain and vegetables. Two Local Government Areas (LGA) being Akure south and Ifedore were randomly selected. Two communities were randomly chosen from each LGA (Ibule Soro and Ilara Mokin from Ifedore LGA and Akure and Oda from Akure South). Five wards were randomly sampled from each selected community giving a total of 20 wards. Furthermore, five respondents were randomly selected from each ward making a total of 100 respondents for the study.

The data used in the study were primary and secondary data. The primary data were collected through the use of well structured questionnaire while the secondary data were from published works and the ministry of commerce and industry. Information was collected on demographic and non-demographic characteristics of the respondents. The data collected were analyzed using descriptive statistics such as frequencies, tables and percentages to show the socio-economic characteristics of the Local rice producers and consumers, their sources of input and consumption level, while correlation matrix was used to show the relationship that existed between the study variables.

Description of variables

Socio-economic characteristics

These villages include age, sex, marital status, and household size, level of education, occupation and income. Questions that elicited the information from the respondents were asked and the responses were coded as:

Age: Less than 25 years, 26-35 years, 36-45 years, 46-55 years and greater than 55 years
Sex: male or female
Marital status; whether single, married, divorced, or widowed
Religion; whether Christianity, Islam or others
Level of Education: Categorized as "No formal education", "Non-formal education", "Primary school education", "Secondary school education" and "Tertiary education"
Attitude: Attitudinal statements were measured on a five-point Likert scale of "very much preferred", "much preferred", "just preferred" "Not preferred" and "very much not preferred". The five-point Likert scale was also as strongly. Agreed", "agreed", "undecided", "disagreed" and "strongly disagreed", values were assigned to each category as 5, 4, 3, 2 and 1 respectively but were reversed (1, 2, 3, 4 and 5) for negative statements. The mean score of response for each statement was obtained and classified as 0.1-1.14, 1.5-2.4, 2.5 – 3.4, 3.5 – 4.4 and > 4.4 and which were interpreted as "strongly disagreed", "disagreed", "undecided", "agreed" and "strongly agreed" respectively. Further interpretation of the classification was also made as "strongly disagreed" and "disagreed" were regarded as "unfavorable attitude", "undecided" as "indifference" while "agreed" and "strongly agreed" were regarded as "favorable attitude"

RESULTS AND DISCUSSION

Table 1 shows the result of the respondents' socio-economic characteristics. It was found that only 17% of the respondents were above 55 years of age while the rest were 55 years and below. It is therefore, inferred that the majority of the respondents ranged between being young and middle aged. These ages could predict what their tastes, likes or dislikes were over some matter, including consumption pattern such as for local rice.

The findings also indicate that both sexes were well represented in the surveys as the males were 58% while the females were 42%. Inferences from this study would therefore be gender friendly. Furthermore, while 22% were single, 70% were married. Consumption of local rice could therefore, be regarded as a family or household decision which would further validate the outcome of the

Table 1. Socio-economic characteristics of t respondents.

Variable	Frequency	Percentage
Age		
1-25	16	16.0
26-35	18	18.0
36-45	23	23.0
46-55	26	26.0
55-80	17	17.0
Total	100	100.0
Sex		
Male	58	58.0
Female	42	42.0
Total	100	100.0
Marital status		
Single	22	22.0
Married	70	70.0
Divorce	1	1.0
Widowed	7	7.0
Total	100	100.0
Religion		
Christianity	77	77.0
Islam	20	20.0
Other	3	3.0
Total	100	100.0
Level of education		
No formal education	13	13.0
Non-formal education	5	5.0
Primary school education	25	25.0
Secondary school education	24	24.0
Tertiary education	33	33.0
Total	100	100.0
Total family size		
1-5	30	30.0
6 – 10	62	62.0
Above 10	8	8.0
Total	100	100.0
Monthly income		
Below 5000	8	8.0
5,000-15,999	31	31.0
16,000-25,999	10	10.0
26,000-35,999	25	25.0
36,000.45,999	16	16.0
Above 46,000	10	10.0
Total	**100**	**100.0**

study. It was further found that, 87% of the respondents attained one level of education or the other. The high literacy rate, could be the fact that, the people in the study area were found in the political region that had early exposure to education consequent of the efforts of their political leader, before and immediate after

Table 2. Rice consumption patterns of the respondents.

Variable	Frequency	Percentage
Consumes local rice		
Yes	93	93.0
No	7	7.0
Total	100	100.0
Frequency of local rice consumption		
Daily	11	11.0
Weekly	34	34.0
Fortnightly	24	24.0
Monthly	18	18.0
Occasional	6	6.0
Nil	7	7.0
Total	100	100.0
Consumes local rice more than imported rice		
Yes	26	26.0
No	74	74.0
Total	100	100.0
Consumers local rice because of its Relatively cheap price:		
Yes	42	42.0
No	46	46.0
Others	12	12.0
Total	**100**	**100.0**

independence and who had as a focus, compulsory and free education. It was therefore, expected that, the information obtained was of valid judgment as they were conscious of the bases of their responses. The finding, on level of education, might equally be the reason while the level of the respondent's income, with a mean of about N27,000 could be regarded as average. With this, they were expected to be able to afford their preferred food item.

Results, as shown in Table 2, revealed that 93% of the respondents actually consume local rice while 1% consumes local rice daily, 34, 24, 18 and 6% consume it weekly, fortnightly, monthly and occasionally respectively. These findings indicate that local rice is still important as staple food in the study area corroborating Odulari (2010) who reported that rice form a staple in the diet of the people. Despite the fact that, imported rice seems to be more appreciated among Nigerian populace, the findings showed that, 74% of the respondents consume more of the rice than the imported. The reason why more of local rice was being consumed than the imported was because of its relatively cheaper price (42%).

Local rice is being sold at a relativity cheaper price so as to encourage patronage and consumption by the producers. This might be in the wisdom of the producers who might have believed that, the processing of local rice has more defects, in terms of dirt content, than the imported corroborating Daramola (2005), FAO (2004) and Bamidele et al. (2010) who observed that local rice is non competitive with the imported rice due to high cost of production and poor processing technique. It is therefore hypothesized that if farmers could adopt more efficient methods of processing, the price of local rice can compete favorably with the imported ones.

Table 3 shows the respondents' perception of local rice. It was found that, even at equal price for both the imported and local rice, 48% still preferred to consume local rice just as 73% agreed to be getting real value for the money expended on local rice. However, the respondents (100%) acknowledged the need for improvement in local rice production, possibly, the processing stage. Furthermore, the findings indicated that, respondents had access to local rice through self production (7%), market contact (80%), as a gift (6%), and cooperative societies (2%) and through other means that were unspecified (5%). This implies that majority of the people buy local rice for consumption. Market forces could therefore, be said to be the determinant of the rate of local rice consumption.

In addition, the results revealed that, as a reason for

Table 3. Respondent's perception of local rice consumption.

Variable	Percentage
At the same price which would you prefer?	
Local rice	48.0
Imported rice	52.0
Are you getting real value for your money?	
Yes	73.0
No	27.0
Would you like improvement in local rice production?	
Yes	100.0
Source of local rice	
Self producing	7.0
Market	80.0
At Gift	6.0
Cooperative	2.0
Others	5.0
Reason for local rice consumption	
More palatable	73.0
Cheaper	21.0
More available	1.0
Others	5.0
Percentage of income spent	
On Local Rice	
10 – 24%	82.0
25 – 40%	16.0
41 – 54%	1.0
Above 54%	1.0
Informal source	90.0
Formal source	10.0

local rice consumption, 73% of the respondents found it to be more palatable, 21% found it cheaper while 1% found it to be more available. Majority of the respondents (82%) was found to be spending between 24% of their income on local rice consumption while 16% spend between 25 and 40% of their income on it. This result could therefore, be said to have confirmed local rice consumption as a staple food in the study area confirming the postulation of Daramola (2005), that the Nigeria's per caput consumption is about 30 kg.

Table 4 shows the attitudes of respondents towards local rice. The findings showed that the mean score of the respondent' responses on their preference for either local or imported rice was 3.32, implying that, respondents were indifferent (undecided) on whch of the two types they preferred. The outcome of the study also showed that 67% of the respondents agreed with the statement that local rice production would soon become a thing of the past", while 36% disagreed. Since 39% was earlier found to have preferred local rice to imported rice, that percentage could not have been part of those (57%) who wished for local rice production terminated, rather, that percentage could likely have included the 22% that never preferred local rice and the 39% who were indifferent to preference for local rice consumption.

The wishing away of local rice production could also be attributed to the taste of the Nigerian populace who has preference for imported commodities generally, and which could be the reason why local rice production is getting dwindling. Giving support to this assertion, further finding showed that, 81% of the respondents supported the fact that, with Nigeria's resources, she should not have been importing rice. This implies that, the neglect of the support for local rice production by the various

Table 4. Attitudes of respondents towards local rice.

Variable	Frequency	Percentage	Means
To what extent do you prefer local rice to imported rice?			
Very much preferred	21	210	1.05
Much preferred	18	18.0	0.72
Undecided	39	39.0	1.17
Not preferred	16	16.0	0.32
Very much un-preferred	6	6.0	0.6
Local rice production would become a thing of the past			
Strongly agreed	34	34.0	1.70
Agreed	23	23.0	0.92
Undecided	7	7.0	0.21
Disagreed	23	23.0	0.46
Strongly disagreed	13	13.0	0.31
With our resources should we import rice?			
Very much supported	57	57.0	2.85
Much supported	24	24.0	0.96
Undecided	11	11.0	0.33
Not Supported	6	6.0	0.12
Very much unsupported	2	2.0	0.02
Total	100	100	4.28

government administrations was likely responsible for the decline in rice production rather than because the people much more prefer the imported rice.

Conclusion

With the abundance of resources; manpower, ecological and other natural resources at the disposal of Nigeria as a nation, it is ironic for her to be largely dependent on importation of rice, to the extent of expending a large proportion of her GDP on it. Nigeria too grows rice, which is normally referred to as local rice. However, some people have greater taste for the imported while some for the local rice. The study has shown that a large percentage of the people consumes local rice, their attitude, however, shows that they were indifferent in their preference of local rice to imported rice. The people also showed unfavorable attitudes towards rice importation at the expense of the locally produced one, given the favorable natural resources at the nation's disposal. Those who consume local rice purchase it from the markets. Some consumers also believed that local rice tastes and cooks better than the imported rice but wished for improvement in the processing techniques. The wish might not be unconnected with the unattractiveness of the local rice in comparison with the imported as a result of foreign objects usually found in it. It is therefore recommended that more effective processing techniques should be adopted to get rid of the foreign contents in local rice to bring it as part with the imported one. In addition, with a large proportion developing good taste for local rice consumption market exits for it, therefore, increase in production should be encouraged by the various levels of government administrations and private investors should also be encouraged.

Conflict of Interests

The author(s) have not declared any conflict of interests.

REFERENCES

Bamidele FS, Abayomi OO, Esther OA (2010). Economic Analysis of Rice Consumption Patterns in Nigeria. J. Agric. Sci. Technol. 12:1-11.

Daramola AG (2005). Government Policies and Competitiveness of Nigerian Rice Economy. A paper presented at the Workshop on Rice Policy and food Security in Sub-Saharan Africa, Organized by WARDA, Cotonou, Republic of Benin.

Food and Agriculture Organization (FAO) (2004). Marketing Integration, Price Transmission and Import Surges. FAO Import Surge Working Paper 6:1-9.

Food and Agriculture Organization (FAO) (2006). Food Production in the Sub- Saharan Africa. FAO of UNO, Rome, P. 23.

Kassali R, Kareem RO, Oluwasola O, Ohaegbulam OM (2010). Analysis of Demand for Rice in Ile-Ife, Osun State, Nigeria. J. Sustain. Develop. Afr. 12(2):63-78.

Lancon F, Ereinstein O, Akande SO, Titilola OG, Ogundele O (2003). Imported rice retailing and purchasing in Nigeria: A Survey. WARDA, Abdijan, Cote D'voire.

Odularu GO (2010). Rice trade policy options in an open developing economy: The Nigerian Case Study. J. Develop. Agric. Econ. 2(5):166-177.

Odusina OA (2008). Urban rice demand analysis: A case study of Ijebu ode township. Middle-East of Scientific Res. 3(2):62-66.

Ogazi CG (2009). Rice output supply response to the changes in real prices in Nigeria: An Autoregressive distributed Lag model Approach. J. Sustain. Develop. Afr. 4(11): 83-100.

West African Rice Development Association (WARDA) (2003). Imported rice retailing and purchasing in Nigeria: A Survey, in "The Nigerian Rice Economy in a competitive World: Constraints, Opportunities and Strategic Choices". Abidjan, Cote D'voire.

Determinants of adoption of upland rice varieties in Fogera district, South Gondar, Ethiopia

Yemane Asmelash

Bahir Dar University, P. O. Box 873, Ethiopia.

The study was conducted in Fogera district, South Gondar zone, Amhara National Regional State (ANRS). Three Peasant Associations (PAs) among the fifteen PAs of upland growers of the district were selected purposively and a total of adopters and non-adopters of 120 selected households were interviewed to generate primary data. Descriptive statistics were used for analyzing quantitative data. Econometric model (binary logit) was employed to determine factors that influence the adoption behavior of farmers respectively. A sum of fifteen explanatory variables for the binary logit model were used, out of which nine variables were found to significantly affect the adoption of improved upland rice variety. These are: sex of the household head, agricultural organization membership, household heads' participation in field days related to improved upland rice, household head contact with extension agents, participation in social organizations, achievement motivation, attitude towards improved upland rice variety, distance to input and output market from residence of the household and active labor force of the household with 15 to 64 age group. Any effort in promoting improved upland rice variety, however, should consider the social, economic, institutional and psychological characteristics for better adoption of the technology. Therefore, attempts to promote transformation in agriculture through improved upland rice variety could be successful if these evaluation criteria and determinants are taken into consideration.

Key words: Adoption, upland rice, binary logit model, New Rice for Africa (NERICA), Fogera district.

INTRODUCTION

Many economic activities depend on agriculture, including marketing, processing, and export of agricultural products. However, agricultural development in Ethiopia is characterized by low productivity due to a number of constraints including: Inadequate production incentives farmers; inadequate generation and application of new technologies; weakness in support service to farmers; and lack of physical infrastructure (FDRE, 1998).

Rice was introduced to Ethiopia in the 1970s and has since been cultivated in small pockets of the country. It is a staple food in the country's east, where rice is imported through Somalia on the black market. Recent surges in demand, especially from city dwellers, are forcing the government to spend large amounts of money on importing rice. If successful, initiatives to boost rice production in Ethiopia can help the country achieve food security. The rice production system in the country has focused mainly on the introduction of improved varieties from a range of different sources, including the International Rice Research Institute (IRRI), the Africa Rice Center (WARDA), Guinea, and Madagascar. Federal and regional research centers are concentrating on the evaluation and release of new varieties for local producers. Three improved irrigated varieties from IRRI and four "New Rice for Africa" (NERICA) Varieties grown in the rain-fed uplands, where farmers do not have access to irrigation systems registered yields of 3 to 6 tons per hectare (IRRI, 2006).

The IRRI varieties, grown in lowland irrigated conditions, achieved 6 to 8 tons/ha. The Sasakawa Africa Association (SAA) through its Sasakawa Global 2000 (SG 2000) program has played a key role in promoting NERICA and other cultivated varieties to the country. In Addition, the Japan International Cooperation Agency (JICA) and SAA have supported the introduction of essential post harvest (storing, milling, and drying) technologies and processing machinery to rice-producing areas.

Despite several efforts to promote these technologies, adoption levels remain low. Therefore, the study aimed at analysis the social-economic determinants of adoption of improved rice varieties by small-scale farming households.

RESEARCH METHODOLOGY

The study area

The study was conducted in Amhara National Regional State (ANRS), South Gondar zone, Fogera district (Ethiopia). The District is geographically located 11° 58'N latitude and 37° 34'E longitudes.

Sampling and data collection

In this study, both primary and secondary data sources were used. Quantitative and qualitative data were gathered from different sources of the study area. Discussion with group of farmers and agricultural extension staff was done to generate information. Key informants were also used as information source from different actors. A two stage sampling procedure was used to select the Peasant Associations (PAs) and sample households. In the first stage, three PAs were selected purposively from 15 PAs where upland rice was producing, based on their potential of improved upland rice production and their relative access to past extension support. In the second stage after lists of farmers were obtained from the district Agricultural and rural development office, farmers who were cultivating upland rice in three PAs that were all 60 adopter sample household heads were taken as respondent using probability proportional to size. Equal numbers of non-adopter respondents were selected using simple random sampling method based on their proportion. The data was collected from September 2010 up to January 2011 for four months.

Analytical framework

In this study, descriptive statistics (percentage and mean) were mainly used. The descriptive analysis was conducted using Statistical Package for Social Science (SPPS). The econometric model applied for analyzing determinants of adoption of improved rice variety is the logit. This model was chosen because it has advantage over other model (LPM, Tobit and Probit) in that, it is an extremely flexible and easily used function. As a result, only the logit model was reported in the paper even if both models were estimated for the purpose of comparison.

In this study, the logit model was used to estimate the probability of adoption of improved upland rice varieties that takes either one of the two values of Y = 1 for adoption and Y = 0 for non-adoption of upland rice varieties. The functional form of the model is presented as follows:

$$Prob(Y=1) = \ln\left(\frac{p_i}{1-p_i}\right) = \beta_o + \beta_1 x_1 + \beta_2 x_2 + \beta_3 x_3 ... + \beta_i X_i + U_i$$

Where, β_o is the constant, $\beta_i = i = 1,2,...n$ are the coefficients of the independent variables to be estimated. x_i is a vector of independent variables; U_i is the error term with zero mean and constant variance. Farmers' decision to adopt or reject new technology is influenced by the combined effects of factors related to farmers' objectives and constraints such as farmer specific variables (age, education, farm size, etc).

Variables in the model

The logit model discussed above suggests many important hypotheses related to the relationship between the adoptions of improved upland rice variety; personal, economic, institutional and socio-psychological factors. The model assumes that the dependent variable which is adoption of improved upland rice varieties depends on the independent variables of age, sex, educational status, farm size, livestock ownership, active labor force (15 to 64 age), access to credit, distance to the nearest market, extension agent contact, field day participation, agricultural organization membership, social participation, cosmopoliteness, achievement motivation and attitude. A list of all the hypothesized independent variables therefore were decided to be included in the model for analysis along with their measurements and hypothesized signs are given in Table 1.

RESULT AND DISCUSSION

Effects of explanatory variables on adoption of improved upland rice variety

As portrayed from Table 2, mean age of the adopters was 44 as compared to 45 years for the non-adopter household heads. The result of t-test showed that there was no significant mean difference (t-value= 0.674, P = 0.502) among farmers category indicating the absence of significant relationship.

The average total land holding of the sample households were 2.28 hectare for adopters and 1.90 hectare for non-adopters (Table 2). Statistical analysis illustrate that mean difference is statistically significant at 10% significance level that is (t-test = 1.887, P-value = 0.062). The results of the study concur with the past research finding of Yishak (2005).

Table 1. Independent variables, hypothesized signs and their measurements.

Independent variable	Expected sign	Variable measurement unit
Age of the household head	Indeterminate	Years
Sex of the respondents	+	Binary variable: 1= male, 0 = female
Educational status	+	Binary variable: 1 = literate, 0 = illiterate
Farm size	Indeterminate	Hectare
Livestock ownership	+	Number of livestock ownership per household in Tropical Livestock Unit (TLU)
Active labor force	+	Man equivalent
Access to credit	+	Binary variable: 1=literate, 0=illiterate
Distance to the nearest market	-	Kilo meter
Extension agent contact	+	Binary variable: 0 = no, 1= yes
Field day participation	+	Binary variable: 0 = no, 1= yes
Agricultural organization membership	+	Binary variable: 0 = no,1= yes
Social participation	+	Binary variable: 0 =no, 1= yes
Cosmopoliteness	+	Binary variable: 0 =no, 1= yes
Achievement motivation	+	Scale:1 = disagree, 2 = undecided and 3 = agree.
Attitude	+	Likert Scale: 1-strongly disagree, 2-disagree, 3-undicided, 4-agree and 5-strongly agree

Table 2. Relationship between continuous variables and mean adoption of improved upland rice variety.

Continuous variables	Adopters (n = 60)	Non-adopters (n = 60)	T-test	
			T-value	P-value
Age of the household head (year)	43	44	0.674	0.502
Total household farm size (hectare)	2.28	1.90	1.887	0.062
Livestock ownership (tropical livestock unit)	9.95	8.55	1.527	0.129
Household family size (number)	3.75	3.22	2.258	0.026
Distance from market center (km)	12.250	9.500	-1.897	0.060
Achievement motivation (scale)	14.550	1.600	2.384	0.019
Attitude (Likert Scale)	31.333	29.150	3.585	0.000

To know whether there is a significance mean difference between adopters and non-adopters in average livestock ownership, t-test statistic was conducted. The result of the t-test (t-value = 1.527, P-value = 0.129) revealed that there is no significant difference in average livestock ownership between adopters and non-adopters as indicated in Table 2.

As shown in Table 2, the average active labor force of the respondents was 3.48 members. The result shows there is positive significance mean difference among farmers category (t-value = 2.258, P-value = 0.026) and active labor force has been found to be positively and significantly influence at 5% significant level with upland rice adoption.

Regarding the distance taken to travel from home to the nearest market place, mean distance to travel to the nearest market center by adopters and non-adopters was 12.250 and 9.500 Km, respectively. Results of t-value (t-value = -1.897, P-value = 0.060) reveals that there is significant mean difference among adopters and non-adopters of farmers (Table 2).

As presented in Table 2, evident from significant mean difference between adopters and non-adopters in achievement motivation of sampled households in which it was found to be 15.183 and 14.550 mean variation respectively. The t-test assured that there is significant difference at 5% significance level.

According to the result of the study from Table 2, there is high significant variation between adopters and non adopters at less than 1% significant level and attitude was found to be influential factor on adoption of improved upland rice variety. This study concedes with the research result obtained by Tesfaye (2009).

As described in the Table 3, out of 120 respondents, 82.5% were males and the rest 17.5% were females. The majority of females were found in non-adopter category of farmers which indicates they are not efficient and able to adopt upland rice variety as compared to their male counterpart. The result of chi-square test $x^2 = 9.755, P-value = 0.002$ revealed that there is significant relationship between sex and the adoption of upland rice variety at less than 1% significance level.

Table 3. Relationship between categorical variables and adoption of improved upland rice variety.

Categorical variables	Percentage proportion across farmers category		Chi-square	
	Adopters	Non-adopters	χ^2	P-value
Sex of the respondents				
Male	93.3	71.7	9.755	0.002
Female	6.7	28.3		
Educational status				
Illiterate	41.3	79.6	10.258	0.016
Literate	58.7	21.4		
Access to credit				
Yes	61.7	45.0	3.348	0.067
No	38.3	55.0		
Contact to extension agent				
Yes	76.7	45.0	12.626	0.000
No	23.3	55.0		
Participation in field day				
Yes	83.3	38.3	25.497	0.000
No	16.7	61.7		
Agricultural organization member				
Yes	88.3	46.7	23.742	0.000
No	11.7	53.3		
Cosmopoliteness				
Yes	83.1	74.6	8.009	0.091
No	16.9	25.4		
Social participation				
Yes	86.7	41.7	26.421	0.000
No	13.3	58.3		

Out of the adopters of improved upland rice varieties, 41% were illiterate, and 59% were literate (Table 3). The existence of a literate in the household enhances the technology adoption. The result revealed that there is difference between education and adoption of improved upland rice variety.

This study revealed that the hypothesized sign is similar with the findings (Table 1). So the significant relationship (χ^2 = 3.348, P-value = 0.067) between credit and adoption of upland rice variety is supported by the mentioned hypothesis and it is significant at 10% significance level as shown in Table 3.

The relationship between extension contacts and adoption of improved upland rice variety was found to be significant as shown in Table 3. From the total sample households 39% were reporting not having contact with extension agent. From the non-adopter categories, 55% of respondents did not have any contact with extension agents. The percentage of respondents not having contact with extension agents; larger shares comes from the non-adopters as compared to the respondents of adopter categories. The chi-square analysis result (χ^2 =12.626, P-value = 0.000) shows significant relationship of contact of extension agent with the adoption of improved rice variety. Earlier studies by Abrhaley (2007) and Tadesse (2008) also revealed similar results.

The participation of respondents in field day with adopters and non-adopters of upland rice variety can be observed in Table 3. To determine the relationship between field days with the adoption of improved upland rice variety chi-square analysis was conducted. The chi-square analysis showed (χ^2 = 25.497, P-value = 0.000) in Table 3 that there existed a significant relationship between them at less than 1% probability level.

As was expected, the membership of cooperative so society had significant relationship (χ^2 = 23.742, P-value =0.000) with the adoption of improved upland rice variety at 5% level of significance (Table 3). The majority (67.5%) of sample households were found to be cooperative members and the rest 32.5% were reported to be non members. From non adopter group, only 46.7% respondents were the member of agricultural organization.

The statistical analysis in Table 3 (χ^2 = 8.009, P-value = 0.091) revealed that there existed a significant relationship between cosmopolitness and adoption of improved upland rice variety at 10% significance level. As indicated in Table 3 the mean scores for sample households were 13.325, while the mean score of social participation for non adopters was 13.100. The results (t-value = 3.585, p-value = 0.001) reveals statistical significant mean difference farmer categories in relation to social participation score at less than 1% probability level. This study is in line with Tesfaye (2006) where he found a positive relationship between adoptions of rain water harvesting and social participation.

Factors determining the adoption of improved upland rice variety

All the hypothesized variables therefore were decided to be included in the model for analysis. In this case, a farmer who used improved upland rice variety is considered to be "an adopter". The dependent variable is either adopter or non-adopter of improved rice variety and logit model was used to estimate the determinants of the hypothesized independent variables on adoption of improved upland rice variety.

The maximum likelihood estimate of the parameters and the effect of independent variables on probability of adoption was analyzed and presented in Table 4.

Various studies showed that adoption of improved technologies have relation with sex of the respondents. The model result showed that a sex difference among the farm households was positively influencing adoption of improved upland rice varieties at 10% significance level. The male-headed households were more likely to adopt improved upland rice varieties than female-headed households. The model revealed that maleness was found to have 26.1% contribution to adoption of improved upland rice varieties. The possible explanation might be male-headed households had better access to information and agricultural inputs. Previous studies reported similar result regarding the effect of sex of the household head on adoption. For example, Minyahil (2008) found that male-headed households are more likely to adopt improved technologies of Triticale than female-headed households.

Participation in agricultural organizations had a positive influence on adoption of improved upland rice variety at 5% level of significance. The model result showed that as the household head involves in agricultural organization, the probability of adoption of improved upland rice variety increase by 81.0%. From this result, it is important that organizing farmers in different agricultural organizations would help farmers to have access to credit, access to extension and to market. This indicates strengthening and expansion of rural organizations is of paramount importance to enhance adoption of improved upland rice variety. The result goes along with the study done by Dereje (2006).

When farmers practically observe a new practice they can weigh the advantage and disadvantages of the new technology. This can facilitate adoption and it can help them to practice it in the right way. It was found that exposure to information due to participation in field days had a positive and significant influence on the probability of adoption of improved upland rice variety at less than 1% significant level. The result of binary logit model in relation to this variable shows that as farmers' exposure to agricultural practical information increases, adopting improved upland rice variety increases by a 67.8%. This implies that farmers who have opportunity to attend field days of improved upland rice variety are more likely to use improved upland rice than those farmers who have no similar opportunity. This result goes along with the study done by Tadesse (2006, 2009).

Extension service as a source of information regarding the importance of upland rice variety has a strong influence on the farmers' adoption decision. This is because, the more agricultural advice a farmer gets and the more closely on using recommended technologies are observed. The variable is positively significant at 5% significance level and the model result showed that if a farmer gets agricultural advice from extension agents, the willingness to adopt improved upland rice varieties increases by 85.4%. Similar results were reported by Bezabih (2000).

Farmers' participation in social organizations was used as important aspect for access to information. This variable is correlated with the probability of adoption positively and is significant in explaining adoption of improved upland rice varieties. Those farmers who participate in social organizations are capable to use the available information related to new technology. The model result shows that as household head involves in social organizations, the probability of adoption of improved upland rice varieties increase by 9.3%. Similar results were reported by Yishak (2005).

Achievement motivation is something that causes a person to make an effort to become successful and be goal oriented. The model output result revealed that a one unit variation in the achievement motivation increases the adoption of improved upland rice varieties by 76.8%. This implies that a strong need for achievement leads for technology adoption. Similar result was reported by Yitayal (2004).

Farmers attitude towards the improved upland rice

Table 4. Binary logit estimates of adoption of improved upland rice variety.

Independent variables	Coefficient	Wald statistics	Exp(β)	Significance
Sex	0.261	3.432	0.104	0.064
Credit	-0.375	0.217	0.687	0.641
Agricultural organization membership	0.810	6.176	0.060	0.013
Education level	1.044	0.361	0.048	0.548
Age	-0.040	0.798	0.960	0.372
Cosmopoliteness	1.060	6.119	1.024	0.130
Participation in field day	0.678	7.960	0.069	0.005
Contact to extension agent	0.854	5.129	0.157	0.024
Social participation	0.093	4.846	0.123	0.028
Total livestock	-0.050	0.345	0.951	0557
Achievement motivation	0.768	5.621	2.155	0.018
Attitude	0.538	9.921	1.712	0.002
Farm size	0.067	0.025	1.069	0.874
Distance to market	-0.190	6.474	1.209	0.011
Active labor force	0.842	4.981	2.321	0.026

varieties was found too positively and significantly affect the adoption of improved upland rice varieties at 1% significant level. The result showed that, those farmers who had higher attitude score towards the improved upland rice varieties have adopted than those farmers who had lower attitude score towards the technology. According to the result of econometric model, the variable accounted 53.8% variation in the adoption of improved upland rice varieties. Similar result was reported by Yitayal (2004) after conducting a study on determinants of use of soil conservation measures by smallholder farmers in Jima Zone.

The model result shows that this variable influences the probability of adoption of improved upland rice variety negatively and significantly at less than 5% significant level. This implies that farmers who are distant from input and output market have less likelihood to adopt the improved upland rice variety. The model result indicated that as input and output market distance from the residence of the household head increases by one unit adoption of upland rice variety decreases by 19.0%. The probable reason could be farmers far away from the market center were not motivated to produce upland rice due to poor access to information, infrastructures and extension services. This result goes along with the study done by Abrhaley (2007). This result is contradictory with the finding of Yishak (2005).

Households with high labor availability are more likely to adopt improved upland rice variety than households with low labor availability. The model outputs showed that labor availability was found to be statistically significant at 5% probability level with the expected value positively related with the adoption of improved upland rice variety. Model results confirm that adoption of improved upland rice variety increase by 84.2% for a unit increase of total household in man days. The probable reason for this result might be due to the fact that upland rice seed production is labor intensive crop particularly at ploughing and weeding periods. Similar result was reported by Tesfaye (2009).

CONCLUSION AND RECOMMENDATION

This study has identified key socio-economic factors that influence adoption process in the study area. According to the findings of this research, it is necessary to establish appropriate extension strategy to bring those non-adopters to adopt improved upland rice varieties. In this regard, attention should be given; to encourage, support and motivate female farmers to achieve their adoption decision behavior.

Being a member of agricultural organization was also positively and significantly related with adoption of improved upland rice varieties. Members of agricultural organization have got credit, seed, fertilizer and other farm inputs at reasonable price from the organization's shop. So strengthening and expansion of agricultural organizations is one means to enhance improved upland rice production in the area. Participation of farmers in extension event, especially in field day has to be strengthened so as to improve farmers' access to information and extension advice.

Contact to extension agents has positively influenced adoption of improved upland rice varieties. This implies that the need for more targeted and continued extension services. Thus, the extension system operating in the area, needed to be strengthened further to increase the flow of information for overall development of the rural area.

Members in different social organizations are more likely to be aware of new practices as they are easily exposed to information. The social organizations found around the study area should be strengthened and much effort should undergo to capacitate their human and financial resources so that they can be the center for technology transformation as well as for the overall development of the agricultural sector. Farmers with strong need for achievement require encouragement and special support in terms of rewards and financial aspects. So the concerned bodies should formulate a strategy for rewarding and recognition for those farmers who are genuinely successful and be goal oriented.

Result of this study indicated that there was significant difference in adoption among farmers, who perceived the improved upland rice varieties positively and not positively, hence emphasis has to be given to create awareness with tangible evidence is important for adoption of improved upland rice varieties.

Distance to the input and output market from residence of the household head was found to be the determinant of adoption of improved upland rice varieties in the study area. One way forward to increase market access of rural people is to improve the infrastructures and to increase the number of roads in each PAs to close average distance gap and as a result vehicles could be easily accessible so that rural people visit to nearby market. Therefore, attention should be given to developing infrastructure and transportation.

The study result also revealed that one of the most important constraints and showed significant influence on adoption of improved upland rice variety is labor shortage at weeding which calls for further research. Even though the researcher considers that efforts are undergoing to develop upland and low land rice production packages/practices at MOARD level, this should have to be conducted with the involvement of farmers.

Farmer- to-farmer seed exchange should be promoted. Farmers should be selected and trained in quality seed production. Knowledge sharing also should be encouraged through informal organizations.

Generally, the result of this study indicates that adoption of improved upland rice variety is a result of an interplay of several factors, which need due attention.

Conflict of Interest

The author(s) have not declared any conflict of interest.

REFERENCES

Abrhaley G (2007). Farmers' perception and adoption of integrated striga management technology in Tahtay Adiabo Woreda, Tigray, Ethiopia. M.Sc Thesis Submitted to School of Graduate Studies, Haramaya University.

Bezabih E (2000). The role of new varieties and chemical fertilizer under risk: The case of small holders in Eastern Oromia, Ethiopia. PhD dissertation, University of Hannover, Shaker Verlung. Germany.

Dereje H (2006). Assessment of farmers' evaluation criteria and adoption of improved bread wheat varieties in Akaki, central Ethiopia. M.Sc Thesis submitted to school of graduate studies of Haramaya University.

FDRE (Federal Democratic Republic of Ethiopia) (1998). PADEP (Peasant Agricultural Development Project), Addis Ababa, Ethiopia.

IRRI (International Rice Research Institute), (2006). Major research in upland rice. Los Baiios, Philippines.

Tadesse A (2008). Farmers evaluation and adoption of improved onion production package in fogera district, South Gonder. M.Sc Thesis (Unpublished) presented to school of graduate studies of Alemaya University. P. 79.

Tesfaye A (2009). Effectiveness of upland rice seed change farmer-to-farmer in Fogera district, South Gonder, Ethiopia. M.Sc. Thesis (Unpublished) Presented to School of Graduate Studies of Alemaya University, P. 79.

Tesfaye W (2006). Analysing factors affecting adoption of rain water harvesting technology in Dugda Bora woreda, Ethiopia. M.Sc. Thesis (Unpublished) Presented to School of Graduate Studies of Alemaya University. P. 79.

Yishak G (2005). Determinants of adoption of improved maize technology in Damote Gale Woreda, Wolaita, Ethiopia. M.Sc. Thesis Presented to School of Graduate Study of Alemaya University.

Yitayal A (2004). Determinants of use of soil conservation measures by smallholder farmers in Jima Zone. The case of Dedo district. Ethiopia. M.Sc Thesis, Alemaya University.

Utilization of soyabean products by farm-families in Abia State, Nigeria

S.O. Olatunji* A. C. Agumagu and O. M. Adesope

Department of Agricultural Economics and Extension, University of Port-Harcourt, Nigeria.

Soyabean has been nicknamed "the poor man's meat" because it contains 40.3% protein, 20% oil and 91.9% total digestible nutrient. This study assessed farmers' awareness, knowledge of preparation and rate of utilization of selected soyabean products. Multistage sampling method was employed to select a sample of 137 farm-families who responded to a 45-item structured interview schedule used for data collection. Data were subjected to descriptive statistical analyses. Among others, it was found that between 73.7 and 93.4% of the farmers are aware of the various selected soyabean products. Friends and neigbours are farmers' most important sources of information about soyabean, followed by television/radio while extension agents ranked third. Between 42.3 and 70.1% of the farmers know how to prepare soya flour, soya milk, soya moinmoin and soya vegetable soup but majority do not possess adequate knowledge for preparing soya meat, soya snacks and soya candies. About 32.9% of infants/children and 61.3% of pregnant women/lactating mothers were utilizing soyabean products at a very low rate of "once a week". It was concluded that majority of the farmers are utilizing soyabean at a lower than adequate rate. It was recommended that more extension efforts be focused on creating awareness of the nutritional values of soyabean and dissemination of preparation technologies for products that most farmers do not know how to prepare. Pregnant women/lactating mothers and infants/children (who are more vulnerable to protein-deficiency diseases) should be persuaded to increase their frequency of utilizing soyabean products.

Key words: Soyabean products, farm-families, utilization, protein, nutritional deficiency diseases, Abia State, Nigeria.

INTRODUCTION

The importance of agriculture to national development cannot be over-emphasized because agriculture is the major source of livelihood for man and animals. Nobody can do anything without food. Generally, all nutrients that are essential for normal growth and development of humans are derived from Agriculture. These include carbohydrates, protein, vitamins and minerals and oil. Carbohydrates refer to any of the various types of food such as cake, bread and potatoes which consist of oxygen, hydrogen and carbon and which provide the body with heat and power (energy). Protein, on the other hand, is any of many substances present in such foods (as meat, eggs, soyabeans, etc) that help to build the body and keep it healthy. Because of its significant roles of body building and maintenance of health, protein has remained a very crucial element in the diet of man. Thus, the need for protein in the diet of man cannot be overemphasized. However, increasing high costs has continued to make protein from animal sources unaffordable to a vast majority of Nigerians, especially farmers. Igbedioh (1990) reported that morbidity and mortality arising from protein calories malnutrition in Nigeria was on the increase. Atinmo (1982) also asserts that malnutrition, especially protein-malnutrition in young children, continues to be the first and most outstanding health problem in Nigeria.

It should be noted that majority of Nigerians in rural areas, are resource-poor and hardly able to afford even their carbohydrates daily requirements talk less of protein. Protein from animal sources is unaffordable by most farm-families, hence the widespread of preventable protein-deficiency diseases in most rural communities in

*Corresponding author. E-mail: sooolatunji@yahoo.com.

Nigeria. It was becoming increasingly necessary to supplement protein requirement of Nigerians (especially the farm-families) with relatively inexpensive plant protein as noted by Weingartner (1987), cited in Tiamiyu et al. (2001). Research evidence has shown that soyabean (glycine max (l) (merrill) has higher and virtually unrivaled protein content than other crops and that it is a rich source of edible vegetable oil, vitamin and minerals. Soyabean, popularly called the "poor man's meat", contains 40.3% protein, 20% oil and as high as 91.9% total digestible nutrient (Fukushima, 2000; Fabiyi, 2007).

Health benefits of soyabean include prevention of heart diseases, cancer, high blood pressure, diabetes-related diseases and many others. Soyabean oil is rich in fatty acids and devoid of cholesterol. It is an excellent source of calcium, iron, and vitamins such as niacin, thiamin and riboflavin (Weingartner, 1992; World's Healthiest Foods, 2004).

Although most farmers in the area of study do not grow soyabeans, it is produced in large quantity in the Middle-belt and Northern States of Nigeria and is readily available in most markets In the Southern and Western States due to virile activities of merchants. Soyabean products and their uses have been promoted in Nigeria (Odebode, 2005) and in the Eastern States of Nigeria through the extension unit of the agricultural development programme (Mathews-Njoku, 2005). The promotions of soyabean by the agricultural extension service (especially the women in agriculture- WIA) have lead to the emergence of several varieties of soyabean products in both rural and urban markets in Nigeria.

Prominent among soyabean products that farm families can readily take advantage of in improving their protein intake are: soya flour, soya milk, soya-pap (ogi or akamu), soya vegetable soup, soya moinmoin, soya snacks, soya meat, soya candies and soya akara. Soya flour or granulated soya is made from dried and grounded soyabeans. To make soya milk, the beans are soaked in water, coats removed, ground and boiled gently, sieved to remove the chaff. Soya palp Is produced from a mixture of maize and soyabean grounded to paste and prepared by adding boiling water. Soya vegetable soup is prepared by grinding the beans to fine paste after removing the coat. The paste is added to vegetable and other ingredients to make soup. Soya moinmoin is made from boiling soyabean paste mixed with other ingredients to taste. Soya snack is made from roasted well-drained beans plus oil, salt and sugar. Soya meat is made from boiling soya milk and magnesium sulphate until it coagulates, and solidified. Then, cook as scrabled eggs or add to stew to supplement meat. Soya candies are made from roasted beans added to sugar, water and lime melted in a frying pan. Soya akara is prepared by mixing beans and soyabeans paste (with ingredients) and deep fry with groundnut oil. It is expected that utilization of these soyabean products will adequately meet protein nutrient needs of the farm families including children, pregnant women and lactating mothers who have been the most vulnerable to protein-deficiency-related diseases and deaths.

There is inadequate information on the extent of adoption of recommended uses of soyabean products in the area of study. The knowledge of farmers' awareness, sources of information, knowledge of preparation and frequency of utilization of these soyabean products would be valuable and pertinent in assessment of the impact of agricultural extension efforts aimed at promoting utilization of soyabean products by farm-families in the recent past. The research will equally provide a basis for re-invigorating efforts at increasing farmers' awareness and utilization of these "life-giving' soyabean products if their utilization is at low ebbs. Indeed, agricultural development in Nigeria would most likely than not depend on existence of healthy work force, especially the farm-families.

In view of the foregoing, answers were provided to the following research questions:

1. What percentage of farmers Is aware of the selected soyabean products?
2. What are farmers' sources of information?
3. What is the percentage of farmers who have knowledge about preparation of selected soyabean products?
4. What is the frequency of utilization of soyabean products by sampled farmers?

METHODOLOGY

The population of study includes all farm families in the three agricultural development programme (ADP) zones of Abia State (Aba, Umuahia and Ohafia) (Appendix, Table 1). Multistage sampling technique was employed to select sample, considering that Abia State have been delineated into zones, blocks, circles and sub-circles. Two zones (Umuahia and Ohafia) were purposively selected from the three ADP zones in the State. Using simple random sampling technique at each stage, 3 blocks were selected from each zone, 2 circles from each block, 2 sub-circles from each circle and six (6) farm-families from each sub-circle. Thus, samples of 72 farm-families each were selected from Umuahia and Ohafia zones respectively. Most farm families in the area of study are made up of 5 to 7 persons. These include father, mother, and 3 to 5 children. In each of the sampled farm families, the father, mother or a grown up child (above 18years of age) who was accessible was interviewed and his or her responses taken as data for that farm-family. The sample size is 144 but only 137 subjects who responded to the entire interview schedule were included in the study. A 45-item structured interview schedule was developed by the researcher, validated by test and measurement experts and pre-tested on similar, but not the same, sample. The reliability coefficient of 0.71 obtained was considered adequate as a justification for using the instrument. Relevant data were subjected to descriptive statistical analysis (mean, percentage and ranking).

RESULTS AND DISCUSSION

From Tables 1 to 5, results of data analysis are presented

Table 1. Percentage distribution of farmers on the basis of their awareness of selected soyabean products.

S/N	Products	Farmers not aware	Percentage	No of farmers who are aware	Percentage of farmers who are aware	Ranks of percentage awareness
1	Granulated soya	14	10.2	123	89.8	2
2	Soya milk	9	6.6	128	93.4	1
3	Soya candies	24	17.5	113	82.5	6
4	Soya ogi (akamu)	16	11.7	121	88.3	3
5	Soyavegetable soup	34	24.8	103	75.2	8
6	Soya meat	36	26.3	101	73.7	9
7	Soya snacks	32	23.4	105	76.6	9
8	Soya Akara	21	15.3	116	84.7	5
9	Soya moinmoin	18	13.1	119	86.9	4

Table 2. Percentage distribution of farmers on the basis of their sources of information about soyabean products.

S/N	Soyabean product	Not aware		EAs		Radio/TV		Print media		Friends/ neighbours	
		No.	%	No.	%	No.	%	No.	%	No.	%
1	Granulated soya	14	10.2	12	8.8	34	24.8	8	5.8	69	50.4
2	Soya milk	09	6.6	07	51	29	21.2	10	5.8	82	59.9
3	Soya candies	24	17.5	30	21.9	39	28.5	6	4.4	38	27.7
4	Soya Ogi (akamu)	16	11.7	25	18.2	61	44.5	5	3.7	30	21.9
5	Soya veg.soup	34	24.8	28	20.4	13	2.5	2	1.5	60	43.8
6	Soya Meat	36	26.3	24	17.5	18	13.1	11	8.0	48	35.0
7	Soya snacks	32	23.4	17	12.4	29	21.2	13	9.5	46	33.6
8	Soya akara	21	15.3	14	10.2	33	24.1	5	3.7	64	46.7
9	Soya moinmoin	18	13.1	32	23.4	19	13.9	6	4.4	62	45.3
	Total	204		189		275		66		494	
	**Multiple ranks			3		2		4		1	

Sources of information (N=137).

Table 3. Percentage distribution of farmers on the basis of knowledge of preparation of soyabean products.

S/N	Product	Knowledge of soyabean product preparation		Rank
		No	%	
1	Soyabean flour	96	70.1	1
2	Soya milk	87	63.5	2
3	Soya candies	41	29.9	9
4	Soya ogi (akamu)	54	39.4	6
5	Soya vegetable soup	59	43.1	3
6	Soya meat	51	37.2	7
7	Soya snacks	49	35.8	8
8	Soya akara	56	40.9	5
9	Soya moinmoin	58	42.3	4

in the serial order of stated research questions.

Percentage distribution of farmers who are aware of selected soyabean products

The implication of the finding in Table 1 is that majority of the selected Farm-families are aware of selected soyabean products. Only about a quarter are not aware of some of the products. This is commendable. It is a first step in the right direction because adoption of innovation usually begins with awareness. The adoption process involves: awareness, interest, evaluation, trial and

Table 4. Percentage distribution of farmers on the basis of frequency of utilization of soyabean products.

S/N	Frequency of soyabean utilization per 3 week									
	Soyabean products	**0**	**%**	**1 to 2**	**%**	**3 to 4**	**%**	**5 to 6**	**%**	**Ranks**
1	Soya flour	15	10.9	45	32.9	30	21.9	47	34.3	2
2	Soya milk	22	16.1	23	16.8	24	17.5	68	49.6	1
3	Soya candies	26	19.0	61	44.5	25	18.3	24	17.5	6
4	Soya ogi	27	27.0	49	35.9	17	12.4	24	24.5	3
5	Soya vegetable	34	24.8	53	38.7	28	20.4	22	16.10	8
6	Soya meat	36	26.3	58	42.3	20	14.6	23	16.8	9
7	Soya snacks	34	24.8	68	49.6	22	16.1	13	9.9	9
8	Soya akara	29	21.1	59	43.1	24	17.5	25	18.5	5
9	Soya moinmoin	35	25.6	49	35.8	27	19.7	26	19.0	4

Table 5. Frequency of utilization of soyabean by infants/children and pregnant women/lactating mothers.

Respondent	Frequency of utilizing soyabean product per week (N=137)							
	0	**%**	**1 to 2**	**%**	**3 to 4**	**%**	**5 and above**	**%**
Infant/children	48	35.1	45	32.9	22	16.1	20	14.6
Pregnant women/ lactating mothers	23	16.8	84	61.3	11	8.0	19	13.9

adoption/use (Orebiyi et al., 2005). No one is expected to adopt a technology of which he is not aware. This high rate of awareness may be due to effective extension activities, enlightenment, education or effective use of communication channels. The result corroborates the findings of a study in which Mathews-Njoku (2005) found that 66.6% rank of her study sample from Owerri North Local Government Area of Imo State Nigeria were aware of and consumes soyabean flour regularly. Higher percentage of awareness is noted in this study than Mathews-Njokus'. The policy implication of this is that extension activities in the study area may be adjusted effective. In the order of magnitude, most farmers are aware of soya milk (ranked first), followed by soya flour (ranked second) soya moinmoin (ranked third) and soya akara (fourth rank). Soyabean products that farmers are least aware of include: soya snacks (7th rank), soya vegetable soup (8th rank) and soya meat (9th rank). This finding suggests that more awareness needs to be created for soya candies, soya snacks, soya vegetable soup and soya meat (Table 1).

Farmers' major sources of information about soyabean products

The percentage distribution of farmers on the basis of their sources of information about soyabean is as shown in Table 2. Friends and neighbours were the most frequently used information source about soyabean product, followed by radio/TV. Extension agents ranked third (3rd) while print media was the least used source of information on soya bean products. Ogunlade (2004) recorded a higher percentage (85%) of his study using extension agents as the major source of information on soya bean in Kwara State, Nigeria than noted in the current study (for example, 23.4% for Soya moinmoin). Most of this study sample used friends and neighbours as the main source of information. For example, 59.9% used friends and neighbours soya milk. The implication could be that extension activities in the area of study are not so effective. The findings that print media ranked last is expected since majority of the respondents are illiterates who cannot read printed extension information.

Percentages of the sampled farmers with adequate knowledge about preparation of selected soyabean products

Relevant data were analyzed using descriptive statistics and the finding are as presented in Table 3.

The products which most of the farmers cannot prepare are soya meat, soya snacks and soya candies. Thus, soya flour, soya milk, soya vegetable soup and soya moinmoin are the products which most farmers have adequate knowledge of preparation. Extension efforts that focus on dissemination of technologies for preparation of soya meat, soya snacks and soya candies require more attention among farmers in Umuahia and Ohafia ADP zones of Abia State.

Frequency of utilization of soyabean products by the sampled farmers

As shown in Table 4, between 10.9 and 26.3% of

sampled farm-families do not adopt selected soya bean products at all. This comprised the group of farmers who are not aware of the products and those who were aware but did not adopt. Considering the fact that most of the farmers in the area of study are resource-poor and cannot afford to source required level of protein for healthy living from animal sources only, the finding that up to 26.3% of respondent do not utilize soyabean product becomes a subject of concern. The implication is that a third of the farmers are likely to be susceptible to protein-deficiency related diseases. This situation will further aggravate the rate of protein calories malnutrition-related morbidity and mortality in Nigeria which Igbedioh (1990) already noted was on the increase.

The frequencies of respondents who utilize soyabean vary with different products. With the exception of soya milk, most of the respondents utilize soyabean products at a rate of 'once-per-week'. Soya milk is the product that farmers are utilizing more frequently than other products. In ranking of products farmers utilize more frequently, the products would appear in this order: (1) soya milk (2) soya flour (3) soya ogi (4) soya moinmoin (5) soya akara (6) soya candies (7) soya meat (8) soya vegetable soup (9) soya snacks.

Data in Table 5 specifically focus on utilization of soyabean products by infant/children and pregnant/lactating mothers. The table reveals that 35.1 and 16.8% of infants/children and pregnant women/lactating mothers, respectively, do not utilize soyabean products. While it is commendable that only 16.8% of pregnant women/lactating mothers do not use soyabean products, it is worrisome that as high as 35% of infants and children do not use them. This finding give credence to the assertion of Atinmo (1982) that malnutrition, (especially protein-energy malnutrition in young children continues to be first and most outstanding health problems in Nigeria. Should this trend continue, there may be astronomic increase in the rates of morbidity and mortality arising from protein calories malnutrition in Nigeria which Igbedioh had reported was on the increase in 1990.

Table 5 also reveals low rates of soyabean product intake among infant/children and pregnant women/lactating mothers. As high as 61.3 and 32.9% of women/lactating mothers and infants/children respectively take soyabean products at a rate just once a week. Only 14.6% of infant/children and 13.9% of women/lactating mothers take soyabean 5 or more times a week.

Although studies are needed to determine the quantity of soyabean product that is adequate for every individual, yet, the finding of this study represents gross insufficiency in the rate of utilization. The implication of eating balanced diet as stressed by WHF (2004) and Fabiyi (2007) would mean that each day's food should be rich in protein. Thus, it seems that with utilization of soyabean products in at least once-a-day would represent the minimum dosage for all farmers, but more especially for infants, children, pregnant women and lactating mothers. Most farmers are very far below that minimum. The findings are as indicated in Tables 4 and 5.

Conclusion

The study revealed high rate of awareness of respondents on the selected soyabean products. Awareness was as high as 93.4% for soya milk while the product that farmers are least aware of is soya meat (73.7%). The highest percentage of sampled farmers who are not aware of some soyabean product is 26.3%. This percentage is considered high enough to generate anxiety because of its health implications. The soyabean products that respondents are least aware of are soya candies, soya snacks, soya vegetable soup and soya meat. Most of the sampled farmers got most information about soyabean products from friends and neighbours, followed by radio/TV, while extension agents ranked third (3rd)˙ The finding suggests that extension activities in the area of study may not be as effective as expected, even as farmers are using other information sources more than the extension agents. Most farmers are aware of processes for preparing soya flour, soya milk, soya vegetable soup and soya moinmoin but do not posses adequate knowledge for preparing soya meat, soya snacks and soya candies.

It was also found that the rate at which most respondents utilize soyabean product is about once-a-week. Only a few utilize soyabean products at five or more times per week. Less than 15% infants, children, pregnant women and lactating mothers utilize soyabean products at the rate of five or more times a week. Majority of the respondents utilize soyabean products at the rate of once per week. The rate is considered grossly inadequate to meet the protein calories requirement of farmers, infants, children, pregnant women and lactating mothers to check increases in morbidity and mortality which literature have established are on the increase in Nigeria (Kennedy and Alderman, 1987; WHF, 2004 and Fabiyi, 2007).

RECOMMENDATIONS

Based on the findings, it was recommended that the current status of awareness be sustained and improved upon. Efforts should be made to reach the 26.3% of farmers who are not aware of some soyabean products. More attention should be focused on soya candies, soya snacks, soya meat and soya vegetable soup which farmers are not so much aware of. Lack of awareness on soyabean processing technologies may have resulted in observed low adoption among the respondents.

Therefore, promotion activities are very necessary to create adequate awareness of existing soyabean foods. Processing knowledge will, more likely than not, increase the rate of utilization of soya meat, soya candies, soya snacks and soya vegetable soup which most farmers do not know how to prepare. These products that farmers tend to use only sparingly (soya snacks, soya vegetable soup, soya meat and soya candies) should be popularized through increased extension activities. The women in agriculture (WIA) could be involved in a special outreach to women in the villages with a view to informing and teaching women on the nutritional benefits, how to prepare and use soyabean products.

There should be widespread campaigns emphasizing the nutritional values of soyabean utilization especially among vulnerable groups such as children, infants, pregnant women and lactating mothers. The current average rate of adoption (once-per-week) should be greatly improved upon.

REFERENCES

Atinmo T (1982). "The role of school meal in combacting malnutrition among school children in Nigeria". Nig. J. Nutr. Sci., 3(2): 81- 86.

Fabiyi EF (2007). An assessment of improved soyabean processing adoption in Dass Local Government Area of Bauchi State, Nigeria. Global J. Agric. Sci., 6(1): 73-76.

Fukushima D (2000). Recent progress in Research and Technology for processing and utilization of soyabeans. The Third International Soyabean Processing and utilization Conference (ISPUC-III) 2000 of the innovative Era for soyabeans. October 15-20, Tsukuba, Ibaraki, Japan, pp. 11-16.

Igbedioh SO (1990). "Macro-economic Adjustment, Food Availability and Nutritional Status in Nigeria: A look at the 1990s. Food Pol., 16(40): 518-524.

Kennedy ET, Alderman HR (1987). Comparative analysis of nutritional effectiveness of food subsidies and other food–related interventions. Washington D.C. Int. Food Pol. Res. Inst., 169-179.

Mathews-Njoku EC (2005). Adoption of soyabean products in Owerri North Local Government Area of Imo State, Nigeria. Agrosearch, 7(1): 17-22.

Odebode SO (2005). The Soybeans Utilization Programme of the Agricultural Development Programme in Nigeria: A Situation Appraisal of Oyo State. Eur. J. Sci. Res., 4(1): 37-45, Austria.

Ogunlade I (2004). Information Sources used and constraints in adoption of soyabean production and utilization technologies in Kwara State, Nigeria. Agrosearch, 6 (1): 23-28.

Orebiyi JS, Benchendo NG, Onyeka UP (2005). Determinants of Contact Farmers' adoption of improved cassava production technologies in Imo State, Nigeria. Agrosearch, 7(1): 7-15.

Tiamiyu SA, Idowu AA, Misari SN (2001)."Determinants of soyabean adoption in Niger State, Nigeria". Nig. Agric. J., 32: 152-161.

Weigartner KE (1987). "Processing, Nutrition and Utilization of soyabean. In: Singh SR, Rachie KO, Dashiel KE (Eds) (1987). Soyabean for the Tropics. New York: John Wiley and Sons, pp. 149-174.

World's Healthiest Foods (WHF) (2004). Soyabean internel: file: 11f:// New Folder \ the World's Healthiest Foods. htm. pp. 1-16 Retrieved on April 14th, 2006.

APPENDIX

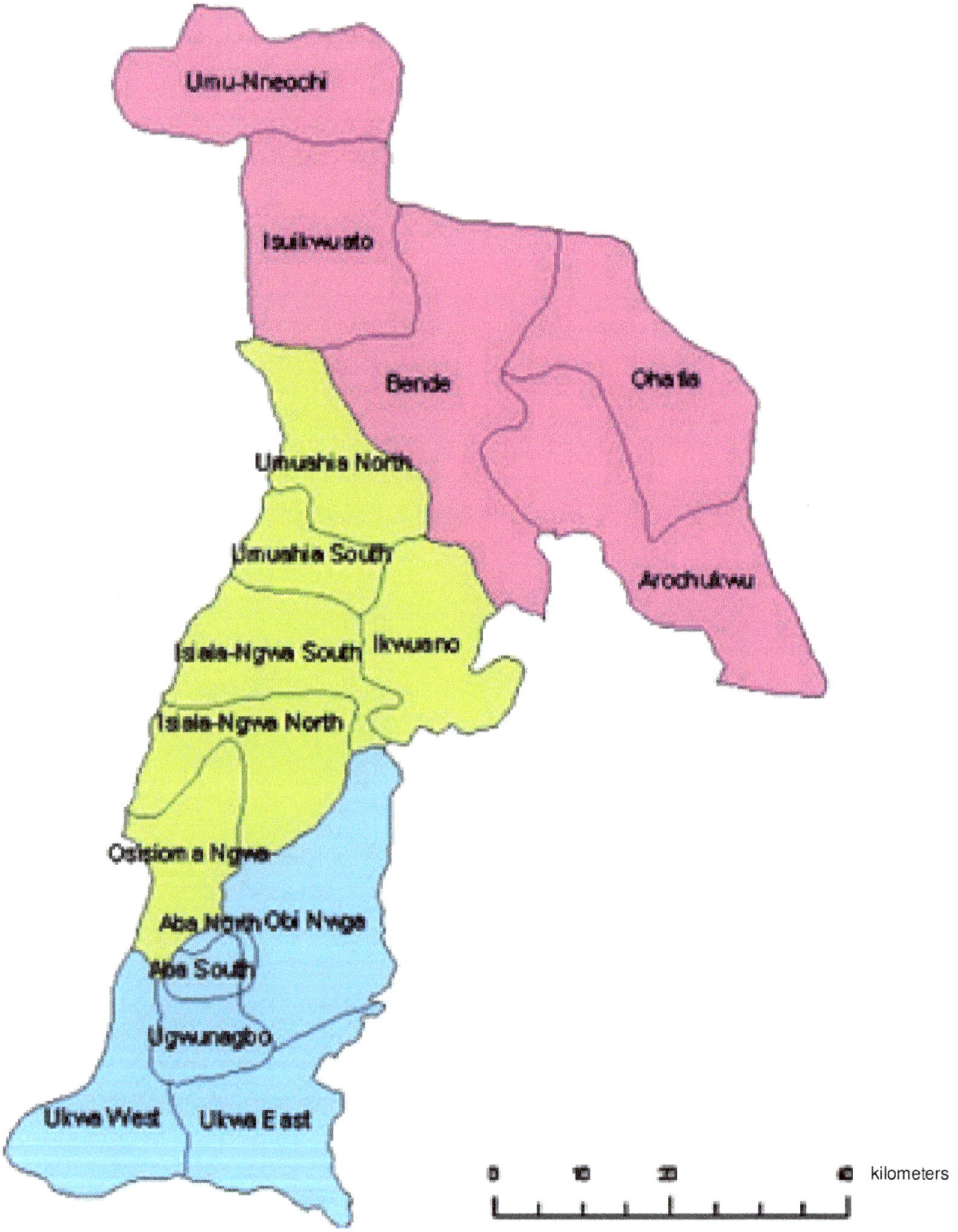

Table 1. Map of Abia State, Nigeria indicating the study area.

Characterization of crop production and marketing to improve food security in Arsi Zone, Ethiopia

Bedada Begna, Messay Yami, Eshetu Lemma, Tesfaye Solomom and Tarekegn Etana

Ethiopian Institute of Agricultural Research (EIAR), Kulumsa Agricultural Research Center (KARC), P. O. Box 489, Assela, Ethiopia.

Crop production provides food and feed; the community generates income from it to cover household expenses and savings. To improve the productivity of the sector, identification of crop production and marketing constraints in Arsi Zone is important to enhance crop production as intended in the Growth and Transformation Plan. This article aims to characterize the crop production, marketing systems and constraints so as to set proper development plans to improve food security in the future. A study was conducted in six randomly selected Peasant Associations from two randomly selected districts among 26 districts in the zone. A total of 120 farm households were selected for the study. The information collected from them was analyzed by descriptive statistics with the help of SPSS computer software. On average, farmers owned 2.645 ha of land, although there were variations from 0.25 and 9.25 ha. Barley, wheat, faba bean and field pea were the major crops, grown by 87.5, 75.8, 74.2 and 34.2% of the respondents respectively. Farmers used the crops produced mainly for family consumption and marketing. December to May was the peak period in which farmers sold their crop grains. District market was the main market where farmers sold their crops. Improving the timely availability of production inputs and improving credit access to farmers are important to enhance crop productivity and food security in the area. Strengthening service-providing institutions and bringing them into function through the provision of required facilities are crucial points which need consideration in the future.

Key words: Crop production, marketing, crop production constraints, food security, Ethiopia.

INTRODUCTION

Most crop production in Ethiopia comes from the peasant sector, whose production technologies are primarily traditional (CSA, 2011). Production and productivity under such traditional systems, however, have been considerably low, due mainly to the inherently low productivity of the crop varieties used, declines in soil fertility, and the effects of crop pests (Firdu and Tsedeke, 2007). But, contrary to this, much increase of crop production in the past decade has been observed due to increase in area cultivated in the country (Alemayehu et al., 2010); however, to what extent the area cultivated can continue to expand remains an important question. It seems that in the highland areas, expansion of cultivated area will have to come almost exclusively from reduction in pasture land.

According to neoclassical economic theory, the development of favorable marketing institutions, trade and specialization are crucial for successful agricultural business.

In line with this, Ethiopian government in recent years has engaged in the establishment of basic development infrastructure (roads, schools, farmer training centers, health centers, and so on) in the whole country emphasizing rural parts, even though farmers are still weakly connected to factor and product markets both domestically and internationally. Only 30% of Ethiopian agricultural output is marketed (Bernard et al., 2006, as cited in Tesfalem, 2008). Better integrating such farmers into factor and product markets is part of the overall development endeavor.

Ethiopian economy has been largely based on agriculture for many past years and some years (5-10) in the future according to Growth and Transformation Plan (GTP) (MoFED, 2010). In transforming agriculture-based economy to one that is industry-based, the agricultural sector should be strengthened including the crop production sector. The country set strategy to increase crop production two-fold, as indicated in the GTP. In doing so, the driving force of the sector, the subsistence farmers, should be supported and accessed to development institutions which are fully equipped with facilities to serve the smallholders closely to enhance their productivity to attain the targets of GTP.

In order to improve crop production and productivity in the zone as well in the country, the rural service-providing institutions need to be strengthened and in line with this the status and situation of the institutions have to be known. Understanding the crop production and marketing constraints of the area is fundamental to improve the crop production and improve the smallholders' livelihoods in general. Hence, the study was designed to characterize the crop production and marketing systems in the Arsi zone and to assess constraints of crop production and marketing systems in the zone.

METHODOLOGY

Description of Arsi Zone

The survey was conducted in Arsi zone located in Oromia Regional State in the South Eastern part of Ethiopia in 2012. It shares boundaries with West Arsi, East Shewa, Bale and West Hararge Zones. Assela, the capital town of the zone, is located 175 km from Finfinne or Addis Ababa on the Finfinne-Adama-Bale Robe main road, 75 km south of Adama.

Arsi zone has four agro-climatic zones, differing mainly in altitude, whose great variation in temperature provides wide opportunities for the production of different types of crop. It is dominantly characterized by moderately cool (about 40%) followed by cool (about 34%) annual temperature. Cool/cold type of thermal zone is found in the highland areas of Chilalo, Bada, Gugu, and Enkolo and Kaka Mountains. The category of moderately warm temperature is found in the lowland areas of Gololcha, Amigna, Seru and Merti districts. It is also found in the Wabi Shabele river valleys and Awash Gorges according to the report of zone BoA and Natural Water Resource Development. The mean annual temperature of the zone is between 20 to 25°C in the lowlands and 10 to 15°C in the central highlands. However, there is a slight variation of temperature by months. February to May are the hottest months while October to January are the coldest months.

The main agricultural feature of the zone is a mixed farming system in which crop production and livestock husbandry complement each other. Crop residues mainly straw used as feed for livestock. On the other hand, the wastes from livestock used in form of compost for fertilizing the soil and livestock supply draught power during land preparation and threshing. The major food crops produced in the Arsi zone are cereals, pulses, oil seeds and others. Among cereals barley and wheat are the pre-dominant and among pulses horse beans and field peas are grown widely.

The zone has high potentialities for livestock rearing. Cattle, draught animals, small ruminants and poultry are the dominant livestock species, which contributing largely to the livelihoods of the community of the area (Messay et al., 2012). Local cattle are the pre-dominant breeds reared in the area; however, improved breeds are being substituted, due to reduced grazing and pasture lands. Improved breeds are productive than local once if properly managed. Keeping few improved breeds on smaller plots than many local breeds is being preferred by the smallholders because of reduction of grazing lands and pastures.

Sampling procedures

Multi-stage sampling techniques were employed for the data collection. Specifically, the study was conducted in Lemu-bilbilo and Munesa districts which were selected randomly at the first stage among 26 districts in Arsi zone. In the second stage, all accessible Peasant Associations (PAs) were listed and three PAs, namely Bokoji Negeso, Chiba Mikael, and Qoma Katare were also selected randomly from the accessible PAs in Lemu-bilbilo. Similarly Chefa, Gumguma and Choba were the PAs selected randomly among accessible PAs in Munesa district. From Munesa district, 14.2, 15.8 and 19.2% of the respondents were selected randomly from Chefa, Gumguma and Choba PAs respectively (Table 1). The number of the respondents involved in the study from each selected PAs was determined in accordance with population as shown in Table 1.

Data analysis

With respect to the expected output of the study, the quantitative data obtained were analyzed using descriptive statistics. Qualitative data collected during the study were used to interpret and conceptually generalize the findings and used to ensure the validity of the results from descriptive statistics. For quantitative data, descriptive statistics were employed, and all findings were combined to give meaningful picture and reliable information. SPSS software was used to analyze descriptive statistics. Means, percentages (shares), ranges, t-tests and χ^2-tests were the main descriptive statistics used in the analysis of quantitative data collected through the formal survey.

RESULTS AND DISCUSSION

Socio-economic characteristics

In the survey areas, the age of respondents (household heads) varied from 20 to 86; the duration of their farming experience and the time they had lived in the area varied from 1 to 63 years and from 10 to 75 years respectively. The overall means of age, farming experience and years lived in the area were 48, 23, and 43 years respectively. The mean ages of respondents in Lemubilbilo and

Table 1. Peasant associations and the proportions of the respondents.

Peasant associations (PAs)	Frequency	Percent
Bokoji Negeso	22	18.3
Chiba Mikael	19	15.8
Qoma Ketare	20	16.7
Cheffa	17	14.2
Gumguma	19	15.8
Choba	23	19.2
Total	120	100.0

Munesa were 46.33 and 49.80 years respectively. The respondents mean farming experiences in Lemubilbilo and Munesa were 24.87 and 30.67 years respectively, whereas the years lived in the areas by the respondents were 38.85 and 46.35 years respectively (Table 2).

In regard to age, a t-test showed the absence of a statistically significant difference between the respondents in the two districts. However, there is a statistically significant difference at the 5% probability level between the two districts with regard to years lived in the areas by the respondents, and another at the 1% probability level between their farming experiences.

Occupation

Crop and livestock farming is the main occupation for more than 89% of the respondents and household chores are also the main activities for about 5% respondents. The majority of farm households in the study districts practiced both crop production and livestock rearing as a mixed farming system. While crop production is practiced together with livestock husbandry, one supports the other in a variety of ways. Residues from crop production serve mainly as feed for the livestock, and the manure is used as crop fertilizer.

In addition to their main occupation, the respondents are practicing a range of activities in order to diversify their sources of income and enhance their livelihoods. Crop and livestock farming, salaried employment, self-employment on farms, casual work on other farms, off-farm casual work, herding and household chores were the secondary occupations of 5, 3.3, 6.7, 3.3, 2.5 and 5.8% of the respondents respectively.

Among 113 respondents involved in the study, crop and livestock farming and household chores were the main occupations for more than 94 and 5% of them respectively (Table 3). Crop and livestock farming was the main occupation for 90.6 and 98.3% of the respondents in Lemubilbilo and Munesa districts respectively (Table 3). The chi-square test shows the presence of significant mean differences between the two districts in referring to the respondents' main occupation statistically.

Walling and roofing materials of main resident house

The walling and roofing materials of the main resident house indicate the living standards of the household. The house of wealthy households is constructed by bricks, stone and iron sheet where as small grass thatched and wooden (timber) mud plastered house belongs to the resource poor households. Wooden mud plastered and earth were the main walling materials of main resident house in rural areas in the past years; however, significant changes are in place currently. Similarly, grass thatching was the common roofing material in many rural parts of Ethiopia.

Wood plastered with mud is main walling material of the main resident house for more than 96% in the study districts. Similarly, for 52 and 41% of respondents, grass thatch and iron sheet are the roofing materials of the main resident houses of the respondents. The result of a chi-square test shows that there is a statistically significant mean difference between the two districts with regard to the walling materials used to construct the main resident house in the areas by the respondents at the 5% probability level and between the roofing materials at the 1% probability level (Table 4). Furthermore, the result indicates the variation in living standards among rural households in Arsi zone.

Land holding

Farmers in the study areas owned land for farming, even though they also accessed land through renting and sharing with other farmers in addition to their own when there is a periodic shortage. Older and resource poor farmers rent out land to youth and wealthier farmers. More than 99% of the respondent farmers owned 2.645 ha on average, varying from 0.25 to 9.25 ha. About 98 and 63% of the respondents allocate 1.77 and 0.76 ha of their own land for cultivation and fallow respectively on average. Among the respondents who used their own land for fallowing, 63 and 37% were Lemubilbilo and Munesa farmers respectively. Fallowing is practised mainly to restore soil fertility and to reduce the incidence of grass weeds, and in the fallowing year the land is used for grazing.

In case of land shortage and based on their capability (finance to cover production cost and oxen ownership), farmers in the study areas accessed land through renting in and sharing in. In contrary to this, due to financial problem and lack of oxen for ploughing, some farmers rented out (0.488ha) and shared out (0.825 ha) some of their land (Table 5).

Land allocation

Barley, wheat, faba bean and field pea are the major

Table 2. Respondents' socio-economic characteristics variables.

Statistic	Age of respondent		Farming experience		Years lived in the area	
	Lemubilbilo	Munessa	Lemubilbilo	Munessa	Lemubilbilo	Munessa
N	60	60	60	60	60	60
Mean	46.33	49.80	24.87	30.67	38.85	46.35
Std. Dev.	11.96	13.87	11.99	14.68	14.12	15.15
t-value	-1.467^{NS}		-2.371**		-2.806***	
P-value	0.147		0.019		0.006	

Source: Analysis of own survey data, January 2013: NS= non-significant, ** significant at less than 5% level of significance, *** Significant at less than 1% level of significance

Table 3. Main occupations, by district.

Main occupation	Lemu-bilbilo		Munesa		Combined		χ^2-value	p-value
	No.	%	No.	%	No.	%		
Crop and livestock farming	48	90.6	59	98.3	107	94.7	5.591**	0.061
Household chores	5	9.4	1	1.7	6	5.3		
Total	53	100.0	60	100.0	113	100.0		

Source: Survey data, January, 2013; ** Significant at less than 5% level of significance.

Table 4. Walling and roofing materials, by district.

Walling and roofing materials	Lemubilbilo		Munesa		Combined		χ^2-value	p-value
	No.	%	No.	%	No.	%		
Earth	4	6.7	0	0	4	3.3	4.138**	0.042
Wooden	56	93.3	60	100	116	96.7		
Total	60	100	60	100	120	100		
Grass thatch	25	41.7	37	61.7	62	51.7	11.506***	0.003
Iron sheet	26	43.3	23	38.3	49	40.8		
Grass thatch and iron sheet	9	15	0	0	9	7.5		
Total	60	100	60	100	120	100		

Table 5. Mean land tenure in the 2012 cropping year.

Land ownership	Mean (ha)	Cultivated land (ha)	Fallow land (ha)	Rented out (ha)	Shared out (ha)
Own	2.645 (99.2)	1.77 (98.3)	0.76 (63.3)	0.488 (8.3)	0.825 (7.5)
Rented in	1.00 (44.2)	0.85 (40.0)	1.17 (7.5)		
Shared in	3.00 (29.2)	0.70 (28.3)	0.67 (2.5)		

Numbers in parentheses are the percentages of the respondents.

crops, grown by 87.5, 75.8, 74.2, and 34.2% of the respondents, who allocate 0.87, 0.57, 0.37 and 0.33 ha of own land to the crops respectively. Farmers rent in land to cultivate crops besides their own land and farmers more than 34% rent in land to wheat and barley among the respondents included in the study. In the study area, more than 87% of the respondent farmers allocate more own land to barley production than other crops (Table 6). However, more land is allocated for crop production, farmers use land for pasture, fallowing and non-agricultural uses in the study area. On average, farmers of the area allocate more than one plot for specific crop or grow a specific crop on different plots located at different places, which indicates a high level of land fragmentation.

Table 6. Land allocation to crops.

Crops	Own			Rented in			No. of plots
	No.	%	Mean (ha)	No.	%	Mean (ha)	
Tef	11	9.2	0.22	1	0.8	0.5	1.08
Wheat	91	75.8	0.57	41	34.2	0.44	2.17
Barley	105	87.5	0.87	41	34.2	0.58	1.38
Faba bean	89	74.2	0.37	21	17.5	0.56	1.22
Field pea	41	34.2	0.33	9	7.5	0.44	1.10
Linseed	38	31.7	0.37	16	13.3	0.38	1.36
Maize	33	27.5	0.25	6	5	0.21	1.22
Land for other crops	12	10	0.65	4	3.3	0.75	1.67
Land for pasture and tree crops	41	34.2	0.56	1	0.8	1.00	1.74
Fallow farmland	46	38.3	0.76	8	6.7	1.47	2.40
Land for non-agricultural uses	24	20	0.71	2	1.7	2.63	2.00

Table 7. Crop production and utilization (averages per household) (N=120).

Crop type	Mean output produced (kg per household)		Consumption (kg)			Mean output sold (kg per household)		
	n_1	Mean	n_2	Mean	n_2/n_1(%)	n_3	Mean	n_3/n_1 (%)
Wheat	95	1273.45	95	336.06	100	68	962.50	72
Barley	115	2892.08	115	464.83	100	62	1562.10	54
Faba bean	96	567.84	93	126.68	97	66	309.02	69
Field pea	38	451.67	32	97.52	84	32	334.37	84
Linseed	37	438.51	5	51.00	14	35	381.58	95
Maize	34	741.10	23	263.91	68	11	413.63	32

Crop production trends

As indicated in the figure, farmers obtain the highest yields from barley; wheat is the next important crop, which gave better yield and competing with barley until 2009. Next to barley and wheat, faba bean is an important cash crop that shows a yield increase from 5 to about 10qt[1] since 2009. Farmers in the study areas produced crops primarily for family consumption and sale to earn cash to cover family expenditure. In general, farmers have obtained higher yields from the major crops in recent five years since 2009 (Figure 1) due to the utilization of improved agricultural technologies according to explanations of the respondents. Due to land scarcity among the farmers, specifically young ones, previously grazing, pasture and forest lands have entered into crop cultivation in recent years. Farmers started farming on hill sides and even on mountain tops in the Arsi zone as well in other parts of the country. Although this has increased production, it will have a high impact on the environment.

Crop production and utilization (averages per household)

Barley and wheat are the dominant food crops produced and utilized by a large number of farmers for consumption and sales (Table 7). Households of the study area used significant amounts of produced crops for seed, and for in-kind payment for land and labor.

As also reported by Alemayehu et al. (2010), farmers use crops produced mainly for both family consumption and marketing. For different crops, the amounts sold and consumed varies across households, and depend on family size, quantity produced, crop type and market demand. For instance, among 37 farmers who produced linseed, only 5 (14%) farmers used it for consumption, and about 35 (94.5%) used the crop for marketing; more than 87% of the linseed produced was sold (Table 7).

If the produced crops are not sufficient, households cope mainly through purchases and food aid or gifts received. Only about 17, 23, 7 and 7% of the respondents bought wheat, barley, faba bean and maize in 2010 to cope with the crop shortage of that year. An insignificant number of the respondents received wheat and barley by food aids/gifts (Table 11).

Marketing of crops

The demand and supply of agricultural produce varies from time to time, leading to fluctuations of price over time. Nowadays, the participation of many actors makes agricultural marketing complex than in previous periods, when market actors and intermediaries were fewer. The increase of market actors and intermediaries could be interlinked with the increasing population and the benefits fetched from grain trading. The farmers sell their crops to different buyers such as farmer groups, farmers' unions, rural assemblers, brokers and rural grain traders; the potential buyers of the crops are urban grain traders and consumers, and other farmers in the study areas (Table 8). Urban grain traders and consumers were the potential market actors purchased wheat from more than 20 and 14% respondent farmers respectively. Similarly, more than 24 and 10% of farmers sold barley for urban grain traders and consumers respectively as depicted in Table 8.

Farmers' trust in traders

Respondent farmers replied differently to the trust statement provided, as shown in Table 9. In order to improve the smallholder farmers living conditions, market actors like traders should buy the product of farmers at a reasonable price at least compensating their costs of production. On the other hand, market crop price could vary overtime and the farmers' access to the correct market information varies among market actors. The correct market price that traders pay to the farmers' product and the return they fetch from it indicates the levels of farmers' trust in traders. Accordingly, farmers indicated their level of trust in traders according to the statement says "most traders can be trusted". Out of 120 respondents, more than 48% were indifferent or farmers were in between to trust or not to trust, 30% disagreed and about 16% agreed with the statement (Table 9). The Likert scale result was 2.84 which is near to 3 (in-different), which indicates the level of farmers trust on traders.

In crop marketing, where the farmers sell the crop or market type is another important determinant that facilitate agricultural marketing. For farmers, selling crop grain at farm gates and village markets reduces ups and downs of travelling far distances and saves much time to accomplish other agricultural or development tasks if just and fair prices are paid. Farmers in the study area sold their crop produce at farm gate, village market and the main district market (Table 10). Main or district market is the place where many farmers sold the crops in the study area. Some farmers sold crops at village markets, and few of them did so at farm gates (Table 9) to rural grain traders or to other village farmers who took the grains to the district market or to secondary markets. Bokoji is the main or district market of Lemubilbilo, and Kersa is the main market of Munesa district. Among 120 respondent farmers, 44.2, 50.0, 42.5, 20.8 and 24.2% sold wheat, barley, faba bean, field pea and linseed respectively at their respective district markets, as depicted in Table 9.

With regard to market type, it is crucial to consider the distance to the nearest village and main markets. Among the respondents, there were farmers who are 12 km from their nearest village market; the average distance was 4.81 km. Similarly, farmers in the study areas on average could travel 6.87 km to reach the main respective district markets which could consume more than one hour in average for a single trip (Table 12). The study result also shows the presence of farmers who travelled up to 24 km (needs 3 h) to arrive at the district main market.

To evaluate the living standards of a society in a given area, considering modes of transport is important. Farmers transport the crops to the village markets or district markets mainly by donkey; however, some farmers used oxen/ horse carts. About 53, 50 and 46% of the respondents used donkey to transport barley, wheat, and faba bean respectively to the markets.

In crop marketing, understanding the farmers' crop selling period or months is also an indispensable factor like market type, market actors, transport mode and market distance. Farmers' crop selling period could be influenced by a variety of factors such as personal, social, economic and cultural ones. As indicated in Figure 2, higher percentages of farmers sold the crop commodities between November and May. January, February and March were the peak crop selling months in the study areas during the survey year in 2012. November and December are harvesting months in study areas. However, the crop selling months vary from year to year and from farmer to farmer depending on the level of production and crop prices in the market; generally many farmers sell major proportion of crops from December to May (Figure 2).

Access to market information and means of access

Access to market information before deciding to sell their crop commodities helps the farmers to make decisions whether to sell or not, and to sell at a fair market price. Among the 120 farmers in this study, 73.3, 75.8, 65, 45 and 23.3% sold wheat, barley, faba bean, field pea and maize respectively in the survey year. Out of the respondents who sold these crops, some were unable to access market information before deciding to sell their crops (Table 11).

Some days before selling, observation of the market and discussion with other village farmers and relatives were the major means of accessing to market information for the farmers in the study area. According to the farmers' explanation, some farmers also have access to market information through telephone discussion with traders and farmers of in other areas. Cell phones are playing a great role in accessing market information.

Food security and coping mechanisms

Crop production of Arsi zone mainly comes from the

Table 8. Buyers of crops in the study areas.

Crop type	Farmer group		Farmer union/ cooperatives		Consumer or other farmer		Rural assembler		Broker/ middle men		Rural grain trader		Urban grain trader	
	N	%	N	%	N	%	N	%	N	%	N	%	N	%
Wheat	3	2.5	0	0	17	14.2	3	2.5	3	2.5	15	12.5	25	20.8
Barley	1	0.8	9	7.5	13	10.8	1	0.8	1	0.8	12	10	29	24.2
Faba bean	1	0.8	1	0.8	18	15.0	2	1.7	3	2.5	10	8.3	25	20.8
Field pea	2	1.7	2	1.7	10	8.3	2	1.7	1	0.8	7	5.8	9	7.5
Linseed	0	0	0	0	5	4.2	0	0	2	1.7	4	3.3	26	21.7
Maize	0	0	0	0	3	2.5	1	0.8	0	0	5	4.2	1	0.8

Table 9. Level of farmers' trust on traders.

Most traders can be trusted	Strongly disagree (1)	Disagree (2)	Indifferent (3)	Agree (4)	Strongly agree (5)	Total
Frequency	4	36	58	19	3	120
Percent	3.33	30.00	48.33	15.83	2.50	100.00
Scale * Frequency	4	72	174	76	15	341

Weighted mean= 341 ÷120=2.84.

Table 10. Farmers selling crops, by market types and crop sold.

Crops	Farm gate		Village market		Main/ District market	
	No.	%	No.	%	No.	%
Wheat	4	3.3	10	8.3	53	44.2
Barley	1	0.8	5	4.2	60	50.0
Faba bean	2	1.7	8	6.7	51	42.5
Field pea	1	0.8	8	6.7	25	20.8
Linseed	0	0	8	6.7	29	24.2
Maize	0	0	2	1.7	9	7.5

traditional farming system based on animal power and rain-fed peasant sector; a limited amount comes from state farms. The peasant sector accounted for about 96.45% of the total cultivated land in the Zone, and 96.2% of the total crop production. The zone is considered as food self-sufficient zone. However, there are households having food insecurity problems, whose severity varies from year to year depending on rainfall, and from district to district, because all districts are not equally vulnerable to drought.

Among the 120 farm households included in the

Table 11. Proportion of farmers selling crops and their levels of access to market information, by crop type.

Crop type	Proportion of farmers selling the crop		Access to market information before deciding to sell the crops			
			Accessed		Not accessed	
	N	%	N	%	N	%
Wheat	88	73.3	79	89.77	9	10.33
Barley	91	75.8	84	92.31	7	7.69
Faba bean	78	65.0	72	92.31	6	7.69
Field pea	54	45.0	47	87.04	7	12.96
Maize	28	23.3	22	78.57	6	21.43

Table 12. Proportion of farmers and amount of crops bought.

Crops	Amount bought (kg)				Amount of food aid or gifts received (kg)			
	N	%	Mean	Sta. dev.	N	%	Mean	Sta. dev.
Wheat	20	16.7	147.5	108.2	2	1.67	80	28.3
Barley	27	22.5	241.6	275.6	1	0.83	300	
Faba bean	8	6.7	46.88	8.84	0			
Field pea	1	0.83	50		0			
Linseed	7	5.83	63.57	104.27	0			
Maize	8	6.7	219.88	205.56	0			

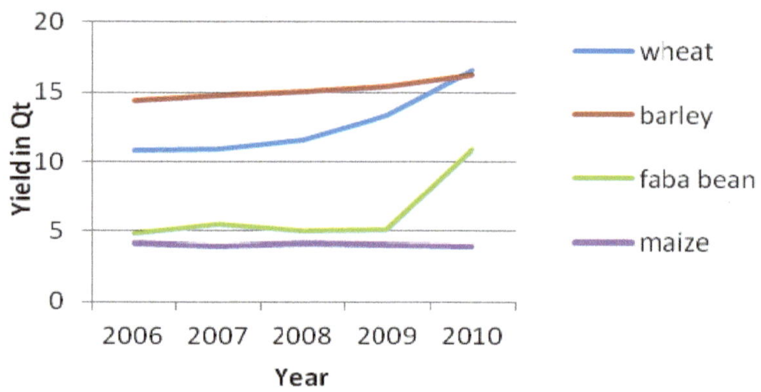

Figure 1. Major crop production trends (Yield in qt.).

study, only a few faced a shortage of food crops for consumption in 2011.Wheat and barley sustained the livelihoods of millions of people residence in highlands of the study areas. These crops were the main food crops consumed almost daily in different forms such *as injera, bread, ganfo and kinche.* Farmers in the study area utilize their crops both for consumption and marketing. About 16.7 and 22.5% of the respondent farmers bought wheat and barley respectively in 2011 due to the shortage of the crops (Table 12). Faba bean, field pea, linseed and maize were the crops cultivated by a few farmers on smaller plots in the study area as indicated in Table 7 above. These crops were consumed proportionately in smaller quantities compared to wheat and barley, and the majority of the farmers used these crops for marketing. Among the sampled farmers, only a few of them bought faba bean, field pea, linseed and maize to overcome their shortage (Table 12). Even though the quantities of food aid/gifts from government and non-government bodies are unknown, an insignificant number of households received wheat and barley as food aid or gifts from relatives, neighbors or friends.

As indicated in many reports, 'food security is defined as physical and economic access by all people at all times to sufficient food to meet their dietary requirements for a reproductive and healthy life'. By this definition, even though some households in the study area faced a food shortage in some periods of the year, they may not

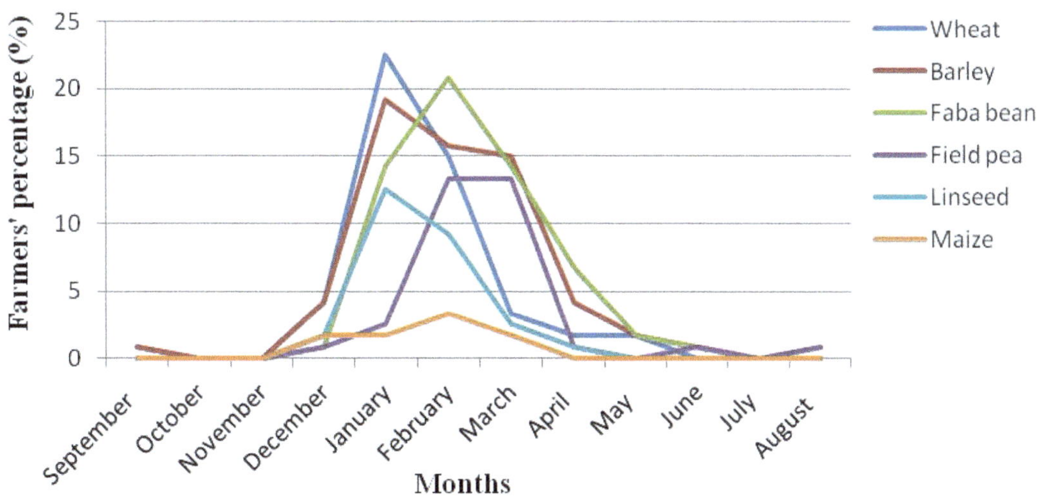

Figure 2. Patterns of crop selling months and farmers' proportion.

Table 13. Average Distance of the nearest institutions from farmer's home.

Distance to the nearest	N	Minimum	Maximum	Mean	Std. Dev.
Village market (km)	91	.001	12.000	4.80847	2.694861
Main market (km)	118	.30	24.00	6.8742	5.59383
Source of seed (km)	120	.001	18.000	4.52643	2.841906
Source of fertilizer (km)	119	.001	21.000	4.53085	2.975629
Source of farm chemicals (km)	110	.001	18.000	5.14228	3.736420
Farmer cooperative (km)	118	.001	12.000	4.29397	2.565329
Agricultural extension DA office (km)	119	.001	11.000	2.82329	1.952660

have been food-insecure, because adequate food crops were available and they could access to these food crops.

Households of the study area used different means to cope when a food crop shortage exists. These coping mechanisms are adopted depending on how bad the crisis is, and what is available to them to manage the situation. According to the explanations of some key informants, farm households used livestock sales (specifically small ruminants), certain types of off farm employment, requesting grain loans, sales of fire wood or charcoal, reduction of crops to be marketed and reduction of consumption were some of the coping mechanisms exercised by households in the study area.

Availability of supportive institutions in the farmers' area

Gebremedhin et al. (2009) showed that the expansion of the agricultural services particularly extension had a significant impact on the intensity of input use, agricultural productivity and market participation of Ethiopian smallholders. The availability of service-providing institutions at a near distance is desirable for the farmers in order to access agricultural technologies and information. Not only availability of these institutions but also their level in facility and capacity to support farmers in the provision of required services is crucial. Even though there are farmers who are far from supportive institutions, the distances seem acceptable (Table 13). The improvements observed in the availability of supportive institutions in the farmers' area attained by the focuses given to farmers and the efforts made by the government in recent years. For instance, on average, farmers are 2.82km distant from an agricultural extension (Development Agents) office. Sources of seed, fertilizer and farm chemicals, and farmer cooperatives, are situated on average at distances of 4.53, 4.53, 5.14 and 4.30 km respectively (Table 13).

Crop production and marketing constraints

Farmers in the study areas are confronted with a variety of crop production and marketing challenges and difficulties as indicated in Table 14. Mesay and Tolosa (2011) reported rather similar wheat production constraints. The price of fertilizer and of improved seed and the

Table 14. Production and marketing constraints for selected major crops.

Constraints	Wheat		Barley		Faba bean		Linseed		Teff	
	N	%	N	%	N	N	%	N	%	N
Socio-economic										
Timely availability of improved seed	64	53.3	57	47.5	68	56.7	39	32.5	13	10.8
Price of improved seed	73	60.8	64	53.3	67	55.8	37	30.8	12	10
Quality of seed	55	45.8	53	44.2	50	41.7	31	25.8	8	6.7
Availability of credit to buy seed	45	37.5	40	33.3	37	30.8	28	23.3	10	8.3
Timely availability of fertilizer	43	35.8	50	41.7	40	33.3	13	12.5	7	5.8
Price of fertilizer	86	71.7	73	60.8	70	58.3	24	20	12	10
Availability of credit to buy fertilizer	44	36.7	36	30.0	34	28.3	18	15	4	3.3
Access to markets and information	32	26.7	31	25.8	31	25.8	23	19.2	3	2.5
Reasonable grain prices	32	26.7	37	30.8	35	29.2	20	16.7	5	4.2
Biological										
Drought	12	10.0	8	6.7	8	6.7	4	3.3	0	0
Floods	31	25.8	19	15.8	18	15.0	10	8.3	4	3.3
Pests	69	57.5	67	55.8	65	54.2	40	33.3	12	10
Diseases	67	55.8	62	51.7	78	65.0	42	35.0	12	10
Soil fertility	54	45.0	43	35.8	41	38.2	33	27.5	12	10

and the timely availability and quality of improved seed, were the main socio-economic problems reported by 71.7, 60.8, 53.3 and 45.8% of the respondent farmers respectively in wheat production. Pests and diseases, as biological/ natural wheat production constraints, were indicated by 57.5% and 55.8 farmers respectively. With regard to barley the price of fertilizer, price of improved and timely availability of improved seed were the principal socio-economic challenges reported by 60.8, 53.3 and 47.5% of farmers respectively while pests and diseases are the major biological constraints of barley, as reported by 55.8 and 51.7% of the respondents. Based on the study results, the price of fertilizer and improved seed, timely availability of improved seed and fertilizer, availability of credit to buy fertilizer and seed and access to market and agricultural information are the socio-economic crop production constraints of the farmers in the study area (Table 13). Further, pests, diseases, soil fertility and floods are important biological crop production constraints in the areas.

CONCLUSIONS AND RECOMMENDATIONS

Crop production plays a significant role in providing food and income for rural society. Wheat, barley, faba bean, field pea, and linseed are the major crops grown by most farmers through the allocation of more proportion of cultivable land. Through the utilization of improved agricultural technologies - particularly improved seeds and by increasing the extent of cropped lands, farmers in the area have obtained higher production levels since 2008. Arsi zone is considered as a food-secure region; however, some rural farm households faced food insecurity during the study period, and coped via bought grain and via food aids or gifts.

Crop marketing is a dynamic and complex phenomenon in which crop demand and supply vary from time to time, and farmers sell their grains at different markets to a great variety of market actors. Urban traders, consumers and rural traders are the main actors who bought crops from farmers in the study area. Most of the farmers sold their crop products from December to May. In recent years, village and main markets are located at relatively acceptable distances from farm households in the study areas although there were farmers who are far from the markets.

Besides to markets, other service-providing institutions also seemed situated at acceptable distances from the farmers. Only the availability of service-providing institutions in the farmers' area without adequate facility is less satisfactory, so strengthening the institutions required facilities is more important and needs emphasis.

The high cost of fertilizer and improved seeds, timely unavailability of inputs, unavailability of credits to buy fertilizers and improved seeds, crop pests and diseases are the major socio-economic and biological problems which are significantly contributing to reduce productivity of crops in Arsi zone. Hence, improving timely availability of production inputs and improving credit accessing ways to farmers are important in the area. Furthermore, strengthening service providing institutions in the area through provision of required facilities are crucial points need to be considered in the future.

Conflict of Interest

The authors have not declared any conflict of interest.

REFERENCES

Alemayehu ST, Dorosh P, Sinafikeh A (2010). Ethiopia Strategy Support Program (ESSP II). Crop Production in Ethiopia: Regional Patterns and Trends Summary of ESSP II Working Paper 16, Research note 11. Available at: http://essp.ifpri.info. Accessed on 14/03/2013

Central Statistics Agency (CSA) (2011). Federal Democratic Republic of Ethiopia.Central Statistical Agency (CSA). Large and Medium Scale Commercial Farms Sample Survey 2010/12. Statistical report on Area Production of Crops and Farm Management Practices. Addis Ababa, Ethiopia.

Firdu A, Tsedeke A (2007). IPM Research Project. Phase I Country Back Ground Papers. Pesticide Action Network, UK.

Gebremedhin B, Jaleta M, Hoekstra D (2009). Smallholders, Institutional Services and Commercial transformation in Ethiopia. Agric. Econ. 40(S):773-787.

Mesay Y, Tolesa A (2011). Baseline Survey in Selected Wheat Producing Areas of Ethiopia: The Case of Arsi Zone. Eastern Africa Agricultural Productivity Programme (EAAPP) – Wheat Region Center of Excellence.

MoFED (Ministry of Finance and Economic Development) (2010). Performance Evaluation of the First Five Years Development Plan (2006-2010) and Growth and Transformation Planning (GTP) Next Five Years (2011-20015). A draft document for discussion with the Regional/ City administrations. Addis Ababa, Ethiopia.

Tesfalem HA (2008). Linking Small Farmers to Markets. Household Determinants of Smallholder Commercialization in Northern Ethiopia. Wageningen, Netherlands.

Foraging and pollination activities of *Xylocopa olivacea* (Hymenoptera, Apidae) on *Phaseolus vulgaris* (Fabaceae) flowers at Dang (Ngaoundere-Cameroon)

Bernice Mireille Tekombo Kingha[1]*, Fernand-Nestor Tchuenguem Fohouo[1], Albert Ngakou[1] and Dorothea Brückner[2]

[1]Department of Biological Sciences, Faculty of Sciences, University of Ngaoundéré, P.O. Box 454 Ngaoundere, Cameroon.
[2]Forschungsstelle für Bienenkunde, Universität Bremen, FB2, Postfach 33 04 40, 28334 Bremen, Germany.

To evaluate the impact of the carpenter bee (*Xylocopa olivacea*) on pod and seed sets of *Phaseolus vulgaris* (black seed outlets), its foraging and pollinating activities were studied in Ngaoundéré, during the June-July 2009 and 2010 cropping seasons. Treatments included unlimited floral access by all visitors, bagged flowers to avoid all visits, and limited visits of *X. olivacea*. Observations were made on 120 flowers per treatment of which all flower visitors were recorded. The carpenter bee seasonal rhythm of activity, its foraging behaviour on flowers, and its pollination efficiency (fruiting rate, number of seeds/pod and percentage of normal or well developed seeds) were recorded. Twenty-four insect species visit *P. vulgaris* flowers. *X. olivacea* was the most frequent visitor and they intensely and exclusively foraged nectar. The foraging speed was 9.94 flowers/min. The foraging activity of *X. olivacea* resulted in a significant increment in fruiting rate by 48.43 and 78.18%, the number of seeds/pod by 19.38 and 18.58% and the normal seeds/pod by 15.67 and 38.25%, respectively in 2009 and 2010. Hence, conservation of *X. olivacea* nests close to *P. vulgaris* crop fields should be recommended to improve pod and seed production in the region.

Key words: *Xylocopa olivacea*, *Phaseolus vulgaris*, foraging activity, nectar, pollination, yield.

INTRODUCTION

Very little information exists on the relationships between flowering insects and many plant species in Cameroon. It is well, known that anthophilous insects including bees usually increase fruit and seed yields of many plant species, through pollination provision (Keller and Waller, 2002; Fluri and Frick, 2005; Sabbahi et al., 2005; Klein et al., 2007; Tchuenguem Fohouo et al., 2009a). Up to date, no detailed work has been investigated on *Phaseolus vulgaris*.

P. vulgaris is an annual plant originated from South and Central America (Graham et al., 1997). The flower is pink, but can vary from white to purple depending on the different varieties (Debouck, 1991) and produces nectar/pollen which attract insects (Ibarra-Perez et al., 1999). In Cameroon, *P. vulgaris* is cultivated as vegetable and can be consumed raw or cooked; its pods are sold fresh (green beans), or transformed into flour, while the stems and leaves are used to feed livestock (Debouck, 1991). Despite the high seeds demand in the country (203053 tones/year), the quantity of *P. vulgaris* available to consumers is very low (147553 tones/years for 124746 hectares), as a result of low pod and seed yields (MINADER/DESA, 2010). Therefore, it is important to investigate on the possibilities of increasing the production of this plant in Cameroon. The plant is allogam/autogam (Ibarra-Perez et al., 1999) and its cross pollination is ensured by insects (Mackie and Smith, 1935;

*Corresponding author. E-mail: bernice2905@yahoo.fr.

Barrons, 1939; Wells et al., 1988; Ibarra-Perez et al., 1997).

Previous researches have shown no set of seeds from undisturbed bean flowers, whereas, flowers manipulated by wing petals have a great number of seeds (Darwin, 1858). Bumblebees were indicated to be the most effective pollinator of beans when visiting blossoms to collect pollen and nectar in South Africa (Palmer, 1967). Genus *Bombus* was associated with pollination activity in common beans in North America (McGregor, 1976). *P. vulgaris* flowers were reported to produce fewer seeds per pod in the absence of efficient pollinators in the United States of America (Ibarra-Perez et al., 1999). Recent research conducted in Kenya has revealed *Apis mellifera* to be the most abundant insect species visiting *P. vulgaris* flowers, followed by *X. calens* (currently *Xylocopa olivacea*) and *X. inconstans* (Kasina et al., 2009).

Prior to these studies, no previous research has been reported on the relationships between *P. vulgaris* and its anthophilous insects, although, the activity and diversity of flowering insects of a plant species vary with regions (Roubik, 2000).

X. olivacea is one of the common carpenter bees in Cameroon. During preliminary investigations on flower-insect relationships in Ngaoundere before 2009 (unpublished data), *X. olivacea* has been seen intensively visiting flowers of *P. vulgaris*.

The main objective of this research was to gather more data on the relationships between *P. vulgaris* and flower visiting insects for optimal management of pollination services. Specific objectives were the registration of the activity of *X. olivacea* on *P. vulgaris* flowers, the evaluation of the impact of visiting insects on pollination, pods and seeds yields of this Fabaceae, and the estimation of pollination efficiency of *X. olivacea* on this plant.

MATERIALS AND METHODS

Study site, experimental plot and biological material

The experiment was carried from June to July, 2009, and from June to July, 2010 at Dang (7°25.372' N, 13°32.566' E and 1092 m above sea level), Ngaoundere, Adamaoua Cameroon. This region is within the high altitude of Guinean savannah agro-ecological zone (Tchuenguem et al., 2007). The climate is characterized by a distinct rainy season (April to October) and dry season (November to March), with an annual rainfall about of 1500 mm. The mean annual temperature is 22°C, while the mean annual relative humidity is 70% (Tchuenguem et al., 2007). The experimental plot measured 437 m^2, in which *P. vulgaris of* small black seeds purchased from a local market was sown. During the study period, 31 to 51 *A. m. adansonii* Latreille (Hymenoptera: Apidae) colonies were located at 3 km in diameter around the experimental site. The vegetation was represented by crops, ornamental plants, hedge plants and native plants of savannah and gallery forests. The vegetation near the *P. vulgaris* field had various unmanaged and cultivated species including bee *X. olivacea* which nests in tree trunks and branches under natural conditions.

Sowing and weeding

On May 8, 2009 and May 5, 2010, the experimental plot was prepared and divided into 8 subplots, each measuring 8 × 4.5 m. Two seeds were sown in 6 lines per subplot, each of which had 30 holes per line. Holes were separated 25 cm from each other, while lines were 75 cm apart. Weeding was performed manually as necessary to maintain plots weed-free.

Determination of the reproduction system of *P. vulgaris*

On 28th June 2009, 30 *P. vulgaris* flowers at the bud stage were labelled on each subplot, giving a total of 240 flowers. One hundred and twenty of the total flowers were allowed to be open pollinated (treatment 1) whilst the other 120 were bagged with 1 mm square gauze bag to prevent visitors or external pollinating agents (treatment 2). On 28st June 2010, the experiment was repeated. Twelve days after shading of the last flower, the numbers of pods were assessed in each treatment. The podding index (*Pi*) was then calculated as described by Tchuenguem et al. (2001):

$Pi = F2/F1$

Where *F2* is the number of pods formed and *F1* the number of viable flowers initially set.

The allogamy rate (*Alr*) from which autogamy rate (*Atr*) was derived was expressed as the difference in podding indexes between unprotected flowers (treatment 1) and protected flowers (treatment 2) as follows (Demarly, 1977):

$Alr = [(Pi1 - Pi2) / Pi1] \times 100$

Where *Pi1* and *Pi2* are respectively the podding average indexes of treatments 1 and 2.

$Atr = 100 - Alr$

Assessment of foraging activity of *X. olivacea* on *P. vulgaris* flowers

Observations were conducted on 120 individual opened pollinated flowers of treatment 1 each day from 29th June to 12th July, 2009 and from 29th June to 10th July, 2010 at 1 h interval from 6.00 to 17.00 h. In a slow walk along all labelled flowers of treatment 1, the identity of all insects that visited *P. vulgaris* flowers was recorded. Specimens of all insect taxa were caught with an insect net on unlabelled flowers; for each species 2 to 10 specimens were captured. These specimens were preserved in 70% ethanol (except butterflies that were preserved dry) for subsequent taxonomic identification. All insects encountered on flowers were recorded and the cumulated results expressed in number of visits to determine the relative frequency of *X. olivacea* in the anthophilous entomofauna of *P. vulgaris*.

In addition to the determination of the floral insect's frequency, direct observations of the foraging activity on flowers were made on insect pollinator fauna in the experimental field. The floral rewards (nectar or pollen) harvested by *X. olivacea* during each floral visit were registered based on its foraging behavior. Nectar foragers were expected to extend their proboscis to the base of the corolla and the stigma, while pollen gatherers were expected to scratch the anthers with their mandibles or legs (Jean-Prost, 1987).

In the morning of each sampling day, the number of opened flowers was counted. The same days as for the frequency of visits, the duration of individual flower visits was recorded (using a stopwatch) at least three times at hourly intervals between 07.00 and 18.00 h. Moreover, the number of pollinating visits which was defined as contact between the bees and stigma upon a visit (Jacob-Remacle, 1989; Freitas, 1997); the abundance of foragers defined as the highest number of individuals simultaneously foraging on a flower/or 1000 flowers (A_{1000}) (Tchuenguem et al., 2004) and the foraging speed, which is the number of flowers visited by a bee per minute (Jacob-Remacle, 1989) were measured. The abundance of insects per flower was recorded following the direct counting, during the same dates and daily periods as for the registration of the duration of visits. The foraging speed (*Vb*) was calculated according to Tchuenguem et al. (2004).

The disruption of the activity of foragers by competitors or predators and the attractiveness exerted by other plant species on *X. olivacea* was assessed by direct observations. The temperature and relative humidity in the station were also registered every 30 min using a mobile thermo-hygrometer (techno WS-7018, Germany) during all sampling periods.

Evaluation of the effect of *X. olivacea* and other insects on *P. vulgaris* yields

This evaluation was based on the impact of visiting insects on pollination, the impact of pollination on fructification of *P. vulgaris*, and the comparison of yields (fruiting rate, mean number of seed per pod and percentage of normal or well developed seeds) of treatments 1 and 2 (open and bagged pollinated flowers). The fruiting rate due to the influence of foraging insects (*Fri*) was calculated by the formula:

$$Fri = \{[(Fr1 - Fr2) / Fr1] \times 100\}$$

Where *Fr1* and *Fr2* are the fruiting rate in treatments 1 and 2. The fruiting rate (*Fr*) is:

$$Fr = [(F2/F1) \times 100]$$

Where *F2* is the number of pods formed and *F1* the number of opened flowers initially set.

At maturity, pods were harvested from each treatment and the mean number of seeds per pod and the percentage of normal seeds were then calculated for each treatment.

Assessment of the pollination efficiency of *X. olivacea* on *P. vulgaris*

To evaluate of the pollination efficiency of *X. olivacea*, 120 and 150 flowers were isolated (treatment 3) respectively in 2009 and 2010. Between 10 a.m. and 1 p.m. of each observation date and the gauze bag was delicately removed from each opened flower and this flower observed for up to 10 min. The flowers visited by *X. olivacea* were labeled after this manipulation. The contribution (*Frx*) of *X. olivacea* to fruiting was calculated by the formula:

$$Frx = \{[(Fr3 - Fr2) / Fr3] \times 100\}$$

Where *Fr3* and *Fr2* are the fruiting rates in treatment 3 (protected flowers visited exclusively by *X. olivacea*) and treatment 2 (protected flowers).

At maturity, pods were harvested from treatment 3 on which the number of seeds per pod were counted. The mean number of seeds per pod and the percentage of normal seeds were then calculated for each treatment.

Data analysis

Data were analyzed using descriptive statistics, Student's *t*-test for the comparison of means of two samples, Correlation coefficient (*r*) for the study of the association between two variables, Chi - Square (χ^2) for the comparison of percentages with Microsoft Excel 2007.

RESULTS

Reproduction system of *P. vulgaris*

The podding indexes of *P. vulgaris* were 0.83, 0.27, 0.95 and 0.2, respectively for treatments 1, 2 in 2009 and 2010. The allogamy and autogamy rates in 2009 and 2010 were 67.47 and 32.53% against 78.95 and 21.05%, respectively. For the two accumulated years, the allogamy rate (*Alr*) was 73.21%, while that of autogamy (*Atr*) was 26.79%. The variety of *P. vulgaris* (small black seed) used in our experiments had a mixed reproduction regime (allogamous-autogamous), with the predominance of allogamy.

Frequency of floral entomofauna of *P. vulgaris*

Amongst the 177 visits of 15 insects species in 2009 and 157 visits of 16 insects species in 2010 recorded on *P. vulgaris* flowers, *X. olivacea* was the most frequent insect with 61 visits (34.46 %) and 36 visits (22.93 %), in 2009 and 2010, respectively (Table 1). The difference between these two percentages is significant ($\chi^2 = 5.37$, $df = 1$, $p < 0.05$).

Activity of *X. olivacea* on *P. vulgaris* flowers

Floral rewards harvested

During each of the two flowering periods, *X. olivacea* was found to intensively and regularly collect nectar from *P. vulgaris* (Figure 3) but no pollen collection was observed.

Relationship between visits and flowering stages

From Figure 2, a positive and significant correlation was found between the number of *P. vulgaris* opened flowers and the number of *X. olivacea* visits in 2009 ($r = 0.81$; $df = 10$; $p < 0.001$) as well as, in 2010 ($r = 0.15$; $df = 8$; $p < 0.05$).

Diurnal flower visits

X. olivacea foraged on *P. vulgaris* flowers throughout the day, with a peak activity between 10.00 and 13.00 h (Figure 1). The activity of *X. olivacea* was influenced by climatic conditions. In 2009, the correlation between the number of *X. olivacea* visits on *P. vulgaris* flowers and the temperature was positive and highly significant (*r* =

Table 1. Diversity of floral insects on *P. vulgaris* flowers in 2009 and 2010, number and percentage of visits of different insects.

Insects			2009		2010	
Order	**Family**	**Genus, species, sub-species**	n_1	P_1 (%)	n_2	P_2 (%)
Hymenoptera	Apidae	*Xylocopa olivacea* (nectar)	61	34.46	36	22.93
		Xylocopa sp. (nectar)	5	2.82	3	1.91
		Apis mellifera adansonii (nectar)	14	7.91	3	1.91
		Amegila sp. 1 (nectar)	2	1.13	13	8.28
		Amegila sp. 2 (nectar)	-	-	2	1.27
		Braunsapis sp. (nectar)	11	6.21	-	-
		Ceratina sp. 1 (nectar+ pollen)	-	-	24	15.29
		Ceratina sp. 2 (nectar)	-	-	11	7.01
	Halictidae	*Lasioglossum* sp. (nectar)	1	0.56	-	-
	Megachilidae	*Chalicodoma rufipes* (nectar)	28	15.82	23	14.65
		Megachile sp. 1 (nectar)	1	0.56	-	-
		Megachile sp. 2 (nectar)	-	-	1	0.64
		Megachile sp. 3 (nectar)	-	-	1	0.64
		Megachile sp. 4 (nectar)	-	-	5	3.18
		Megachile sp. 5 (nectar)	-	-	2	1.27
	Formicidae	*Camponotus flavomarginatus* (nectar)	2	1.13	-	-
		Camponotus sp. (nectar)	1	0.56	-	-
	Total Hymenoptera		126	71.19	124	78.98
Lepidoptera	Pieridae	*Eurema* sp. 1 (nectar)	1	0.56	8	5.1
		Eurema sp. 2 (nectar)	2	1.13	-	-
	Lycaenidae	(sp. 1) (nectar)	-	-	2	1.27
		(sp. 2) (nectar)	-	-	4	2.55
	Hesperiidae	*Lambrix* sp. (nectar)	2	1.13	-	-
	Total Lepidoptera		5	2.82	14	8.92
Diptera	Syrphidae	*Episyrphus* sp. (nectar)	4	2.26	-	-
Coleoptera	Meloidae	*Coryna* sp. (eat flowers)	42	23.73	19	12.1
Total			177	100	157	100

n_1: number of visits on 120 flowers in 11 days, n_2: number of visits on 120 flowers in 09 days, p_1 and p_2: percentages of visits, $p_1 = (n_1/177) \times 100$, $p_2 = (n_2/157) \times 100$. Comparison of percentages of *X. olivacea* visits for two years: $\chi^2 = 5.37$; $t = 2.32$; $p < 10^{-2}$; sp: unidentified species.

0.92; $df = 4$; $p < 0.001$), but was negative and significant at ($r = 0.85$; $df = 4$; $p < 0.05$) between the number of *X. olivacea* visits and relative humidity. In 2010, the correlation was positive and significant at ($r = 0.55$; $df = 4$; $p < 0.05$) between the number of *X. olivacea* visits on *P. vulgaris* flowers and the temperature, positive and significant at ($r = 0.36$; $df = 4$; $p < 0.05$) between the number of *X. olivacea* visits and relative humidity.

Abundance of X. olivacea

In 2009, the highest mean number of *X. olivacea* simultaneously in activity was 1 per flower (n = 394; s = 0.63) and 256 per 1000 flowers ($n = 106$; $s = 98.97$; maxi = 333). In 2010, the corresponding figures were 1 ($n = 128$; $s = 0$) and 42 ($n = 128$; $s = 15.00$; maxi = 67). The difference between the mean number of foragers per 1000 flowers in 2009 and 2010 was highly significant ($t = 21.99$; $p < 0.001$).

Duration of visits per flower

In 2009, the mean duration of a flower visit was 8.46 s ($n = 425$; $s = 8.06$; maxi = 60 s), while in 2010, the

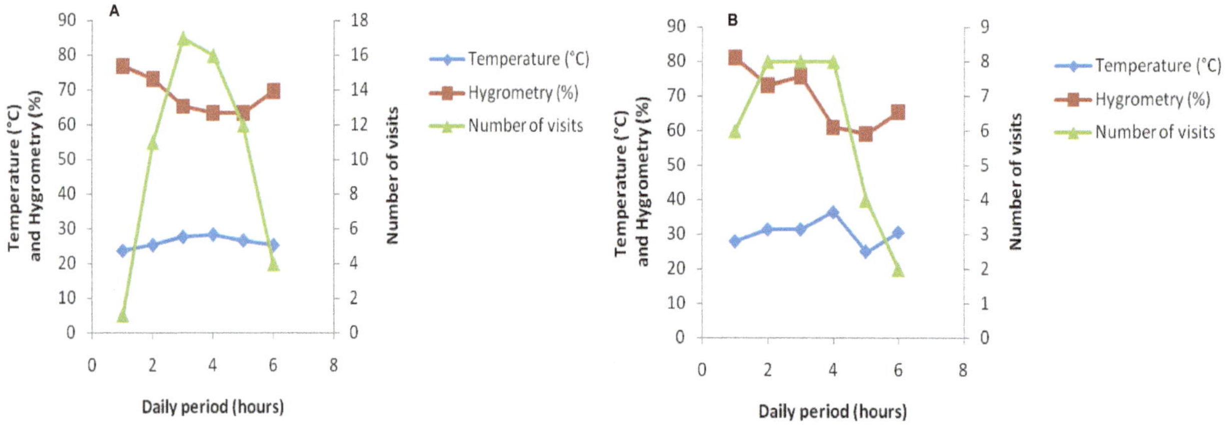

Figure 1: Daily distribution of *X. olivacea* visits on 120 and 120 *P. vulgaris* flowers over 11 days in 2009 (A) and 09 days in 2010 (B) respectively, mean temperature and mean humidity of the study site.

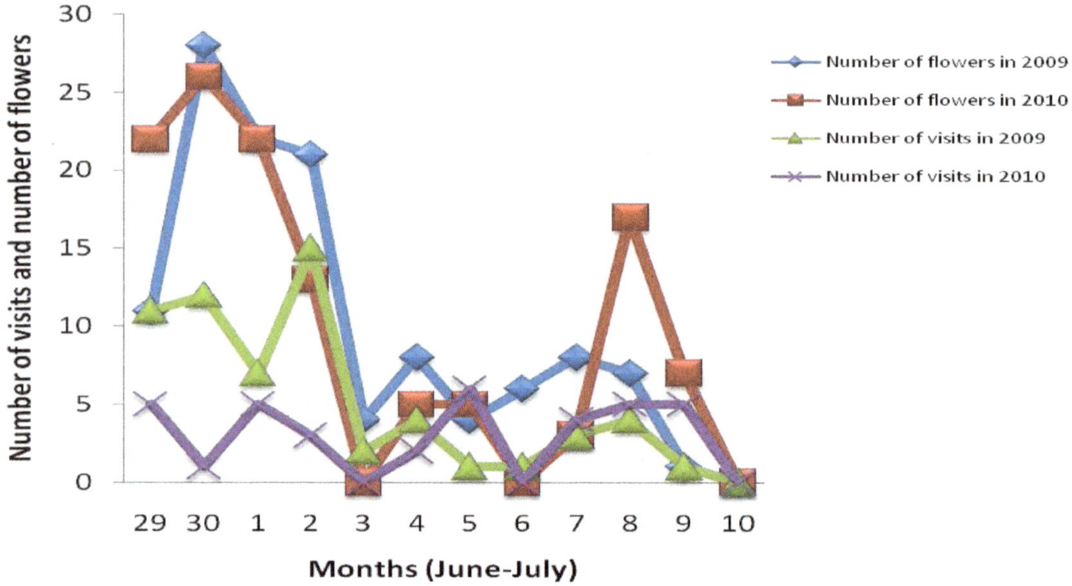

Figure 2. Seasonal distribution of the number of *P. vulgaris* opened flowers and the number of *X. olivacea* visits in 2009 and 2010.

corresponding figure was 5.12 s (n = 425; s = 2.98; maxi = 19 s), giving a highly significant difference of (t = 8.01; p < 0.001) between the two sample years. For the two cumulated years, the mean duration of a flower visit was 6.79 s.

Foraging speed of X. olivacea on *P. vulgaris* flowers

On the experimental plot, *X. olivacea* visited *P. vulgaris* corresponding figure was 5.12 s (n = 425; s = 2.98; maxi = 19 s), giving a highly significant difference of (t = 8.01; p < 0.001) between the two sample years. For the two cumulated years, the mean duration of a flower visit was 6.79 s. between 2 and 28 flowers/min in 2009 and between 4 and 20 flowers/min in 2010. The mean foraging speed was 10.45 flowers/min (n = 184; s = 12.67) in 2009 and 9.42 flowers/min (n = 134; s = 3.26) in 2010. The difference between these means was not significant (t = 1.06; p > 0.05). For the two cumulated years, the mean foraging speed was 9.94 flowers/min.

Influence of neighboring flora

During the observation period, flowers of many other plant species growing in the study area were visited by *X. olivacea* individuals, for nectar (ne) and/o pollen (po).

Figure 3. Flower of *P. vulgaris* plant showing *X. olivacea* collecting nectar on opened flower.

Among these plants were: *Tithonia diversifolia* (Asteraceae; ne and po); *Mimosa invisa* (Mimosaceae; po), *Bidens pilosa* (Asteraceae; ne and po), *Phaseolus coccineus* (Fabaceae; ne), *Cajanus cajan* (Fabaceae; ne), *Vigna unguiculata* (Fabaceae; ne and po), *Psidium guajava* (Myrtaceae; ne and po), *Senna mimosoides* (Mimosaceae; po) and *Gossypium hirsutum* (Malvaceae; ne and po). During the whole observation period individual bees foraging on *P. vulgaris* were not observed moving to a neighboring plant species and vice versa.

Impact of anthophilous insects on pod/set formation and seed yields of *P. vulgaris*

During nectar and pollen harvest on *P. vulgaris*, some foraging insects always shake flowers and contact anthers and stigma increasing the cross pollination possibility of *P. vulgaris*. When the fruiting rate was compared (Table 2), the differences observed were highly significant between free opened flowers (treatment 1) and bagged flowers (treatment 2), the first year at (χ^2 = 75.45; t = 8.69; p < 0.001) and the second year at (χ^2 = 138.11; t = 11.75; p < 0.001). The difference between the two years as far as treatment 1 is concerned was highly significant (χ^2 = 9.39; t = 3.06; p < 0.001). Consequently, the fruiting rate of the unprotected flowers was higher than that of protected flowers in 2009 and in 2010. The fruiting rate due to the action of flowering insects was 67.67% in 2009 and 78.95% in 2010. For all of the flowers studied, the fructification rate attributed to the influence of insects was 73.31%.

There was a highly significant difference between treatments 1 and 2 (t=3.49; p<0.001) the first year and the second year at (t=3.27; p<0.001) as far as, the mean number of seeds per pod is concerned (Table 2). For treatment 1, the difference between the two studied years were not significant at (t=0.35; p>0.05). Consequently, a high mean number of seeds per pod in opened flowers (treatment 1) were noticed compared to bagged flowers

Table 2. *P. vulgaris* mean yields under different pollination treatments.

Treatments	Years	Flowers	Pod	Fruiting rate	Seeds/pod		Total seeds	Normal seeds	% normal seeds
					mean	Sd			
Ff (unlimited visits)	2009	120	99	82.5	6.33	1.25	627	572	91.22
Pf (bagged flowers)	2009	120	32	26.67	5.16	1.76	165	129	78.18
Ff (unlimited visits)	2010	120	114	95	6.39	1.22	728	678	93.13
Pf (bagged flowers)	2010	120	24	20	5.17	1.74	124	70	56.45
Fva (*X. olivacea* flowers)	2009	29	15	51.72	6.4	1.18	96	89	92.71
Fva (*X. olivacea* flowers)	2010	144	132	91.67	6.4	1.34	851	778	91.42

Ff: free flower, Pf: protected flowers, Fva: flowers visited exclusively by *X. olivacea*.

(treatments 2). The number of seeds per pod attributed to the activity of flowering insects was 18.48% in 2009 and 19.09% in 2010, giving an overall mean of 18.79%.

The comparison of the percentage of normal seeds (Table 2) indicate that there were highly significant difference between free opened flowers (treatment 1) and bagged flowers (treatment 2) the first year (χ^2=21.86; p<0.001) and the second year (χ^2=133.02; p<0.001). For treatment 1, the difference between the two studied years was not significant (χ^2=1.71; p>0.05). Thus, the percentage of normal seeds in opened flowers was higher than that of protected flowers in 2009 and 2010. The percentage of the normal seeds due to the action of insects was 14.3% in 2009 and 39.39% in 2010. For all the flowers studied, the percentage of the normal seeds due to flowering insects was 26.85%.

Pollination efficiency of X. olivacea on P. vulgaris

During the nectar harvest from flowers, foragers were always in contact with the stigma and the anthers. The total number of visits expressed as percentage during which foragers bees came into contact with anthers and stigma was 100% during nectar harvest. Thus, this carpenter bee highly increased the pollination of *P. vulgaris* flowers.

The comparison of the fruiting rate (Table 2) shows that the differences observed were highly significant between treatments 2 and 3 (χ^2 = 6.79; p < 0.01) in 2009, treatments 2 and 3 (χ^2 = 139.07; p < 0.001) in 2010 and treatment 3 of the two years (χ^2 = 30.16; p < 0.001). The fruiting rate of flowers exclusively visited by *X. olivacea* (treatment 3) was significantly higher than that of flowers bagged during their flowering period (treatment 2) in 2009 and in 2010. The fruiting rate due to *X. olivacea* activity was 48.43% in 2009 and 78.18% in 2010. For all the flowers studied, the fruiting rate attributed to the influence of *X. olivacea* was 63.31%.

The comparison of the mean number of seeds per pod (Table 2) revealed that the differences observed were highly significant between treatments 2 and 3 (t = 2.85; p < 0.01) in 2009 and treatments 2 and 3 at (t = 3.13; p < 0.001) in 2010. The difference between treatment 3 in the two studied years was not significant (t = 0.03; p > 0.05). Therefore, high mean number of seeds per pod of flowers visited exclusively by *X. olivacea* (treatment 3) when compared to bagged flowers (treatment 2). The percentage of the number of seeds per pod due to *X. olivacea* was 19.38% in 2009 and 18.58% in 2010. For all the flowers studied, the percentage of the number of seeds per pod attributed to the influence of *X. olivacea* was 18.98%.

The normal seeds expressed as percentage (Table 2) demonstrates that the differences were highly significant between bagged flowers and flowers exclusively visited by *X. olivacea* (χ^2 = 9.31; p < 0.001) in 2009, and in 2010 at (χ^2 = 115.96; p < 0.001), and non significant between flowers exclusively visited by *X. olivacea* in the two studied years (χ^2 = 0.19; p > 0.05). Hence, the percentage of normal seeds of bagged flowers and those exclusively visited by *X. olivacea* was higher than that of protected flowers in 2009 and 2010. The percentage of the normal seeds due to *X. olivacea* was 15.67% in 2009 and 38.25% in 2010. For all the flowers studied, the percentage of the number of seeds per pod attributed to the influence of *X. olivacea* was 26.96%.

DISCUSSION

X. olivacea was the main floral visitor of *P. vulgaris* during the observation period. Elsewhere, bumblebees in South Africa (Palmer, 1967), *A. mellifera* in Western Kenya (Kasina et al., 2009) have respectively been reported as the main floral visitor of this crop. This could be due to the absence or low abundance of this bee in those countries. *X. olivacea* was shown to be the most abundant floral visitors of *P. coccineus* in Yaoundé (Pando et al., 2011a) and *Luffa aegyptiaca* in Cape Coast site (Mensah and Kudom, 2011). The significant difference between the percentage visits of *X. olivacea* within studied years could be explained by the presence of several nests of *X. olivacea* near the experimental plot in 2009 when compared to that of 2010. This could also be attributed to the experimental site variation.

The peak activity of *X. olivacea* on *P. vulgaris* flowers was located between 10.00 and 13.00 h, which correlated with the highest availability period of nectar on *P. vulgaris* flowers. However, this decreased activity after 16.00 to 17.00 h could be related to decreased temperature in the experimental field. Although, foragers preferred warm or sunny days for good floral activity (Kasper et al., 2008), the enhanced temperature positively influenced the insect activity on foraged flowers. Similarly, rainfall has been documented as an environmental factor that can disrupt the floral insect activity (McGregor, 1976). The abundance of *X. olivacea* foragers on 1000 flowers and the positive and highly significant correlation between the number of *P. vulgaris* flowers at bloom, as well as, the number of *X. olivacea* visits indicates the attractiveness of *P. vulgaris* nectar with respect to this bee. In fact, weather during bloom was demonstrated to affect the abundance and foraging of pollinator insects (Bramel et al., 2004, Julianna and Rufus, 2010). Among the 24 insect species visiting *P. vulgaris* flowers, *X. olivacea* was the most abundant (28.7%), followed by *Coryna* sp. (17.92%), *Chalicodoma rufipes* (15.24%) and *Ceratina* sp.1 (7.65%).

The significant difference between the duration of visits in 2009 and 2010 could be attributed to the availability of floral products or the variation of diversity of flowering insects from one year to another. During each of the two flowering periods of *P. vulgaris*, *X. olivacea* intensely and regularly harvested nectar. This could be attributed to the needs of individuals at flowering period. The disruptions of visits by other insects reduced the time frame visits of certain *X. olivacea*. This obliged some carpenter bees to visit more flowers for a foraging trip in order to maximize their nectar loads. Similar observations were made for *A. mellifera adansonii* workers foraging on *Entada africana* (Fabaceae) flowers, *P. guajava* (Myrtaceae) flowers (Tchuenguem et al., 2007), *Croton macrostachyus* (Euphorbiaceae) flowers, *Syzygium guineense* var. *guineense* (Myrtaceae) flowers (Tchuenguem et al., 2008a), *Persea americana* (Lauraceae) flowers, *Vitellaria paradoxa* (Sapotaceae) flowers (Tchuenguem et al., 2008b), *V. unguiculata* (L.) (Fabaceae) flowers (Tchuenguem et al., 2009b), *Combretum nigricans*, *Erythrina sigmoidea*, *Lannea kerstingii*, *Vernonia amygdalina* flowers (Tchuenguem et al., 2010) and for *Chalicodoma cincta* (Hymenoptera: Megachilidae) foraging on *C. cajan* (Fabaceae) flowers (Pando et al., 2011b).

The carpenter bee foragers had a high affinity with respect to *P. vulgaris* when compared to the neighboring plant species, indicating their faithfulness to this Fabaceae, a phenomenon known as "floral constancy" (Louveaux, 1984; Backhaus, 1993; Basualdo et al., 2000). Flower constancy is an important aspect in the management of pollination. For this research, it indicates that *X. olivacea* can provide benefits to pollination management of *P. vulgaris*.

During the collection of nectar on each flower, *X. olivacea* foragers regularly come into contact with the stigma. They were also able to carry pollen with their hairs, legs and mouth accessories from a flower of one plant to stigma of another flower of the same plant (geitonogamy), to the same flower (autogamy) or to that of another plant (xenogamy).

The significant contribution of *X. olivacea* in pods and seed yield of *P. vulgaris* is in agreement with similar findings in Britain (Darwin, 1858) and United State of America (Ibarra-Perez et al., 1999) which showed that *P. vulgaris* flowers produce fewer seeds per pod in the absence of efficient pollinators.

The contribution of *X. olivacea* to *P. vulgaris* production through its pollination efficiency was significantly higher than that of all insects on the exposed flowers. The weight of *X. olivacea* played a positive role during nectar collection. *X. olivacea* shook flowers facilitating the liberation of pollen by anthers for the optimal occupation of the stigma. Our results confirmed those of Mensah and Kudom (2011) who revealed that the development of fruits from *L. aegyptiaca* flowers that have received a single visit of *X. olivacea* produced a mean weight of 428.7 g that was 1.5 times heavier than fruits from flowers visited by *A. mellifera* (286.76 g). This phenomenon was also reported by Vanderborght and Rasmont (1987) for *X. bariwal*, an efficient *P. coccineus* pollinator.

Higher productivity of pods and seeds in unlimited visits when compared with bagged flowers showed that insect visits were effective in increasing cross-pollination. Our results confirmed those of Webster et al. (1982), Wells et al. (1988) and Ibarra-Perez et al. (1997) who revealed that *P. vulgaris* flowers set little pods in the absence of insect pollinators. Similar experiments in England (Free, 1966) and in Brazil (Free, 1993) have shown that pollination by insects was not always needed. Darwin (1876) showed that self-pollination of *P. vulgaris* flowers produced as many pods and seeds as exposed plants. Thus, pollination requirements may differ between plant varieties.

Conclusion

This study reveals that *P. vulgaris* black seed outlets is a highly nectariferous bee plant that obtained benefits from the pollination by insects among which *X. olivacea* is of great importance. The comparison of pods and seeds set of unprotected flowers with that of flowers visited exclusively by *X. olivacea* underscores the value of this bee in increasing pods and seed yields as well as seed quality. The installation of *X. olivacea* nests at the proximity of *P. vulgaris* small black seed fields should be recommended for the increase of pods and seed yields of this valuable crop.

ACKNOWLEDGEMENT

The financial support of IFS that contributed to the installation of hives within the University campus is hereby acknowledged.

REFERENCES

Backhaus W (1993). Colour vision and colour choice behaviour of the honey bee. Apidologie, 24: 309-331.

Barrons KC (1939). Natural crossing in beans at different degrees of isolation. Am. Soc. Hort. Sci. Proc., 36: 637-640.

Basualdo M, Bedascarrasbure E, De JD (2000). Africanized honey bees (Hymenoptera: Apidae) have a greater fidelity to sunflowers than European bees. J. Econ. Entomol., 2: 304-307.

Bramel J, Kiran S, Reddy J, Ford-Lloyd B, Chancra S (2004). Degree and distribution of pigeon pea landrace morphological diversity in traditional cropping systems in Andhra Pradesh. In: Bramel PJ, editor. Assessing the risk of losses of biodiversity in traditional cropping systems: A case study of pigeon pea in Andhra Pradesh. International Crop Research Institute for the semi-arid Tropics, 1-45. Pantacheru Andhra Pradesh, India.

Darwin C (1858). On the agency of bees in the fertilization of papilionaceous flowers, and on the crossing of kidney. Gard. Chron., 18: 828-829.

Darwin C (1876). The effects of cross and self fertilization in the vegetable Kingdom. London: John Murray. p. 505.

Debouck D (1991). Systematics and morphology. In: Common beans: research for crop improvement, Van Schoonhoven, A and Voyset, O (eds), Cali, Colombie. pp 55-118.

Demarly Y (1977). Genetic and amelioration of plants. Masson, Paris. p. 577.

Fluri P, Frick R (2005). Apiculture in Sweden: state and perspectives. Rev. Suisse. Agric., 37(2): 81-86.

Free JB (1966). The pollination of the beans *Phaseolus multiflorus* and *Phaseolus vulgaris* by honeybees. J. Apicult. Res., 5: 87-91.

Free JB (1993). Insect pollination of crops. Academic Press, London, UK. pp. 263-270.

Freitas BM (1997). Number and distribution of cashew (*Anacardium occidentale*) pollen grains on the bodies of its pollinators, *Apis mellifera* and *Centris tarsata*. J. Apicul. Res., 36: 15-22.

Graham PH , Ranalli P (1997). Common bean (*Phaseolus vulgaris* L.). Field Crop Res., 53: 131-146.

Ibarra-Perez FJ, Barnhart D, Ehdaie B, Knio KM, Waines JG (1999). Effects of insect tripping on seed yield of common bean. Crop Sci., 39: 425-433.

Ibarra-Perez FJ, Ehdaie B, Waines JG (1997). Estimation of outcrossing rate in common bean. Crop Sci., 37: 60-65.

Jacob RA (1989). Behaviour of domestic and wild forager bees in vergers de pommiers in Belgium. Apidologie, 20: 271-285.

Jean-Prost P (1987). Knowing the bees-Leading the hives, Sixth edition, Lavoisier, Paris, France. p. 579.

Julianna KT. Rufus I (2010). Weather during affects pollination and yield of highbush blueberry. J. Econ. Entomol., 103(3): 557-562.

Kasina M, Kraemer M, Martius C. Wittmann D (2009). Diversity and activity density of bees visiting crop flowers in Kakamega, Western Kenya. J. Apicul. Res.. 48(2): 134-139.

Kasper ML, Reeson AF, Mackay DA Austin AD (2008). Environmental factors influencing daily foraging activity of *Vespula germanica* (Hymenoptera, Vespidae) in Mediterranean Australia. Insect Soc., 55: 288-296.

Keller LF. Waller DM (2002). Inbreeding effects in wild populations. Trends Ecol. Evol., 17: 230-241.

Klein AM, Vaissière BE, Cane JH, Steffan DI, Cunnigham SA, Kremen C, Tscharntke T (2007). Importance of pollinators in changing landscapes for world crops. Proc. Royal Soc., 274: 303-313.

Louveaux J (1984). Domestic bee in its relations with grown plants. In: Pollination and plant production, Pesson P. and Louveaux J. (eds), INRA, Paris, pp. 527-555.

Mackie WW, Smith FL (1935). Evidence of field hybridization in beans. J. Am. Soc. Agron., 27: 903-909.

McGregor SE (1976). Insect pollination of cultivated crop plants. Agricultural Research Service. United States Department of Agriculture, Agric. Handb, Washington, no. 496, p. 411.

Mensah BA, Kudom AA (2011). Foraging dynamics and pollination efficiency of *Apis mellifera* and *Xylocopa olivacea* on *Luffa aegyptiaca* Mill (Cucurbitaceae) in Southern Ghana. J. Pollin. Ecol., 4(5): 34-38.

MINADER/DESA (2010). Annuaire des Statistiques du Secteur Agricole Campagnes 2007 & 2008. AGRI-STAT (16): 98.

Palmer R (1967). Bees and bean pollination. S. Afr. Bee J., 39(4): 12-13.

Pando JB, Tchuenguem FFN, Tamesse JL (2011a). Foraging and pollination behaviour of *Xylocopa calens* (Hymenoptera: Apidae) on *Phaseolus coccineus* L. (Fabaceae) flowers at Yaoundé (Cameroon). Entomol. Res., 41: 185-193.

Pando JB, Tchuenguem FFN, Tamesse JL (2011b). Pollination and yield responses of pigeon pea (*Cajanus cajan* L. Mill sp.) to the foraging activity of *Chalicodoma cincta cincta* (Hymenoptera: Megachilidae) in Yaoundé (Cameroon). J. Anim. Plant Sci., 11(1): 1346-1357.

Roubik DW (2000). Pollination system stability in Tropical America. Conserv. Biol., 14(5): 1235-1236.

Sabbahi R, De OD, Marceau J (2005). Influence of honey bee (Hymenoptera: Apidae) density on the production of canola (Crucifera: Brassicacae). J. Econ. Entomol., 98(2): 367-372.

Tchuenguem FFN (2005). Foraging and pollination activity of *Apis mellifera adansonii* Latreille (Hymenoptera : Apidae, Apinae) on flower of three plants at Ngaoundéré (Cameroon) : *Callistemon rigidus* (Myrtaceae), *Sygygium guineense* var. *macrocarpum* (Myrtaceae) and *Voacanga Africana* (Apocynaceae). State doctorate, University of Yaounde I. p. 103.

Tchuenguem FFN, Djonwangwe D, Brückner D (2008b). Foraging behaviour of the African honey bee (*Apis mellifera adansonii*) on *Annona senegalensis*, *Croton macrostachyus*, *Psorospermum febrifugum* and *Syzygium guineense* var. *guineense* flowers at Ngaoundéré (Cameroon). Pak. J. Biol. Sci., 11: 719-725.

Tchuenguem FFN, Djonwangwé D, Messi J, Brückner D (2007). Exploitation of *Entada africana*, *Eucalyptus camadulensis*, *Psidium guajava* and *Trichillia emetica* flowers by *Apis mellifera adansonii* at Dang (Ngaoundéré, Cameroon). Cameroon J. Exp. Biol., 3: 50-60.

Tchuenguem FFN, Djonwangwe D, Messi J, Brückner D (2008a). Exploitation of *Dichrostachys cinerea*, *Vitellaria paradoxa*, *Persea americana* and *Securidaca longepedunculata* flowers by *Apis mellifera adansonii* Latreille (Hymenoptera: Apidae) at Dang (Ngaoundéré, Cameroon). Inter. J. Trop. Insect Sci., 28: 225-233.

Tchuenguem FFN, Djonwangwé D, Messi J, Brückner D (2009a). Foraging and pollination activities of *Apis mellifera adansonii* (Hymenoptera: Apidae) at Ngaoundéré (Cameroon). Cameroon J. Exp. Biol., 5(1): 1-9.

Tchuenguem FFN, Fameni TS, Mbianda PA, Messi J, Brückner D (2010). Foraging behaviour of *Apis mellifera adansonii* (Hymenoptera: Apidae) on *Combretum nigricans*, *Erythrina sigmoidea*, *Lannea kerstingii* and *Vernonia amygdalina* flowers at Dang (Ngaoundéré, Cameroon). Inter. J. Trop. Insect Sci., 1: 40-47.

Tchuenguem FFN, Messi J, Brückner D, Bouba B, Mbofung G, Hentchoya HJ (2004). Foraging and pollination behaviour of the African honey bee (*Apis mellifera adansonii*) on *Callistemon rigidus* flowers at Ngaoundéré (Cameroon). J. Cam. Acad. Sci., 4(2): 133-140.

Tchuenguem FFN, Messi J, Pauly A (2001). Activity of *Meliponula erythra* on *Dacryodes edulis* flowers and its impact on fructification. Fruits, 56(3): 179-188.

Tchuenguem FFN, Ngakou A, Kengni BS (2009b). Pollination and yield responses of cowpea (*Vigna unguiculata* L. Walp.) to the foraging activity of *Apis mellifera adansonii* (Hymenoptera: Apidae) at Ngaoundéré (Cameroon). Afr. J. Biotechnol., 9: 1988-1996.

Vanderborght I, Rasmont P (1987). Utilization of the domestic bees (*Apis mellifera* L.) and the Xylocope (*Xylocopa bariwal* MAIDL) (Hymenoptera, Apidae) for pollination in cage of *Phaseolus coccineus* L. Bull. Rech. Agron. Gemb., 22(4): 279-284.

Webster BD, Ross RM, Evans T (1982). Nectar and nectary of *Phaseolus vulgaris.* L. J. Am. Soc. Hort. Sci., 107: 497-503.

Wells WC, Isom WH, Waines JG (1988). Outcrossing rates of six common bean lines. Crop Sci., 28: 177-178.

19

Revisiting uncertainty and price forecast indicators in corn and wheat markets

Hélyette Geman[1,2] and Pedro Vergel Eleuterio[1]

[1]Birkbeck, University of London, London, England.
[2]Johns Hopkins University, United States.

The purpose of this paper is twofold: First, we look at the fundamentals of spot prices of corn and wheat and analyse several measures of dispersion, arguing that the use of the standard deviation of prices is more instructive for regulators and world food organisations than volatility, that is, standard deviation of returns. Second, we look at alternative predictors of corn and wheat spot prices and exhibit that the average value of the forward curve introduced by Borovkova and Geman (2006) performs better than individual forward prices to forecast spot prices at future dates.

Key words: Grain markets, volatility of returns, standard deviation of prices, geometric average of forward prices, commodity spot price predictors.

INTRODUCTION

It is usually recognized that Futures markets incorporate information quicker than spot markets due to low transaction costs, liquidity and the feasibility of long and short positions. Price discovery allows the transfer of price information, from commercial merchants that have more accurate information about planting decisions and future harvests to other players that do not have access to this information. If the number of players on the buy and sell sides with full information was large enough, the expectation of futures prices should in principle be an unbiased prediction of spot prices.

In contrast to crude oil prices which started rising in 2002, followed by copper, gold and other metals, agricultural commodity prices were essentially flat (declining in fact if adjusted for inflation) until 2006. In the agricultural year 2006-2007, different weather events around the world sent corn and wheat prices to unprecedented levels. As of that moment, food price risk became a large concern, for government and regulators alike.

Our goal in the first part of this article is to argue that the famous volatility (that is, standard deviation of returns) that is widely discussed, analysed and estimated in the financial markets is not the most informative quantity in the case of commodities, particularly agricultural commodities. We propose instead to focus on the signals provided by the coefficient of variation of prices on one hand, and the standard deviation of price levels on the other hand.

In the second part of the paper, we compare the quality of future price prediction provided by individual forward contracts versus the geometric average of the forward curve introduced in Borovkova and Geman (2006). We show that the latter performs better for corn and wheat spot prices. The measures presented in the first and second parts of the paper are highly related because they all provide important information for participants in the market, including farmers, consumers, and government regulators and policymakers.

Since the measures of dispersion and the predictive power of Futures prices, the two focal points of our study, are directly linked to the realities of the physical markets themselves, we also provide the reader with the essential background needed to understand the corn and wheat markets. Explanations of the physical commodities are vital in order to be able to see how Futures prices are capturing the market.

FUNDAMENTALS OF CORN AND WHEAT MARKETS

Corn

As a feed, corn is the highest valued among the cereal grains for its energy content, consisting of 65% starch, 4% oil and 10% fibre. In temperate climates, corn must be planted in the spring. Corn is a more water efficient crop than soybeans or alfalfa but requires a larger amount of fertilizers. A corn crop producing six to nine tonnes of grains per hectare requires 31 to 50 kg of phosphate fertilizer while a soybean crop requires 20 to 25 kg per hectare (Geman and Vergel Eleuterio, 2013).

The United States is the biggest producer of corn in the world, followed by China, Brazil, and the EU-27. Main world exporters are the United States, Argentina, Brazil, and Ukraine. The biggest importers are Japan, South Korea, the European Union and Mexico. Depending on the government policy and climate conditions, China alternates between years of high exports and high imports, thus making it a source of uncertainty for the world corn market.

In the US, corn production accounts for over 95% of total feed grains production. Of the other feed grains, sorghum accounts for 2.9, barley for 1.5 and oats for 0.5%. The Chicago Mercantile Exchange (CME) offers a wide variety of financial derivatives written on corn including corn Futures, a mini-corn Futures contract, calendar swaps, and a wide variety of options contracts, including new crop options.

Wheat

A fundamental staple of the human diet, wheat comprises approximately 20% of calories and proteins consumed on a global scale. Worldwide, approximately 10% of wheat grain production is used annually for feed. A high starch content of roughly 70% dry matter makes this cereal grain rich in carbohydrates. In addition to having a greater amount of protein (hard and durum wheat have more proteins than soft wheat) than corn or barley, wheat protein is also of a higher calibre, making the grain a valued substitute for corn. It is important to note, however, that the content and quality of crude protein and starch can vary with growing conditions, wheat species, fertilizers, and other factors.

Common (or bread) wheat grain is primarily processed into flour for bread, pastry, and the confectionary industry. The production of bread and similar products is aided by gluten, a protein unique to wheat, which helps dough to rise. Gluten is commonly used to thicken foodstuffs including soup, gravy, and sauces. Durum wheat, which is lower in gluten content, is used in the production of pastas, semolina, couscous, pizza bases, and other flat breads.

The main producers of wheat are EU-27, China, India, and the US. Major exporters are the US, EU-27, and Australia and the largest importers include Egypt, Brazil, Indonesia, Japan, Algeria EU-27 and South Korea. Wheat futures, including a mini-wheat Futures contract, calendar swaps and options contracts as well as Black Sea Wheat Futures, Kansas City Wheat Futures (KC Wheat), options and swaps from the Kansas City Board of trade (KCBT) can be purchased through the CME. Other products include Minneapolis Grain Exchange - Chicago Board of Trade Wheat Spread options (commonly known as MGEX-CBOT Wheat Spread Options) as well as Futures and short-dated new crop options.

Links between corn and wheat

In Crops and Man (1975), renowned botanist Jack Harlan stated the following: "Fully domesticated plants are artefacts produced by man as much as an arrowhead, a clay pot, or a stone axe." This is certainly true of most, if not all, agricultural commodities currently traded and consumed on a global scale. Centuries of research and careful breeding (for example, to create hybrids) together with the development of fertilizers, pesticides, and herbicides have resulted in better quality crops, higher yields, and greater resistance to pests and adverse weather conditions.

Corn and wheat are two central agricultural commodities which have benefited from such developments. With incomes rising in developing countries such as China and India, there has been a shift from basics such as rice and wheat to more expensive foodstuffs including meat, dairy, and vegetable oils. The increase in demand for meat and dairy requires the expansion of the meat production industry which, in turn, requires more feed grains. Although the rate of population growth has slowed significantly with the decline in birth rates, demographers still forecast a rise to ten billion just after 2050. Providing nourishment for everyone will require at least 35% more calories than what is currently produced today. Taking into account the continual increase in meat consumption, the percentage of grains needed is much greater. Animal nutrition, which is ultimately dependent on agricultural commodities, will be crucial in meeting objectives for meat production.

Both corn and wheat are used for animal feed. Corn and

its by-products are valued for their high energy content but have low protein content and need to be supplemented in order to provide the appropriate amount of proteins and amino acids. For this reason, corn is traditionally mixed with soybean meal when used as an animal feed. Wheat that is not fit for human consumption or food processing is used for animal feed. With a higher amount of protein, minerals, oil and fibre than wheat grain, wheat bran - a by-product of the dry milling process used to make flour and produce pasta from durum wheat - is also a major animal feed. The link of corn, wheat and soybeans is apparent in the offering of financial products. The CME offers a wide variety of cross commodity financial products such as wheat-corn spread options and soybean-corn price ratio options.

Wheat and corn also share important connections through crop rotation, double cropping, and intercropping. It has long been shown that crop rotation – a planting method involving the growth of various crops in a strategic, sequential order – of corn, soybean, and wheat in a three year rotation results in improved soil quality, reduced risk of pests and pathogens, better weed control and most importantly, sustained yields. A study by Michigan State University (Lipps et al., 2001) reports that including wheat in the rotation is vital for increasing yields of other crops like corn and soybeans by at least 10%. In addition, crop rotation can be beneficial in reducing risks associated with poor weather conditions; for example, during the drought of 2012, wheat yields were above average while corn was adversely affected. Double-cropping, when a second crop is planted after harvesting the first, is also common in the cases of corn, wheat and soybeans. In the US, it is usual to find double-cropping, for example, with corn and alfalfa rotating in colder areas or corn and soybeans in areas with longer summers. A third crop, such as winter wheat, can also be added to the rotation. In "relay intercropping", two crops are planted in the same field; for example, soybeans can be planted on a field where wheat is currently growing.

Since the start of 2004, a dramatic increase has prevailed in the open interest of CBOT Corn and CBOT Wheat Futures contracts. The biggest rise in open interest was for CBOT wheat, which increased from around 100,000 contracts in 2005 to a high of 550,000 in 2006. Corn followed and by spring 2006, banks accounted for 20% of the total open interest for wheat and 12% for corn. Additionally, from mid-2007 to mid-2008, we observe a widening of the spread between wheat and corn prices. This coincided with a period of in-creased prices due to tight wheat supplies, record wheat prices, human consumption demand being inelastic to prices, and a large amount of trading carried out by index funds. Historically, CBOT Wheat and CBOT Corn have moved together, with the price of wheat typically one to two dollars higher than corn. Only on very rare occasions does this spread invert. However, corn prices reached a record high in April 2011 and surpassed the price of wheat of wheat due to a growing demand for corn-based ethanol and tight level of supplies not seen since the 1930s (Figure 1).

VOLATILITY, COEFFICIENT OF VARIATION AND STANDARD DEVIATION OF PRICES OF CORN AND WHEAT

Traditionally, finance literature has concentrated on the use of returns rather than prices, and volatility rather than standard deviation of prices as a measure of variation. It started with the Portfolio Theory of Markowitz (1958) where the naturally important quantities to analyse are returns on stocks, whatever the initial wealth owned by the investor. It continued with the Black-Scholes-Merton (1973) model, in which the mathematical assumption on the underlying stock price was expressed through the stochastic dynamics of its returns.

Turning to commodities, the monthly volatility of returns - with returns classically approximated by log differences - is computed from monthly data as follows:

$$Volatility = \sqrt{\frac{1}{n-1}\sum_{i=1}^{n}\left(\Delta \ln p_i - \frac{1}{n}\sum_{i=1}^{n}\Delta \ln p_i\right)^2}$$

where p_i is the dollar value of the CBOT closing price of the commodity nearby future contract on month i and n is the number of monthly observations. It is annualized by multiplying the monthly volatility, as calculated above, by $\sqrt{12}$. We calculate the annualized volatility in a similar way when using daily or weekly data, by multiplying volatility by $\sqrt{250}$ or $\sqrt{52}$, respectively .

From an econometric point of view, the use of returns makes sense since the return series is more likely to be stationary, which leads to the use of a wide range of econometric tools. This reasoning is not optimal, however, for agricultural commodities. The repeated crises since 2006 have brought the focus on the right entity of concern, namely the price prevailing in the world market. Price levels are reflected in the cost of food and food security for poorer individuals who have no interest in returns, and in turn for policymakers.

Hence, an investigation of the different measures of dispersion of prices (and not just returns) is, in our view, hard to avoid when dealing with food commodities. Different types of participants may benefit from distinct measures of dispersion. For instance, since farmers are directly affected by price volatility, which can have a long term influence on producers and their production schedule, they should be most concerned with the standard deviation of prices. This measure best captures intra-year volatility of the price level driven by low stock to use ratio, or tight supply. In contrast, investors care about the return of the asset and the risk attached, which is

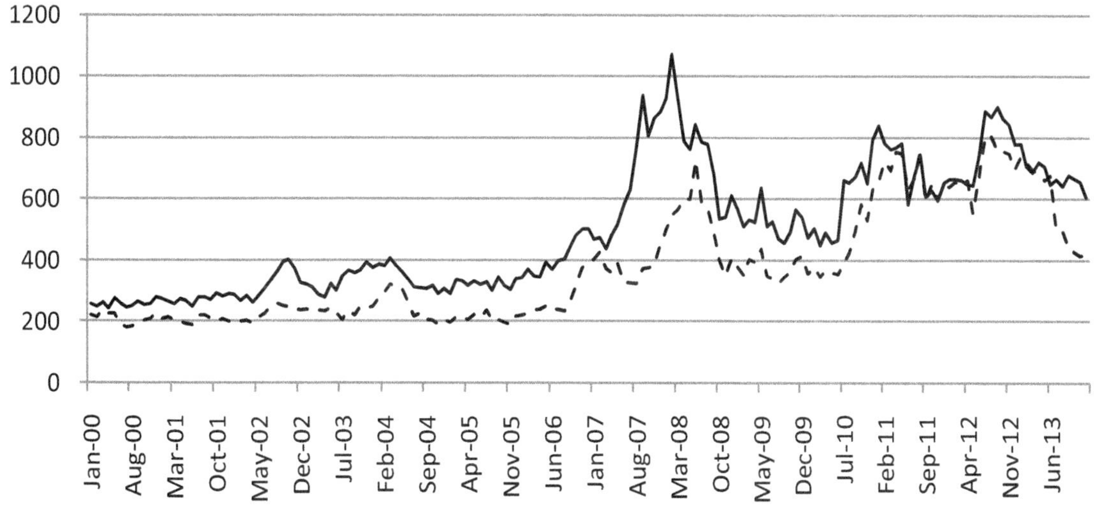

Figure 1. First nearby prices of CBOT Wheat (solid line) and CBOT Corn (dash line) in US cents per Bushel.

best measured by the volatility of returns. Government regulators and policymakers would benefit from using the coefficient of variation since it gives better intra-year information (for example, in 2004 when there was a dramatic increase in the open interest of corn and wheat Futures contracts); this measure could allow them to confront changing conditions more quickly.

The coefficient of variation lies somewhat in the middle of volatility and standard deviation of prices. Like volatility, it is independent of units of measurement; however, its computation bears directly on prices. The annualised coefficient of variation, CV, is usually defined as the ratio of standard deviation σ and the mean μ of the price series:

$$CV = \frac{\sigma}{\mu}$$

It is customary to use the sample standard deviation and the sample mean when calculating the coefficient of variation. In the continuity of our analysis above, we use monthly prices and annualize in the same way. We recall that the standard deviation of the sample of size n is computed as:

$$Standard\ deviation = \sqrt{\frac{1}{n-1}\sum_{i=1}^{n}\left(p_i - \frac{1}{n}\sum_{i=1}^{n} p_i\right)^2}$$

We use in the formulas above monthly data from January 2000 to December 2013. The first nearby, F1, is used as a proxy for the spot price, S_t, throughout the paper. The contracts chosen for each commodity are their respective world benchmarks. In the case of wheat, we use Future contracts on No. 2 Soft Red Winter wheat and for corn, the No. 2 Yellow Corn Future contract, both traded on the CME Futures U.S. Exchange. In all cases, we respect the 'last trading day' rule for consistency.

Corn

From Figures 2 and 3 depicting the results in Table 1, we obtain a clearer picture of the pros and cons of each measure. Starting with volatility, we observe values around 20% until 2003, increasing to around 30% from 2004 to 2007, with a peak of 47% in 2008 and values well over 30% afterwards. The coefficient of variation also shows low values up to 2003, and then starts exhibiting a spiky behavior. Turning to the standard deviation of prices, we observe in the years 2000 to 2003 values lower than 100, then much higher, which should have been a striking signal for regulators and policy makers.

The year 2004 was an exceptional year for corn: Remarkably low levels of carried stock from 2003 and the 2004 harvest the largest one on record. The effects of both scarcity and abundance of grains that year are easily observable in the price movements. Corn prices reached a maximum in April at 316.5 cents per bushel and went down to 192.5 by November to partially recover by the end of year. This intra-year variation is particularly well reflected in the coefficient of variation and the standard deviation of prices, making the case for these two values we propose to bring a large attention to. Regulators should have read in these numbers warnings for the bigger crises to come.

Wheat

In the case of CBOT wheat, we also observe how the alternative measures of dispersion give different views on

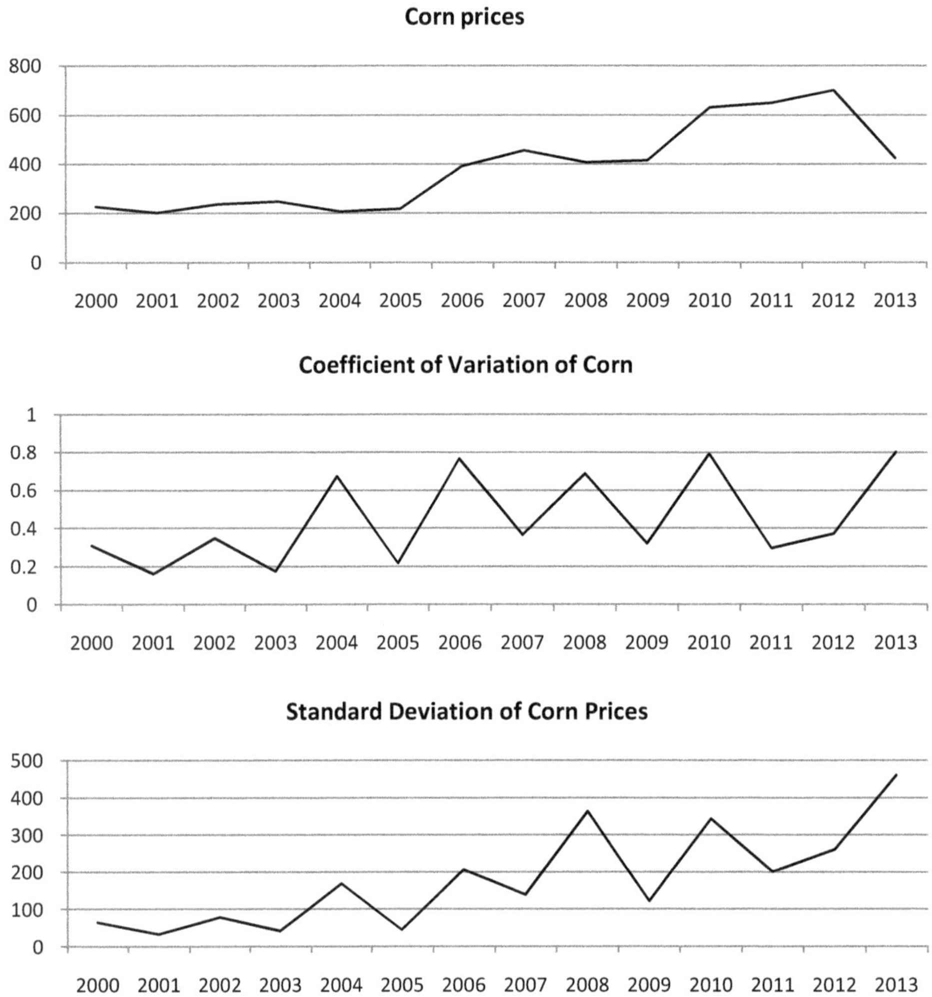

Figure 2. CBOT corn: Closing price, coefficient of variation, and standard deviation of prices per year from 2000 to 2013.

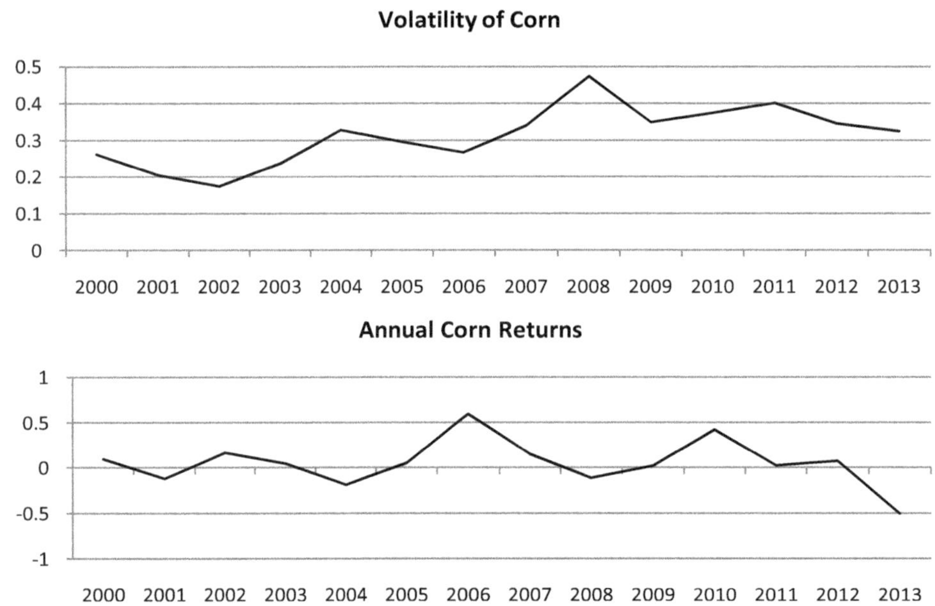

Figure 3. CBOT corn annualized monthly volatility and annual returns from 2000 to 2013.

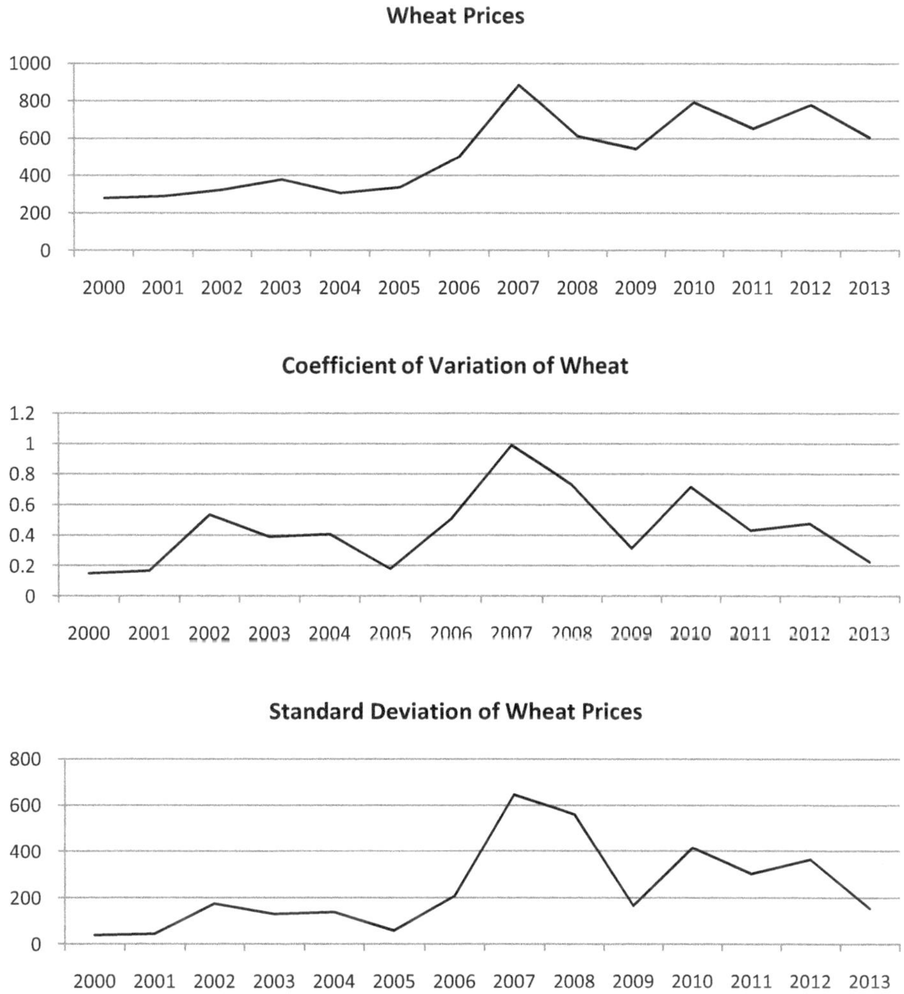

Figure 4. CBOT wheat: Closing price, coefficient of variation, and standard deviation of prices per year from 2000 to 2013.

what happened in recent years. While volatility presents a clear upward trend from 2006 until it started declining in 2011, this decline happens much earlier in both the annualised coefficient of variation and standard deviation (Table 2 and Figures 4 and 5).

In conclusion, for both corn and wheat, the increase of the coefficient of variation and standard deviation of prices was much more dramatic than volatility as of the middle of the year 2005 and could have been a warning signal for governments and regulators.

EXAMINATION OF ALTERNATIVE PREDICTORS OF SPOT PRICES

There is a long history of price expectations models, beginning with Hicks' publications "Value and Capital" (1939, 1946) and "Capital and Growth" (1965). It was Muth (1961) who developed the econometric version of the Rational Expectations Hypothesis. Following his work, numerous studies have been conducted to test the unbiasedness of the forward exchange rate as a predictor of the spot exchange rate in the future; for example, Cornell (1977), Geweke and Feige (1979), Hansen and Hodrick (1980), Longworth (1981), and Frenkel (1981). There also exist empirical studies for commodities, including those of Goss (1983), and Pieroni and Riociarelli (2005).

Lucas (1972) extended the Rational Expectations Hypothesis (REH) to macroeconomics and was awarded a Nobel Prize for his work (Lucas, 1995). According to his perspective, the REH is a conjecture that can be the core of an empirically testable price expectations model. The REH implies that the forward price at date t for maturity date $t+h$, F_t, should be an unbiased predictor of the commodity spot price at $t+h$:

$$f(t, t+h) = E_p(S(t+h)/I_t) + u_{t+h}$$

Where I_t is the filtration incorporating all information until date t, and u is an error term with conditional expected value of 0 and uncorrelated to the information at time t.

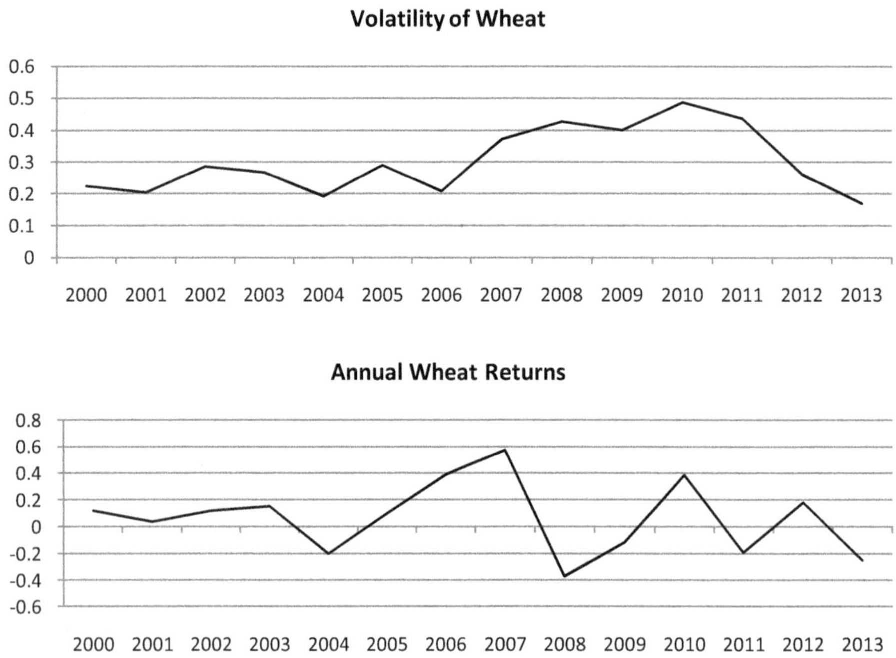

Figure 5. CBOT wheat: Annualised monthly volatility and annual returns from 2000 to 2013.

By adding the no- arbitrage assumption and changing the probability measure, the above relationship can be written as an exact equality.

In this paper, we adapt the tests of the REH in the style of Muth (1961) to find the optimal lag h, if any, for an estimate of the future spot price.

Individual forward contracts as predictors of future spot prices

In order to test the relationship between spot and forward prices, we use log prices of Futures daily data for corn and wheat. Under constant interest rates or absence of correlation of these to the underlying asset, arguably the case for agricultural commodities, forward and Future prices are equal. Both commodities have Future contracts with delivery months in March, May, July, September and December (examples of forward curves in Figure 6) and their last trade date is the last business day prior to the 15th calendar day of the contract month. The first test involves running the regression:

$$\ln S_t = \alpha + \beta \ln f_{t-h} + \varepsilon \quad (1)$$

With h expressed in number of months. In this analysis, we are interested in the results of the F-tests on the intercept of the regression and the slope of the lagged forward price. If $\alpha=0$ and $\beta=1$, even if the series are not stationary, they are cointegrated and hence the forward price is an unbiased predictor of the future spot price. The main drawback of this test is that it could be allocating a predictive power to the forward price that could also be attributed to the lagged spot price. Hence, a second F-test is used based on the regression:

$$\ln S_t - \ln S_{t-h} = \alpha + \rho \quad (\ln f_{t-h} - \ln S_{t-h}) + \varepsilon \quad (2)$$

In this test, the change in the spot price, $\ln S_t - \ln S_{t-h}$,is explained by $\ln f_{t-h} - \ln S_{t-h}$, a quantity that defines the magnitude of backwardation or contango at date *t-h*. This test deals with log differences, which are generally stationary, and subtracts the effect of the lagged spot price from both sides of the equation. When ρ =1, the regression represented by Equation (2) reduces to that of Equation (1) when $\beta = 1$.

We pose two questions. First, do forward prices before the planting period of corn predict spot prices after the harvest in the US? Second, will these estimates be affected by the proximity of the observed forward prices to the contract expiry date? In order to answer these questions, we test the predictive power of each maturity with its corresponding spot price, that is, we compare how the July Future price observed at time *t-h* predict July spot prices, the September Future prices predict September spot prices, etc.

First, we place ourselves at two particular points in time: the beginning of February and the beginning of April. These are the dates when observations of the forward curve are taken; *date t-h* in Table 3, for a length of 10 business days. The corn heading (stage of development of grains where the head pushes its way

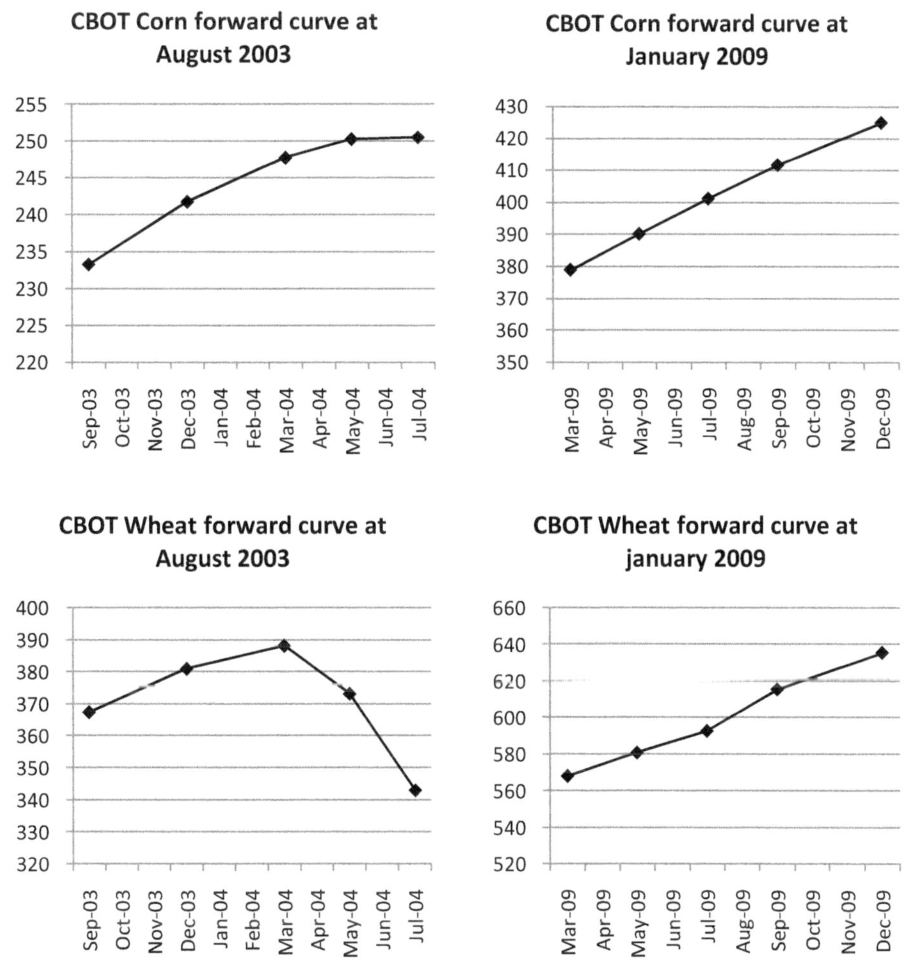

Figure 6. Forward curves of CBOT corn and wheat, August 2003 and January 2009.

through the flag leaf sheath) and harvesting happen during the summer months. The farmer will be interested in inferring information about prices after the harvest, that is, after September, from the forward curve. Accordingly, the October corn option contract is the most traded.

From Table 3, we clearly see that forward prices observed at the beginning of February produce better estimates of corn spot prices than the ones observed in the beginning of April, when planting is already underway (most planting is done in the US from April to May). Although the March Futures contract maturing the following year offers unbiased predictions of future March spot prices in both cases, the earliest contract that provides some post-harvest information is the December Futures contract (Table 3).

We note an important effect on the predictive power of the forward prices as the contracts get closer to expiry. When we compare results from Table 3 with the results from Table 4, where we observe the forward curve during the last 10 trading days before maturity of the contract, in general, the predictive power of the forward prices is greatly reduced.

Similar questions can be asked for wheat, although the case of wheat is markedly different. First, as we did in the case of corn, we compare the results derived from forward curves observed at the beginning of February to those observed at the beginning of April (Table 5). At the beginning of February we only obtain unbiased estimates of spot prices from the September Futures contract (it is important to note that March is a weather/moisture critical period in the heading of wheat in the southern hemisphere), whereas at the beginning of April, we obtain unbiased estimates from all maturities. When we turn our attention to the issue of closeness-to-expiry effects in wheat, we note that the predictive power of many of the maturities is greatly reduced (Table 6). Furthermore, as with corn, short term maturities have little predictive power for future spot prices.

In conclusion, the predictive power of individual forward contracts seems to be most negatively impacted at times of planting, heading and harvesting, and in the days

Table 1. CBOT Corn: Annual returns, annualized monthly volatility, coefficient of variation, standard deviation of prices, and year closing prices from 2000 to 2013.

Years	2000	2001	2002	2003	2004	2005	2006	2007	2008	2009	2010	2011	2012	2013
Annual returns	0.09	-0.12	0.16	0.04	-0.18	0.05	0.59	0.15	-0.11	0.02	0.42	0.03	0.08	-0.50
Volatility	0.26	0.20	0.17	0.24	0.33	0.30	0.27	0.34	0.47	0.35	0.37	0.40	0.35	0.32
CV	0.31	0.16	0.35	0.18	0.67	0.22	0.77	0.37	0.69	0.32	0.79	0.30	0.37	0.80
SD	64.58	33.37	79.32	41.52	169.21	45.32	206.35	138.28	363.16	121.08	343.86	201.34	261.15	459.23
Price	224.75	199.75	235.25	246	204.75	215.75	390.25	455.5	407	414.5	629	646.5	698.25	422

Table 2. CBOT Wheat: Annual returns, annualized monthly volatility, coefficient of variation, standard deviation of prices, and year closing prices from 2000 to 2013.

Years	2000	2001	2002	2003	2004	2005	2006	2007	2008	2009	2010	2011	2012	2013
Annual returns	0.12	0.03	0.12	0.15	-0.20	0.10	0.39	0.57	-0.37	-0.12	0.38	-0.20	0.18	-0.25
Volatility	0.22	0.20	0.28	0.27	0.19	0.29	0.21	0.37	0.43	0.40	0.49	0.44	0.26	0.17
CV	0.15	0.17	0.53	0.39	0.41	0.18	0.51	0.99	0.73	0.31	0.72	0.43	0.48	0.23
SD	38.93	45.64	173	130.64	139.27	57.90	206.43	646.39	560.92	165.98	417.44	303.96	364.05	155.21
Price	279.5	289	325	377	307.5	339.25	501	885	610.75	541.5	794.25	652.75	778	605.25

Table 3. F-test results for equations (1) and (2) for CBOT corn, with forward curves observed at approximately February 1 to 14 and April 1 to 14 respectively.

Observation period of the forward curve:		From: Last trade date of F1 March - 30 days; To: Last trade date of F1 March - 20 days				Observation period of the forward curve		From: Last trade date of F1 May - 30 days; To: Last trade date of F1 May - 20 days			
Approx. Date:		1-14 February				Approx. date		1-14 April			
Maturities	h	Equation (1)		Equation (2)		Maturities	h	Equation (1)		Equation (2)	
		F-test	p-value	F-test	p-value			F-test	p-value	F-test	p-value
May	2	0.38	0.69	0.10	0.91	Jul	2	9.12	<0.01	4.56	0.01
Jul	4	4.18	0.02	0.97	0.38	Sep	4	11.91	<0.01	17.72	<0.01
Sep	6	15.54	<0.01	11.03	<0.01	Dec	7	3.72	0.03	3.42	0.04
Dec	9	1.84	0.16	3.99	0.02	Mar	10	0.45	0.64	0.33	0.72
Mar	12	0.76	0.47	0.24	0.79	May	12	0.43	0.65	0.40	0.67

days when Future contracts are close to expiry.

Optimal lags of prediction of future contracts

A study of forward prices, not by maturity but by their position inside the current forward curve, may better represent traders' daily activities. In order to build consistent monthly data series, we use the individual maturities to create continuous series of deliveries, based on how close they are to the spot price. We refer to them as second nearby Future contracts, that is, F2, third nearby F3, etc. (Table 7). We adapt Equations (1) and (2) to find the optimal lag among the first 24 lags for each nearby, F2 to F7:

$$\ln S_t = \alpha + \beta \ln F_{t-i} + \varepsilon \quad (1)$$

Table 4. F-test results for equations (1) and (2) for CBOT Corn, with forward curves observed between March 1st to 14th and May 1st to 14th, respectively.

Observation period of the forward curve		From: Last trade date of F1 March– 10 days To: Last trade date of F1 March				Observation period of the forward curve		From: Last trade date of F1 May– 10 days; to: Last trade date of F1 May			
Approx. Date:		1-14 March				Approx. date:		1-14 May			
Maturities	h	Equation (1)		Equation (2)		Maturities	h	Equation (1)		Equation (2)	
		F-test	p-value	F-test	p-value			F-test	p-value	F-test	p-value
May	2	3.71	0.03	3.52	0.03	Jul	2	5.37	<0.01	10.44	<0.01
Jul	4	16.26	<0.01	12.14	<0.01	Sep	4	4.13	0.02	10.51	<0.01
Sep	6	13.90	<0.01	10.07	<0.01	Dec	7	6.34	<0.01	3.22	0.04
Dec	9	4.31	0.02	4.75	0.01	Mar	10	1.00	0.37	0.69	0.50
Mar	12	0.65	0.52	0.66	0.52	May	12	0.21	0.81	0.10	0.90

Table 5. F-test results for equations (1) and (2) for CBOT wheat, with forward curves observed at approximately February 1st to 14th and April 1st to 14th, respectively.

Observation period of the forward curve		From: Last trade date of F1 March - 30 days; To: Last trade date of F1 March - 20 days				Observation period of the forward curve:		From: Last trade date of F1 May - 30 days; To: Last trade date of F1 May- 20 days			
Approx. Date:		1-14 February				Approx. Date:		1-14 April			
Maturities	h	Equation (1)		Equation (2)		Maturities	h	Equation (1)		Equation (2)	
		F-test	p-value	F-test	p-value			F-test	p-value	F-test	p-value
May	2	93.74	<0.01	67.17	<0.01	Jul	2	9.26	<0.01	1.18	0.31
Jul	4	38.70	<0.01	19.80	<0.01	Sep	4	2.72	0.07	1.76	0.18
Sep	6	4.00	0.02	0.67	0.52	Dec	7	15.13	<0.01	1.58	0.21
Dec	9	17.64	<0.01	11.67	<0.01	Mar	10	7.34	<0.01	1.96	0.14
Mar	12	7.65	<0.01	7.96	<0.01	May	12	11.91	<0.01	1.29	0.28

Table 6. F-test results for equations (1) and (2) for CBOT wheat, with forward curves observed at approximately March 1st to 14th and May 1st to 14th, respectively.

Observation period of the forward curve		From: Last trade date of F1 March– 10 days; To: Last trade date of F1 March				Observation period of the forward curve		From: Last trade date of F1 May– 10 days; To: Last trade date of F1 May			
Approx. Date:		1-14 March				Approx. date:		1-14 May			
Maturities	h	Equation (1)		Equation (2)		Maturities	h	Equation (1)		Equation (2)	
		F-test	p-value	F-test	p-value			F-test	p-value	F-test	p-value
May	2	38.52	<0.01	15.22	<0.01	Jul	2	0.51	0.60	13.84	<0.01
Jul	4	13.26	<0.01	5.03	<0.01	Sep	4	3.13	0.05	6.07	<0.01
Sep	6	6.21	<0.01	0.40	0.67	Dec	7	7.72	<0.01	1.37	0.26
Dec	9	16.78	<0.01	6.44	<0.01	Mar	10	5.46	<0.01	3.23	0.04
Mar	12	12.37	<0.01	4.05	0.02	May	12	10.38	<0.01	4.28	0.02

Table 7. Individual Future contracts and their position in the forward curve through time.

Nearby	F1	F2	F3	F4	F5	F6	F7
Year 1	Mar	May	Jul	Sep	Dec	Mar	May
	May	Jul	Sep	Dec	Mar	May	Jul
	Jul	Sep	Dec	Mar	May	Jul	Sep
	Sep	Dec	Mar	May	Jul	Sep	Dec
	Dec	Mar	May	Jul	Sep	Dec	Mar
Year 2	Mar	May	Jul	Sep	Dec	Mar	May
...	...	...	...	...	...	...	

Table 8. Relationship between individual contracts, nearby Future contracts and optimal lags.

Location in the curve	Observed prices at t-h and corresponding h for each maturity									
Spot: Nearby	**Mar**	**h**	**May**	**h**	**Jul**	**h**	**Sep**	**h**	**Dec**	**h**
F2	May	2	Jul	2	Sep	2	Dec	3	Mar	3
F3	Jul	4	Sep	4	Dec	5	Mar	6	May	5
F4	Sep	6	Dec	7	Mar	8	May	8	Jul	7
F5	Dec	9	Mar	10	May	10	Jul	10	Sep	9
F6	Mar	12	May	12	Jul	12	Sep	12	Dec	12
F7	May	14	Jul	14	Sep	14	Dec	15	Mar	15

$$\ln S_t - \ln S_{t-i} = \alpha + \rho \ (\ln F_{t-i} - \ln S_{t-i}) + \varepsilon \quad (2)$$

With i = 1,2,3,..., 24 months. Additionally, we extend the methodology proposed by Muth (1961) with a non-parametric test, the Wilcoxon signed-rank test, which allows for the absence of the strict assumptions needed to apply the F-test, such as the absence of serial correlation, which is present in both commodities. This test also takes into account the magnitude of the changes in the comparison of both distributions (Campbell and Dufour, 1991).

By studying each nearby we can infer how the position of the individual maturities on the forward curve affect their predictive power. For example, in Table 7, at the beginning of Year 1, March of year 2 occupies the penultimate point of our forward curve, F6, and as time passes the same March maturity will become the fifth, fourth, third, second and finally at the beginning of year 2, that March contract will be the first nearby and reach expiry.

We depict the distance h for each Future contract from the examples in previous subsections in the first two columns of Table 8, which we extend to all contracts up to the seventh nearby.

When we study each contract, since we have unequally spaced maturities, we can only obtain an interval of lags where the prediction is unbiased and an average lag when each nearby will have the highest predictive power. The 'theoretical' optimal lag will depend on the point in time when we observe the forward curve. For example, in the case of the fourth nearby, if we observe F4 in March, it corresponds to the September Future contract, and the theoretical optimal lag should be six months, that is, the September Future price predicts the September spot price. If we observe F4 in May, it corresponds to the December Future contract and the optimal lag is seven. If we are in July, F4 corresponds to the March Future contract and lag eight, and so on. Hence, a result between six and eight would mean all individual maturities on average give an unbiased prediction at F4. We derive all the theoretical optimal lag intervals with the help of Table 8 for each nearby in the first column of Table 9.

With our tests, we obtain a choice of lags in Equation (2) that includes the theoretical lag interval for each nearby, with the only partial exception of F3 in corn, meaning that, in general, individual maturities keep predictive power as time passes (Table 9). However, Durbin-Watson tests reveal autocorrelation in the residuals across all lags in both commodities. The presence of autocorrelation could be explained by the behaviour of the many market players who use technical analysis, which is based on signals such as moving averages for momentum trading.

When we study the optimal lags in Table 9 (lags with the highest p-values from Equation (2)), in some cases they fall outside their respective theoretical lag interval. In the case of CBOT corn, only the optimal lags for F2 and F3 are inside the theoretical lag interval, while for the rest of the nearby Future contracts the optimal lags are lower. Lower optimal lags indicate that when individual Future contracts are in nearby positions F4 to F7, the best

Table 9. Intervals of theoretical optimal lags, choice of monthly lags in which F-tests and Wilcoxon signed-rank tests present p-values greater than 0.05, and optimal lag, from second to seventh nearby for CBOT corn (top) and CBOT wheat (bottom).

Parameter	Theoretical optimal lag Interval	Equation(1a)	Equation(2a)	Wilcoxon Test	Optimal Lag	
		Choice of lags	Choice of lags	Choice of lags	Highest p-value lag	p-value
CBOT corn						
F2	2 - 3	2 - 5	2 - 3	2 - 11	2	0.34
F3	4 - 6	3 - 7	3 - 5	4 - 13	4	0.30
F4	6 - 8	4 - 11	4 - 14	5 - 15	5	0.29
F5	9 - 10	5 - 16	5 - 16	6 - 17	7	0.45
F6	12	6 - 18	6 - 17	6 - 19	9	0.5
F7	14 - 15	7 - 20	7 - 17	8 - 21	11	0.56
CBOT wheat						
F2	2 - 3	-	2 - 11	3 - 12	7	0.84
F3	4 - 6	-	4 - 12	6 - 15	7	0.78
F4	6 - 8	-	6 - 14	7 - 16	10	0.96
F5	9 - 10	-	8 - 15	8 - 18	12	0.95
F6	12	-	9 - 16	10 - 20	12	0.76
F7	14 - 15	-	10 - 17	11 - 22	13	0.67

Table 10. Choice of monthly lags in which F-test and Wilcoxon signed-rank test present p-values greater than 0.05 and respective individual lags with the highest p-value

Summary results	Eq (1b) F-test		Eq (2b) F-test		Wilcoxon	Optimal Lag	p-values
	Choice of lags	Highest p-value lag	Choice of lags	Highest p-value lag	Choice of lags	Highest p-value lag	Eq (2b) F-test
$\bar{F}$(F2 to F6)							
CBOT corn	4-12	6	4-6	5	4-7	5	0.27
CBOT wheat	-	-	6-12	9	7-16	9	0.9
$\bar{F}$ (F3 to F7)							
CBOT corn	5-16	8	5-12	6	6-17	6	0.21
CBOT wheat	-	-	8-14	11	8-18	11	0.91

predictions do not correspond to the months suggested by the theoretical interval but to future spot prices that happen earlier. In contrast, in the case of CBOT Wheat, only the optimal lags for F6 are inside the theoretical lag interval, while optimal lags obtained for F2 to F5 are higher, that individual Future contracts offer a better prediction of later future spot prices. Consequently, while for CBOT wheat, maturities expiring one year ahead a twelve month prediction(F6 is exactly at lag 12), they only offer a nine month prediction for CBOT Corn.

These features could be explained by a fundamental difference in the global trade of corn and wheat. In the case of corn, world prices are usually set by the US domestic supply-demand forces and farmers in the Southern hemisphere, namely Argentina (generally the second largest exporter of corn), adjust their crop output in reaction to US corn harvests and prices (with US news during the summer curtailing the predictive power of the forward curves). Wheat, on the other hand, is grown in more places and climatic conditions than any other cereal grain, from cold environments near the Arctic Circle to tropical regions close to the Equator; there are up to 20 species and more than 25,000 varieties in existence. Hence, CBOT wheat (it is worth noting that although the CBOT Wheat futures contract is linked to the price of No. 2 Soft Red winter wheat, it is generally used in hedging activities for all kinds of wheat), is less dependent on a single country of production and the arrival of news from all over the world has an influence in the predictive power of the forward curves in the short term, although long term predictions are more reliable than in the case of CBOT corn.

Testing the geometric average $\bar{F}$as a predictor of future spot prices

The seasonal cost-of-carry model for commodity forward curves developed by Borovkova and Geman (2006) introduces the geometric average of the forward prices as an alternative to the spot price for the first state variable when managing a portfolio of seasonal or non-seasonal commodity Futures:

$$\bar{F} = \left(\prod_{T=1}^{N} F(t,T)\right)^{\frac{1}{N}}$$

This quantity has the merit of being less volatile than the noisy spot price and always observable. Two other points are worth noting: first, if the N months encompass an integer number of calendar years, $\bar{F}$ is a measure that is, by construction, devoid of seasonality, as proposed by Borovkova and Geman (2006), who study seasonal energy commodities such as natural gas and heating oil. Second, in our setting, it seems reasonable to expect that the whole forward curve contains more information to build estimators of future spot prices than an individual Futures contract.

In order to study the relationship between $\bar{F}$ and the spot price for each of the commodities, we compute $\bar{F}$ as the average of maturities across twelve months (Figure 7). Equations (1) and (2) have been adjusted to study the relationship between the spot price and $\bar{F}$ through the following regressions:

$$\ln S_t = \alpha + \beta \ln \bar{F}_{t-j} + \varepsilon \qquad (3)$$

$$\ln S_t - \ln S_{t-j} = \alpha + \rho \left(\ln \bar{F}_{t-j} - \ln S_{t-j}\right) + \varepsilon \qquad (4)$$

with j = 1,2,3,…,24 months.

In Equation (4), spot price changes,$\ln S_t - \ln S_{t-j}$, are explained by $\ln \bar{F}_{t-j} - \ln S_{t-j}$ at the optimal lag. In Table 10, we provide two different constructions of $\bar{F}$. In the first one, $\bar{F}$ is built using the second nearby to the sixth, while in the second construction, the second nearby is avoided and $\bar{F}$ includes the third to the seventh nearby future contracts. In practice, it is usual to avoid the second nearby in the construction of$\bar{F}$.

For each commodity, $\bar{F}$ shows intervals of lags where econometric tests have p-values above 0.5. Additionally, each $\bar{F}$ captures the difference in range of prediction found in the nearby futures at their optimal lags, for instance, the $\bar{F}$ of wheat captures the fact that Future contracts relatively further from the spot price offer a better prediction. Hence, in the case of wheat, the optimal lags of $\bar{F}$ are 9 and 11 respectively for each construction of $\bar{F}$, which are much higher than those of corn, 5 and 6 res-pectively. These facts show the reliability of $\bar{F}$throughout time and as a better predictor than individual maturities for both wheat and corn markets.

When we explore the best construction of $\bar{F}$, results are also consistent with previous findings for each nearby. The best $\bar{F}$ for wheat includes the maturities further along the forward curve, from F3 to F7, while $\bar{F}$ for corn clearly benefits from including F2 to F6, which increases its p-value to 0.27.

Conclusion

We argue in this paper in favour of alternative estimators of price dispersion in corn and wheat markets, namely the consideration of the coefficient of variation and standard deviation of prices since their levels are the quantities defining the cost of food supply for population around the world.

We also analyse the performance of several forward measures as predictors of future spot prices. We find that the average value of all liquid forward contracts provides several advantages over the use of forward prices of individual maturities; in particular, during the periods of planting, heading, and harvesting, these have a much lower predictive power.

Individually, these measures provide a limited amount of information; however, together they provide a wealth of information based not only on returns but directly on prices. In effect, these indicators combined provide participants with an important toolbox that can be used to analyse the performance of the corn and wheat markets.

Conflict of Interest

The authors have not declared any conflict of interest.

REFERENCES

Bai J, Perron P (2003). Computation and analysis of multiple structural change models. J. Appl. Econometrics 18: 1-22.

Borovkova S, Geman H (2006). Seasonal and Stochastic Effects in Commodity Forward Curves. Rev. Derivatives Res. 9:167-186.

Campbell B, Dufour JM (1991). Over-rejections in rational expectations models. Econ. Lett.35:285-290.

Cornell B (1977). Spot Rates, Forward Rates, and Exchange Market Efficiency. J. Finan. Econ. 5:55-65.

Frenkel JA (1981). Flexible exchange rates, prices and the role of 'news': lessons from the 1970s. J. Polit. Econ. 89: 665-705.

Geman H, Vergel Eleuterio P (2013). "Investing in fertilizer–mining companies in times of food scarcity. Resour. Policy 38(4):470-480.

Geweke JF, Feige EL (1979). Some Joint Tests of the Efficiency of Markets for Forward Foreign Exchange. Rev. Econ. Stat. 61:334-41.

Goss BA (1983). The semi-strong efficiency of the London Metal Exchange. Appl. Econ. 15:681-698.

Hansen LP, Hodrick RJ (1980). Forward Exchange Rates as Optimal Predictors of Future Spot Rates: An Econometric Analysis. J. Polit. Econ. 88:829-853.

Harlan JR (1975). Crops & Man. American Society of Agronomy & the Crop Science Society of America, Madison, Wisconsin.

Hicks JR (1946). Value and Capital, 2nd edition. Oxford: Clarendon Press.

Hicks JR (1965). Capital and Growth. Oxford: Clarendon Press.

Lipps P, Beuerlein J, Loux M, Wilson H, Eisley B, Lentz E (2001). Improving Wheat Fields in Ohio. The Ohio State University. Available at http://ohioline.osu.edu/iwy/index.html

Longworth D (1981). Testing the Efficiency of the Canadian Dollar - U.S. Exchange Market under the Assumption of No Risk Premium. J. Finan. 36(1): 43-49.

Lucas RE (1972). Expectations and the Neutrality of Money. J. Econ. Theory 4:103-124.

Lucas RE (1995). Monetary Neutrality. Prize Lecture.

Muth JF (1961). Rational Expectations and the Theory of Price Movements. Econometrica 29:315-335.

Pieroni L, Riociarelli M (2005). Testing rational expectations in primary commodities markets. Appl. Econ. 37:1705-1718.

20

Contribution of the "Cassava: Adding Value for Africa" (C:AVA) project to income generation in Nigeria: The case of cassava farmers and processors in Ewekoro area of Ogun State

ODUNAYA Adewale Adeleke

Ogun State Agricultural Development Programme, P. M. B. 2122, Sapon, Abeokuta, Ogun State, Nigeria.

This Research was about the effects of "Cassava: Adding Value for Africa" (C:AVA) project on cassava farmers and processors in the Ewekoro area of Ogun state, Nigeria. Respondent for this study were selected from five groups of processors targeted by the project in the area. They consisted of five pure cassava farmers and fifteen cassava processors because the ratio of men to women in the groups was about 1:3. Five respondents agreed that the new cassava variety had contributed to higher output. Cassava farmers did not report any change in asset as a result of their increased output. Six gari processors reported reduction in cost of cassava grating from one hundred and fifty naira to fifty naira. Five fufu processors reported reduction in cost of hiring drums from two hundred and fifty naira to a hundred naira. Eleven cassava processors reported that the processing equipment received from C:AVA had increased the volume of cassava processed and their capability to meet family expenditures due to reduction in cost. The cassava grits which is an intermediate product of High Quality Cassava flour (HQCF) was not processed in all the villages visited and groups interviewed. Based on these findings, this study concluded that the TME 491 cassava stems distributed to farmers had potential to increase farmers output but the quantity distributed to them was insufficient, the stem cuttings distributed to farmers did not result in acquisition of new assets, the processing equipment had increased the volume of cassava processed and reduced their cost.

Key word: Processors, Cassava: Adding Value for Africa" (C:AVA) project, gari.

INTRODUCTION

This study focused on the evaluation of Cassava: Adding Value for Africa (C:AVA) project in Nigeria supported by a grant from the Bill and Melinda Gates Foundation. The project was involved in development of value chains for High Quality Cassava Flour (HQCF) in Ghana, Tanzania, Uganda, Nigeria and Malawi to improve the livelihoods and incomes of at least 90,000 smallholders' households as direct beneficiaries including women and disadvantaged groups NRI (2011). The project was led by the Natural Resources Institute of the University of Greenwich working closely with University of Agriculture, Abeokuta, Nigeria; Food Research Institute, Ghana; by the Natural Resources Institute of the University of Greenwich working closely with University of Agriculture, Tanzania Food and Nutrition, Tanzania; Africa Innovations Institute, Uganda; and Chancellor College, University of Malawi and a range of other partners. The purpose of C:AVA intervention at the rural level was to ensure that by the end of 2012, at least 90,000 smallholder farmers are earning additional $0.52 per day from each producing 0.8 tons per farmer of consistent quality cassava grits per annum for the HQCF industry across the five countries (Adebayo, 2011). At the farmers' level, the aim of the project was to ensure consistent supply of raw materials namely fresh cassava tubers and processed cassava grits.

The implementers of the project in Ogun state were the Agricultural and Media Resources Centre (AMREC) and Ogun State Agricultural Development (OGADEP) who jointly rendered extension services to farmers and were the links through which all the extension messages, TME 419 cassava variety stem cuttings and processing equipment got to the beneficiaries. These agencies were under the coordination of C:AVA country office at Federal University of Agriculture, Abeokuta (FUNAAB). AMREC is the extension arm of Federal University of Agriculture, Abeokuta and provided extension services to some villages in some parts of Ogun state. OGADEP was the state government extension agency in the state responsible for disseminating information on improved technologies and production recommendations to farmers in the state. At OGADEP, the activities of the project was under the coordination of the C:AVA desk officer supervised by Director of Agro processing unit of OGADEP. This office coordinates training, delivery of cassava stem and processing equipment as well as monitoring the installation of the equipment and implementation of C:AVA project in the villages under their jurisdiction.

Problem definition

The increasing popularity of cassava in production of high quality cassava flour had led to increased income of farmers involved. Diverse uses of cassava flour for food products such as its glucose for pharmaceuticals products as well as food supplements to make alcohol and other beverages, would make farmers smile all the way to the bank (Adebayo, 2010). The Federal government of Nigeria's mandate to include 20% High Quality Cassava flour in all products of wheat flour for bread-making is the needed motivation for cassava farmers, and processing firms. Cassava farmers and processors were supposed to benefit from this policy by processing cassava tubers into cassava grits sold to cassava flour industries thereby earning additional income from value addition but the weak link between cassava farmers and processing groups on one hand and the industries using cassava as raw materials on the other, is depriving farmers and processors from earning the expected income.

Research problem

C:AVA was launched in Nigeria in June, 2008 to develop value chain for high quality cassava flour. Ogun state was one of the two states where the project was implemented in Nigeria. Implementation of C:AVA started in Ewekoro local government area in 2009. Seven cassava processing groups consisting of pure cassava farmers and cassava processors (present in the same group) were selected as beneficiaries of the project in this area. Although recent researches on the impact of C:AVA e.g. Abdulsalam-Saghir (2010) had discovered that incomes of smallholder households in some part of Ogun, Oyo and Ondo states had increased significantly by fifty Euro (ten thousand naira) per month, the effect of the project on the income at the levels of cassava farmers and processors in Ewekoro area was not known.

Purpose of the study

The thesis seeks to identify the effect of C:AVA project on income of cassava farmers and processors involved in the project in Ewekoro area of Ogun state, Nigeria.
To achieve this objective the following research questions were formulated.

Research questions

1. What effect did TME 419 cassava variety distributed to cassava farmers have on income generated by farmers?
Sub-questions:
a. To what extent had the cassava variety received by farmers affected their output?
b. Was there any change in asset and expenditure of cassava farmers that received the TME cassava variety?

2. What effects did the processing equipment distributed to cassava processors have on their income generation?

Sub-questions

a. What type of processing activities were cassava processors in the area involved in?
b. What was the processing equipment given to processors in the area?
c. Was there any change in volume of cassava processed by the processors' groups that received the processing equipment?
d. What changes had occurred in assets and family expenditure of cassava processors that received the processing equipment?

3. What effect did processing equipment donated to groups had on the processing cost?

a. What were the costs incurred by cassava processors in the study area?

b. What were the changes in processing cost that occur as a result of processing equipment received by cassava processors?

The study area

Ewekoro Local Government first came into existence on 22nd of May, 1981, but it was later merged with Ifo local government in 1989 by the military government. However, on 16 December, 1996 Ewekoro Local Government was restored as an autonomous local government by the then Federal Military Government alongside five other in the state, thus increasing the number of Local Government in Ogun State from fifteen to twenty. The Local Government was bounded by Yewa South in the west, Ifo Local Government in the south, Abeokuta North and Obafemi Owode in the north and east respectively and it consisted of ten wards. It has a land area of 63.5 km^2 with an estimated population of about 55,000 people (2006 population census). The indigenous dwellers of Ewekoro Local Government area were mainly the Egbas. The people engaged primarily in farming and trading, the local government area consisted majorly of rural settlements. Cassava, yam, cocoyam and rice are the most cultivated food crops in this area. Processing of the product was done by women who also own personal farms although farm sizes are smaller compared to men's farms in most cases.

Cassava processing groups in the area consisted of pure cassava farmer (men) and cassava processors who were women although they owned individual cassava farms. Ownership of land in this area was by inheritance; men and women had right to inherit land.

Cassava farmers

Cassava farmers in Ewekoro area consist of small scale (0.5 to 2 ha) and medium scale (2 to 4 ha). Cassava was grown as intercrop of maize, rice, melon and vegetables (Olasunkanmi et al., 2012). Cassava cultivation was done by men and women in the study area but male farmers cultivates larger farm than their female counterparts in most cases although there were some women with big farm sizes. Cassava was grown on rented or lease land as well as owned land. It was very easy to rent or lease a land in the area if the farmer was recommended by known indigene or if the farmer resides in the area.

Cassava processors

Processing of cassava in Ewekoro area was mainly women occupation although cassava graters are owned by men and are pushed on a truck around villages. Cassava was processed into gari, fufu and lafun but gari and fufu processing was more prevalent than lafun because lafun requires sun drying for its processing. Processed products were sold in the local market (Wasimi) where traders from Lagos were the major buyers.

Selection of the participants for the project

Selection of cassava farmers and processors that are benefitting from C:AVA project was done by OGADEP. Seven cassava processing groups consisting of cassava farmers and processors were benefiting from the project in the local government although more groups have shown interest in the project. Members of these processing groups are small scale farmers and processors. Selection criteria used for these farmers are based on the recommendation of OGADEP extension component of OGADEP selected these groups based on the possession of registration certificate, regular meetings with good attendance of members and their willingness to participate. Each processing groups consists of 30 members who are pure cassava farmers or cassava processors for easy management and coordination.

Challenges involved in evaluating income effect of projects

Evaluation of income effects of project can be quantitative or qualitative. In quantitative evaluation, data collection and analysis are carried out, while economical models are used for data analysis. Qualitative analysis of income effect is also possible but does not enjoy the level of acceptance and recognition as quantitative evaluation. Income evaluation also has other constraints such as under reporting (for a variety of reasons income is often under reported). Respondents perceive income-related questions as being invasive because of tax related problems. In developing countries, income and expenditure surveys yield estimates of saving that are absurdly low compared to estimates derived from national income. Reported expenditures often exceed reported income.

Conceptual framework

Evaluation of income effect of any project on farmers and processors was a very challenging task, especially when cost and revenues are not recorded down. This research had resolved this issue by looking on the effect of TME419 cassava variety on (a) change in output of cassava farmers in the study area (b) Change in assets and expenditure of cassava farmers that received the TME 419 cassava variety. The effects of the processing equipment donated by C:AVA on (a) Volume of cassava processed (b) Change in processing cost as result of equipment received by processors (c) change in asset and expenditure of processors who received processing equipment from C:AVA were used as indirect indicators of income effect of C:AVA on the benefiting groups in the study area.

Information on change in expenditure (expenses on children education, expenses on food and other household need and expenses on social obligation) was another indicator used because it is easier to recall than income data and it seemed to be less prone to error because expenditures of household on food, children education and other household needs are continuous and recurrent, so it is easily remembered but income is earned at a point in time and the information may be easily forgotten.

Indicators for cassava farmers

Change in output of cassava farmers

Change in output was used to measure income of farmers. This indicator was used because cassava farmers may experience increased in output when they increase land size used for cassava cultivation or when they plant high yielding cassava variety or both. Cassava farmers in the processing groups interviewed were given 100 bundles of TME 419 cassava stems free; the change in income that occurred as a result sale of increased or otherwise of tubers harvested may serve as indicator of the effect of the intervention on the farmers that received the cassava stem cuttings.

Change in assets and expenditure

Increase output and favourable cassava tuber prices will result in increased income. Increased income may lead to acquisition of new assets (e.g. bicycles, motor cycles, houses and livestock) and higher expenditures on food, children education, and other social obligations. These indicators were used to probe the changes of income of cassava farmers that occur as a result of engaging in the project or as a complement of any other on-going project in the area.

Indicators for cassava processors

Change in volume of cassava processed

Cassava processors in the study area received processing equipment from C:AVA. These equipment were distributed to them through Ogun State Agricultural Development Program. Two Steel cassava graters (Figures 1 and 2) and two hydraulic presses (Figure 4) were given to gari processors at Eleyele and Awowo while fufu processors at Asipa-ilaho, Awowo and Gudugba received sixty processing drums (Figure 3) used for soaking fufu before sieving. The increase in volume of cassava processed by these fufu processors will lead to more income because limited numbers of drums were available for rent before the intervention and processors had to use it in rotation. Similarly gari processing groups will be able to increase the volume of cassava processed and may not have to wait for cassava grater owners in the village. This indicator provided a good measure of effect of this project on the activities of the cassava processors.

Change in processing cost

When the cost of processing increase or decrease, it will affect the income derive from processing activities. This indicator was used as a measure of change in income of processors. Cassava processors in the study area has benefitted from processing equipment like steel cassava grater, hydraulic press, processing drums and processing slab. These equipment were distributed by C:AVA through OGADEP to them free. The reduction in processing cost will lead to saving on expenses on processing which is unearned an income. This indicator was used in this research.

Change in asset and expenditure

Changes that occur in the expenditures and asset of cassava processors who received the processing equipment may indicate change in their income unless they are engaged in other income activities. Since the income of cassava processors is no easy to estimate because of poor recording keeping, change in their expenditure on food, children education, assets (pepper milling machine, livestock, houses etc.) and social obligation were used to make fair judgement on their income. This indicator was used in this research to explore the change in income of cassava processors who received the processing equipment in the study area.

Assets in the study area include livestock like sheep and goat, motorcycle, bicycle, landed properties.

Cassava grater, pepper milling machine were some of the asset that farmers and processor strive to acquire when their income increase. This indicator offered reliable measure of income of cassava farmers and processors in the study area although it may not be totally reliable.

RESEARCH METHODOLOGY

Research strategy

Case study was used as strategy for this research because in-depth information is needed to be able to unravel the effects of this

Figure 1. Gari processor at work at Asipa-ilaho.

Figure 2. Steel cassava grater donated by C:AVA to gari processing group.

project on the beneficiaries. This strategy was also chosen because the respondents are mostly illiterate and may not be able to fill structure questionnaire themselves. The research was conducted in five villages in Ewekoro Local government in Ogun state namely Eleyele, Awowo, Asipa-ilaho Osupori and Gudugba. These villages were chosen because C:AVA project was implemented in all these villages and members of these processing groups were active farmers and processors and as such were be able to provide relevant information of the effect of the project. Qualitative data on the effects of cassava variety distributed to cassava groups and processing equipment distributed to processing groups were collected in order to make inference about changes in income that occur to participants of the project.

This strategy was chosen because most small scale farmers in Nigeria hardly keep records of their operations therefore great insight was obtained by involving them in discussion which can generate the needed information. In order to gain access to the study area, the agricultural extension officer was contacted before each interview was conducted. The extension officer assisted the researcher to make appointment based on the availability of the respondents and the appointment date was communicated to him.

Research population

The total population of this research were the beneficiaries of C:AVA in Ewekoro area. These were the two hundred and ten cassava farmers and processor belonging to seven groups in the local government. These groups were selected among existing cassava processing groups in the local government. It was

Figure 3. Dug out well and processing drum donated by C:AVA at Asipa-ilaho village.

Figure 4. Hydraulic press donated by C:AVA to processors (not in use).

observed that all the groups have more women than men in the ratio of 3:1. Respondents of this research were pure cassava farmers and cassava processors who are beneficiaries of the project because they will be able to provide in-depth information on the subject matter.

Selection of processing groups interviewed for data collection was done using purposive sampling. Cassava processors that have participated in the C:AVA project for more than two years were selected because they were able to provide information of the effect of the project on their income. Five out of the seven cassava processing group in the area were thus selected because they have participated in the project for more than two years; the remaining two groups started participating in 2012. Four respondents consisting of three women and one man were chosen per group making a total of twenty respondents for the five groups. Selection of respondents within groups was done by telling the members to pick tallies with "yes" or "no" and respondents with yes were chosen for the interview. This method was chosen to reduce bias and misconception among respondents who are mostly female. The initial plan in the thesis proposal was to choose equal numbers of men and women but the situation on the field revealed that there were more women in these groups than men in the ratio of 3:1 in all the groups interviewed. The researcher had no personal knowledge of respondents before the selection therefore issues of bias were minimal.

RESULTS

In this chapter, the findings of this research were

presented. Twenty respondents consisting of five male and fifteen female belonging to five groups in the study area provided the responses for these findings. In the proposal the respondents were to be chosen on equal basis of male and female for five of the selected groups but the situation in the study area revealed that there are more female than male. All female respondents were involved in cassava cultivation and processing while male respondents were only involved in cassava cultivation. In some of the groups visited, the project just commenced early this year and the impact of the project is yet to fully manifest, hence the reduction in the number of groups selected for the study.

Effect of the cassava variety received by farmers on their output

Six out of twenty respondents interviewed in the study area agreed that the implementation of CAVA project in the study area had contributed to higher output. These farmers consisted of five cassava farmers selected for the interview and one processor who also cultivates cassava farm because she was also given five bundles of the new cassava stems. Fifteen respondents (processors with farms) could not benefit because the cassava stems were not sufficient for all the members.

Change in assets and expenditure of cassava farmers that received the TME 419 cassava variety

The five Cassava farmers interviewed did not report any change in asset as a result of increased output but theircapability to meet their family expenditure on children schooling, food and social obligation increased. One female farmer reported that output from her farm supported her through her time of trial when her husband died. She processed the tuber herself and earned additional income as a result of high starch content of the TME419 variety.

Change in volume of cassava processed by the processors' groups that received the processing equipment

Six out of eight gari processors interviewed reported increase in volume of cassava processed. They said they processed two baskets (500 kg) of cassava tubers per week before but the amount processed has increased to three to four baskets (750 to 1000 kg) of the raw tuber weekly. When asked the reason for change increase in volume, they said they had easy access to the processing equipment donated by C:AVA through OGADEP and also enjoyed benefit to grate their cassava tubers on credit and made payment after the sales of their gari. This privilege was not available before they received the equipment. Two gari processors interviewed said the volume of cassava processed did not change because their houses were located further away from the group's processing shed. Five fufu processors reported increase in volume of fufu processed because of the twenty drums donated by C:AVA which reduce the drudgery experienced before the intervention.

Type of processing activities of cassava processors in the study area

Participants of C:AVA in the study were engaged in processing of fufu and gari, the commonest food of the people of the state. Seven out of the fifteen processors interviewed were involved in fufu processing while eight respondents processed gari (Figure 2). At the Agbedara group in Asipa village, the respondent interviewed said they were formally gari processors but due to the training and encouragement they received from the extension agent working with them on C:AVA coupled with twenty drums
donated to the group by C:AVA, they were motivated to engage in fufu processing which is required less labour and input than gari. At Fagbesoro fufu processing group in Osupori village, one of the respondent engaged in gari and fufu processing said she will continue to process gari although it is labour intensive because it had longer shelf life than fufu (which must be sold immediately it was processed) whether the prevailing price was profitable or not. She also said gari was very important for food security of household with children living in the city because it can be consumed in more ways and can be sold to earn income to support children in school as well.
The fufu processors in the area offered their product for sale as wet fufu paste. The cassava grits which is an intermediate product of High Quality Cassava flour (HQCF), the topmost priority of C:AVA was not processed in all the villages visited and groups interviewed.

Cost of labour for different operation of cassava processors

Cassava processors in the study area incurred various cost on different operation carried in processing activities. Transportation cost of six thousand naira to eight thousand was paid for a pick-up van load (3 ton) of cassava tubers depending on the distance of the farm to the village. Cost of harvesting of tuber was two hundred naira for a basket two hundred and fifty kilograms and this cost is paid by the farmer. For cassava tuber weighing 250 kg (one basket), Gari processor paid two hundred and fifty naira for peeling, six hundred naira for grating, and four hundred naira for frying. Fufu processors paid the same peeling cost and sieving cost of five hundred naira for the same quantity of tuber. In practice, respondents in the study area performed some or all of these operations to be able to obtain some margin on their activities. Additional transportation cost is also paid on the gari and fufu to market centres. One hundred naira is paid on a bag of gari or wet fufu for transportation to the market but lower cost can be paid if large numbers of bags are being transported by a processor through bargaining.

Change in processing cost that occur as a result of processing equipment received by cassava processors

Some of the cassava processors in the study area reported reduction in processing cost in their processing activities as a result of processing equipment they received from C:AVA project. Six out of eight gari processors interviewed reported that the reduction in cost was as result of reduction in grating cost from one hundred naira which was paid previously to fifty naira which they paid into the group purse to provide the running and maintenance cost for the equipment (Table 1). However two respondents could not benefit from the reduction in grating cost because their houses were located further away from the processing shed. Five out of seven fufu processors also reported reduction in processing cost due reduction in cost of hiring drums used for soaking cassava tubers before sieving from two hundred naira paid previously to a hundred naira (Table 2). This reduction occur as a result of twenty processing drums they received from C:AVA project. The remaining two respondents have individual processing drums. The hydraulic press also donated by C:AVA to gari processors were not used in all the villages visited because it cannot contain the type of sac used in the area (Appendix 1) and the processors were not willing to change the sac because it is used as scale of the expected processed gari.

Change in asset and family expenditure as a result of processing equipment received by cassava processors

Eleven out of the fifteen cassava processors interviewed reported that the processing equipment received from C:AVA had increased their capability to meet family expenditures like payment of children school fees, training of their children on crafts, provision of their food needs and social obligation because of reduction in their processing cost. One of the fufu processors interviewed said she benefitted from marketing linkage with a firm producing High Quality Cassava Flour (HQCF) at the commencement of the project but the linkage stop later when the firm halted production because of poor electricity supply. She was able to buy a milling machine and start a small petty trade in addition to fufu processing. All the interviewed members of Fagbesoro processing group at Osupori village unanimously agreed that the implementation of the project in their village had impacted positively on their capability to meet their family expenditure and social obligations. Another respondent, an elderly woman at Osupori village said the project increased her income generation because of higher turn-over as e result of higher starch content of the TME 419 cassava variety and reduction in cost and hence was able to support the completion of her children higher education that was abandoned to her by her polygamous husband.

DISCUSSION

In this chapter, an attempt was made to analyse the findings to see the extent the research question were answered to fulfil the research objective.

Effects of the cassava variety received by farmers on their output

One hundred bundles of TME 491 cassava stems distributed to farmers among the processing groups interviewed had increased their output but the quantity distributed to them was not sufficient for all the farmers. Only five members out of twenty respondents selected for the interviewed were able to derive increased income as a result of increase in output by planting the high yielding, TME 419 cassava variety. The prevailing price in the study area encouraged the planting of the variety but it is not known whether the current price can be sustained because cassava tubers and product usually experience price fluctuation and if the variety was supplied and there was increase output, the marketing of the surplus must be guaranteed for farmers to continue to enjoy increase income by planting the variety.

Changes in assets and expenditure of cassava farmers that received the TME 419 cassava variety

The distribution of TME 419 cassava variety to farmers did not result in acquisition of new assets although the beneficiaries reported increase in their capability to meet expenditure on food, children education, and meeting social obligation like naming ceremonies.

This was an indication that planting of the variety at the moment had not increase farmers income to the extent of having surpluses to acquire new asset because the quantity of cassava stem received was not sufficient.

Change in volume of cassava processed by the processors' groups that received the processing equipment

The processing equipment donated by C:AVA to the

Table 1. Estimation of Reduction in processing cost for 1 ton of cassava tubers for gari processors using the data from cost of labour for different operation of cassava processors subsection.

Items	Beneficiaries	Non-beneficiaries
Transport cost	₦2000	₦2000
Cost of peeling	₦1000	₦1000
Cost of grating	₦800	₦2400
Cost of frying	₦1600	₦1600
Total	₦5400	₦7000
Reduction in processing cost = ₦(7000 - 5400) = ₦1600 = $9.70 ($1 = ₦165.00		

Table 2. Estimation of reduction in cost of processing for 1 ton cassava tubers for fufu processors using the data from cost of labour for different operation of cassava processors subsection.

Items	Beneficiaries	Non-beneficiaries
Transport cost	₦2000	₦2000
Cost of peeling	₦1000	₦1000
Cost of renting drums	₦400	₦1000
Cost of sieving	₦2000	₦2000
Total	₦5600	₦6000
Reduction in processing ₦6000 - ₦5600 = ₦600 = $3.64 ($1 = ₦165.00)		

processors' groups had led to increase in volume of cassava processed. Although some members of these groups were not able to use the facilities, eleven out of fifteen processors interviewed reported increase in volume of their processing activities as a result of the processing equipment they received. The processors reported that they do not have to compete with other villagers on the use of limited number of cassava grater and processing drums in the villages.

Type of processing activities of cassava processors in the study area

Processing of cassava grit used which an intermediate product for cassava flour processing was not observed in all the groups visited. Only one cassava processor reported that she was formally engaged in the processing when she was linked with a cassava flour firm but she stopped the processing when the firm no longer demand the product from her. The remaining fourteen cassava processors did not process cassava grit at all although they reported that had benefitted from training on processing of cassava grits, cassava flour and the utilization of this product for making chin-chin, pies and other recipes. None of these processors were involved in the processing of the chin-chin, pies and sausages because of lack of demand. The cassava grits innovation could be said to be inappropriate in this area at the moment because there was no cassava flour industries in the area to use the product as existing ones are out of operation because of unstable electricity supply. The cassava processors in the study area were mostly involved in gari and fufu processing although they also process lafun when they have sunny weather. Eggleston et al. (1992) as cited in Kehinde (2007), reported that of the main products obtained from processing cassava, gari, is the most common in West Africa agreed with this finding.

Cost of labour for different operation of farmers and processors

Cost of labour for farming operation and processing activities in the study area was expensive compared with the return farmers and processor earned. Although the project provided processing equipment for the processors, there was no provision for farmers on herbicide which can reduce weeding cost as well as enhanced timely weeding. Labour cost also change with prices of cassava tuber and cassava products in the study area. Farmers paid huge labour cost on clearing, heaping and weeding only for them to sell their harvest tuber at a give-away price during harvest when there was glut in the market. Labour saving processing equipment for cassava processing like hand peeler for tuber peeling was not used by cassava processors although this project did not provide such to processors. Donald et al. (2000) also reported similar finding by stating that

Cassava is labour intensive and costs of production can be quite high, compared to other carbohydrate sources. Adeniji et al., and Egyleston et al. (1992) cited in Taiwo (2006) reported similar finding stating "cassava processing by traditional methods is labour-intensive but the application of improved processing technology has reduced processing time and labour and encouraged further production", although most of these cassava processing technologies like mechanised cassava harvester and cassava peeling machine or cassava hand peeler were not in use in the study area.

Change in processing cost that occurred as a result of processing equipment received by cassava processors

One thousand and six hundred naira (per ton of fresh tubers processed) reduction in processing cost was reported by gari processors that benefitted from the two steel cassava grater donated by C:AVA in the study area. Similarly fufu processors reported six hundred naira per ton reduction in cost of hiring processing drums due sixty drums donated by C:AVA to the fufu processing groups. This reduction in cost is calculated based on processing of one ton of cassava tubers using the processing equipment and the prevailing processing cost presented in findings above. The saving on processing cost can be considered as an income though it is not earned. However two out of eight gari processors interviewed could not benefit because their houses were located further away from the processing shed. The sustainability of the donation of processing equipment to cassava processors in the area was not guaranteed because C:AVA project may end in 2013 and that may signify the end of the gesture unless the project is extended in another phase.

Conclusion

This study has analysed the effects of Cassava Adding Value for Africa in Ewekoro area of Ogun state, Nigeria with the aim of investigating whether participation in C.AVA Project has contributed to better Income generation among farmers involved in the project in order to make recommendations for similar intervention in the future. In order to achieve this objective, three research questions were formulated with sub-questions which were used to prepare an interview guide used on the field. The following conclusions can therefore be drawn from the findings presented in the previous chapter of this work.

Cassava farmers

The TME 491 cassava stems distributed to farmers among the five processing groups had potential to increase farmers output in the area but the fact that the quantity distributed to them was insufficient for all the members did not allow the increase output to reach some members.

The bundles of TME 419 cassava variety distributed to farmers in the five processing groups interviewed did not result in acquisition of new assets although the beneficiaries reported increase in their capability to meet expenditure on food, children education, and meeting social obligation like naming ceremonies.

Cassava processors

1. The processing equipment donated to 5 cassava processors' groups in the study area has increased the volume of cassava processed by members.
2. Cassava processors in the study area were not involved in the processing of cassava grits used for High Quality Cassava flour which was the topmost priorities of CAVA project. This was as result of low demand by cassava flour firms caused by poor electricity supply but were rather engaged in gari and fufu processing.
3. The steel cassava grater and processing drumsdistributed to Eleyele, Asipa-ilaho, Osupori, Awowo and Gudugba cassava processing groups had led to reduction in processing cost.
4. The reduction in grating cost represents saving on processing cost which is an indication of increased income.
5. The hydraulic press and jack distributed to Eleyele and Awowo gari processors was not utilised because it cannot contain the sacs used for packaging grated cassava for dewatering and they are not willing to change the size of the sac used because the sac was used as scale of the expected gari processed.
6. Two out of eight gari processors interviewed could not benefit from reduction in grating cost because their houses were located further away from the processing shed.

RECOMMENDATION

Based on these findings and conclusion presented, the author proposes the following recommendations to the implementer of CAVA project in Nigeria, cassava farmers and processors.

Implementer of CAVA project in Nigeria

1. Multiplication of the TME419 cassava variety should be carried out by OGADEP this area either by selecting some of the farmers' farms for multiplication or buying directly from IITA.
2. Linking farmers with multiplication centres for purchase of TME419 cassava stem should be facilitated by OGADEP.

3. Gari and fufu processors in the study area should be linked with credit sources so that they can increase the volume and quality of their product because there is ready market for these products in the local market as well as Lagos which is less than an hour journey to the area
4. Processing of cassava grits should be encouraged in area where cassava processors have easy access to cassava flour industries and good demand for the product because there is no demand for the product in the area
5. The hydraulic press should be withdrawn from processors by OGADEP and distributed to medium scale processors that show willingness to use it because the processors in the area did not show commitment and desire to change the sac used.
6. Gari processors that show interest in the hydraulic press should be encouraged to use sac sizes that are compatible with the press by given them enough information on the sizes of sac that can be used and the minimum quantity of grated cassava that it can contain before distribution..
7. C:AVA project should look into possibility of giving the processing equipment to processors on credit or at subsidized rate to make the project sustainable and allow more groups to benefit.
8. Hand drawn or push truck should also be provided for processing groups to facilitate easy movement of the peeled cassava to the processing shed.

Farmers and processors

1. Farmers must be willing to plant the cassava variety separately to facilitate easy identification during harvesting and to prevent confusion during distribution of the stem cutting to other farmers.
2. Cassava processors must take good care of cassava grater, hydraulic press and drum received from CAVA project. They should also make use of the fund generated from the use of these facilities to obtain other equipment needed in the processing sheds.

Abbreviations: OGADEP, Ogun State Agricultural Development Programme; C:AVA, cassava adding value for Africa; FAO, food and agriculture organisation; HQC, high quality cassava flour; IFAD, International Fund for Agricultural Development; IITA, International Institute of Tropical Agriculture; NEPAD, New Partnership for African Development; CGIAR, Consultative Group on International Agricultural Research; NGO, non-governmental organisation; RTEP, root and tuber expansion programme.

REFERENCES

Adebayo K (2010). Sustainable Inclusion of Smallholders in the Emerging High Quality Cassava Flour Value Chains in Africa: Challenges for Agricultural Extension Services. J. Agric. Exten. 14(1)

Adebayo K (2009). Dynamics of Technology Adoption in Rural-based Cassava Processing Enterprises in South-West Nigeria. Int. J. Agricultural Econ. 2(1).

Adebayo K (2011). Cassava-Adding Value to Rural Farming Systems. Common wealth Scholarship Commission

Donald LP, Truman PP, Robert BK (2000). A Global Development Strategy for Cassava: Transforming a traditional Tropical Root crop.

Natural Resource Institute (2011). Cassava: Adding value for Africa.

Olasunkanmi M, Bamiro MA, Daramola, F (2012). Enterprise Combinations in Cassava Based Food Crop Farming System in Nigeria: Evidence from Ogun State. Greener J. Agric. Sci. ISSN: 2276-7770. 2(1):13-20.

Taiwo KA (2006). Utilization Potentials of Cassava in Nigeria: The Domestic and Industrial Products. Food Rev. Int. 22(1):29-42.

APPENDIX

Appendix 1. Grated cassava in sacs used in the study area.

Knowledge gaps in potato technology adoption: The case of central highlands of Ethiopia

Beliyu Limenih[1] and Tederose Tefera[2]

[1]Ethiopia Institute of Agricultural Research, HARC, P. O. Box, 2003, Addis Ababa, Ethiopia.
[2]College of Agriculture, Hawassa University, P. O. Box, 05, Hawassa, Ethiopia.

The paper intends to identify knowledge gaps in potato technology adoption among farm households in Welmera, Ethiopia. Within the district, four administrative peasant associations were selected based on their potato production potential. From these peasant associations, a total of 112 farm households were selected using a simple random sampling method proportional to population size. The sampled households were interviewed by using a structured interview schedule. Group discussions were undertaken with selected households, development agents and researchers to gather qualitative data. For data analysis, statistical tests like t-tests, chi-squares and one way analysis of variance (ANOVAs) were used to test the variation of means and associations among the sampled households. About 79.5% of households were male headed, whereas the remaining 20.5% of households were female headed. ANOVA results indicated that there was significant variation in the adoption index score in sampled households. Non adopters, low adopters and high adopters account 46.4, 27 and 27% with the mean adoption indices of 0.0000, 0.5233 and 0.74. Moreover, all recommended potato production packages were not implemented by all adopter farmers. Low adopters used seed rates, fertilizer rates, diffused light stores (DLS) and chemical applications below the recommended packages. On the other hand, high adopters used seed rates and DLS below recommended quality and spacing and fertilizer application above the recommended rate. The overall finding of the study underlined the high importance of institutional support in improving seed rate, DLS construction, spacing, fertilizer rate and chemical application. Therefore, research center, development agents and other stakeholders should provide on farm extension training, strengthening cooperative societies, and improving market condition to fill knowledge gaps in adoption of improved potato production packages.

Key words: Potato technology, production, knowledge, market condition.

INTRODUCTION

For household food security, vegetable crops have been contributing highly for small land holder farmers. Horticultural production is usually money spinning as compared to staple crops. The production of fruits and vegetables has a comparative advantage to cereals particularly under conditions where arable land is scarce, labor is abundant and markets are accessible (Lumpkin et al., 2005). Some studies show that horticultural

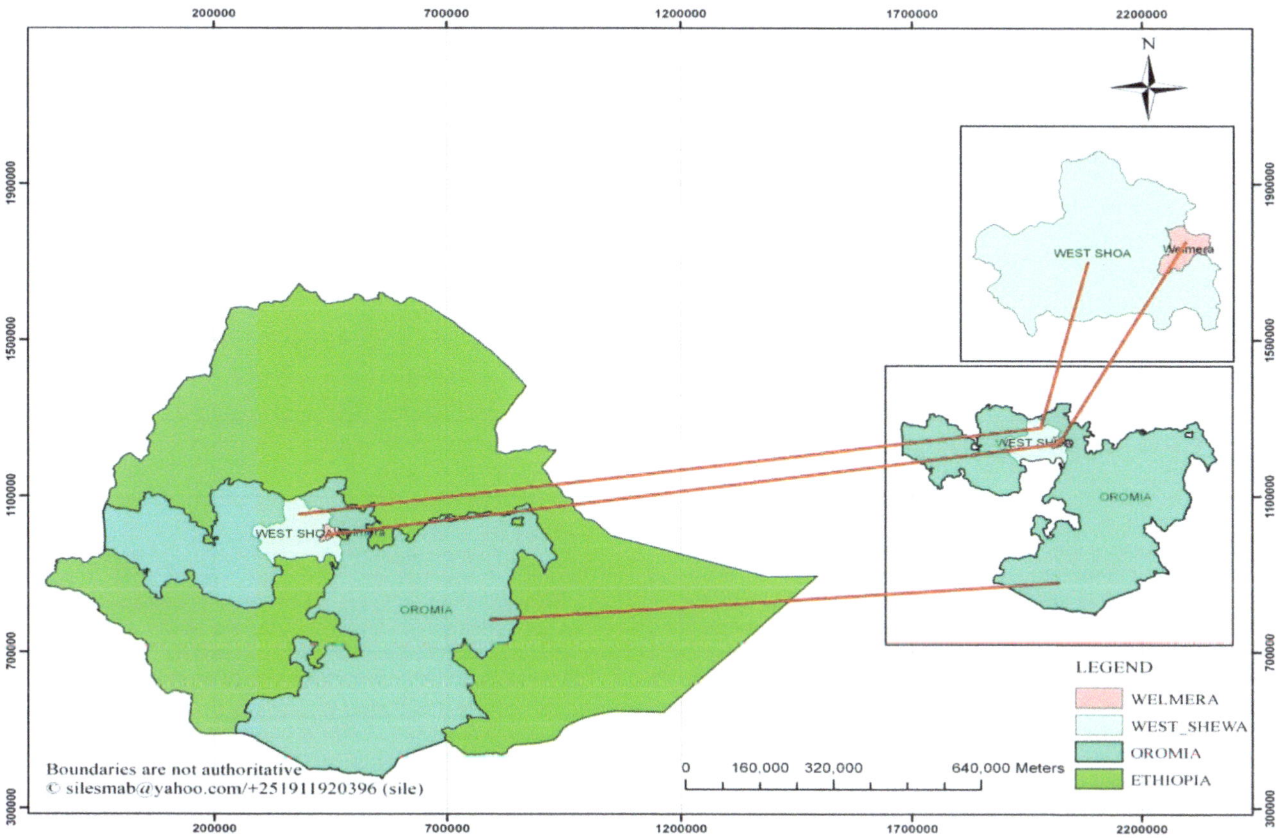

Figure 1. Map of Welmera district.

production needs twice labor forces as compared to stable crops. Increasing horticultural production contributes to commercialization of the rural economy and creates many off-farm jobs (Lumpkin et al., 2005). Potato is one of the horticultural crops, in general, and vegetable crops, in particular, that plays a key role as source of food and cash income for small-land holder producers. As a food crop, it has a great potential to supply high quality food within a relatively short period and is one of the cheapest sources of energy. Potato produce 54% more protein per unit of land area than wheat and 78% more than rice (Stevenson et al., 2001).

Ethiopia has many places where horticultural crops are produced in small scale and commercial scales. The potential attainable average yields of the crop on research and farmers' fields are 45 and 25 tons/ha, while the current national average production is limited to about 10 tons/ha (CSA, 2007). In sub-Saharan African countries the yield is around 8 tons/ha on the continent compared to a world average of 16 tons/ha (FAOSTAT, 2008). Today, more than 160,000 ha are planted. With a population of 93 million and a land size almost double that of Texas, Ethiopia can accommodate growing 3 million hectares of potatoes (Cornell, 2014).

Because of this, Ethiopian Institute of Agricultural Research (EIAR) under its different research centers has been introducing different kinds of improved potato varieties. The released potato varieties along with improved package of technologies have been disseminated through Ministry of Agriculture (MOA) and its regional and grassroots structures. For this reason most farmers in West Shewa zone are able to produce thousands of quintals of potato tuber. However, some potato growing farmers do not apply all the recommended technology packages. Identifying the knowledge gaps on potato technology adoption is critical for professionals engaged in agricultural development, researchers, policy makers and institutions. This study aimed to identify the knowledge gaps in adoption of improved potato technology packages. This was conducted in one of potato grown region of Oromia Special Zone in Welmera district.

METHODOLOGY

The study site

Welmera district (Figure 1) is one of the eight districts in the Oromia Special Zone, covering a total area of 755 km. It is found 40 km east of Addis Ababa and has 24 rural peasant associations. According to 2007 Census data, the district had a total population of

Table 1. Recommended package of improved potato production.

S/N	Name of packages	Abbreviation	Recommended rate
1	Land preparation	LP	Fine seed bed preparation (3-4times)
2	Varieties	VA	Gudena and Jalene
3	Seed rate	SR	20 q/ha
4	Fertilizer rate	FR	60 kg of urea and 60 kg of DAP
5	Chemical application	CA	2 times sprayed 50 - 55, 70 - 75 days
6	Spacing	SP	30 × 75 cm - 30 cm between plant -75 cm between row
7	Seed size	SS	Medium
8	Planting method	PM	Row
9	Weeding /hoeing and hilling frequency	WF	2 hand weeding at 35-40, 60-70 days
10	Seed storage	ST	Diffused light store (DLS)

Source: Holetta Research Center (2011).

83,784. The altitude of the district ranges between 2060 to 3380 m above sea level. Temperature ranges from 0.1 up to 26.9°C, and the mean annual rainfall ranges from 1000 to 1100 mm. The rainfall pattern is usually bimodal from January to June for summer and from June to October for off season time. Livestock is used for meat and milk in addition to that supports crop production as draught power.

Sampling techniques

In this study, two types of sampling techniques were applied. These are purposive and simple random sampling techniques. The district and four administrative peasant associations were selected purposively by reviewing secondary data about potato production experience and area coverage under potato production. After preparing up-to-date list of the sampling frame, households were selected based on probability proportional to size of total potato growing farmers in each peasant associations. Following that 112 sampled farmers were selected by using simple random sampling techniques.

Data type, source, data collection and analysis methods

Individual interviews and four group discussions from each peasant association were conducted to gather qualitative and quantitative data types. Relevant secondary data were collected from annual reports, journals, articles, book in chapters, and manuals. Descriptive statistics were applied for quantitative data type. In addition to that analytical techniques like t-tests, chi-square tests and one way analysis of variance (ANOVA) analyses were carried out.

The Adoption index (AI_i) in this study indicates the extent of adoption of improved potato recommended packages. The farmers who meet the recommended rate in the packages were given 1 and 0 for those who did not. The formula was calculated as follows:

$$AI_i = \frac{LP_i + VA_i + SR_i + FR_i + CA_i + SP_i + SS_i + PM_i + WF_i + ST_i \times 100}{10}$$

Where LP_i, Land preparation of *i* farmer; VA_i, varieties land coverage by *i* farmer farm; SR_i, seed rate usage by *i* farmer; FR_i, fertilizer rate usage by *i* farmer; CA_i, chemical application by *i* farmer; SP_i, spacing usage by *i* farmer; SS_i, seed size usage by *i* farmer; PM_i, planting method by *i* farmer; WF_i, weeding /hoeing and hilling frequency applied by *i* farmer; ST_i, seed storage method by *i* farmer.

This equation includes all of these variables due to the fact that not all sampled households use the entire recommended potato package. On the basis of adoption index respondent farmers were classified in to two categories; that is, low and high adopters.

The above method was applied because of variation in potato production packages. The recommended packages (Table 1) are put in terms of number and word. During group discussion with farmers, development agent, and researchers gave equal weights for each package. The reason was that every package has equal contribution to the production of improved potato varieties.

RESULTS AND DISCUSSION

Extent of adoption of improved potato production package

In order to know the level of adoption of each respondent, the Adoption index score was calculated. Before the calculation the technology packages were listed and weighted. Equal weights were given to all technology packages through discussion with districts experts, development agents and researchers in the study area. The sampled households' Adoption index scores were categorized into three adopter groups; namely non-adopter, low and high adopter. The actual Adoption index score ranges from 0 to 1. An Adoption index score of 0 point implies non-adoption of the overall improved potato production package. Statistical analysis of ANOVA indicated that there was significant variation ($F = 50.796$, $P = 0.000$) among the Adoption index score among the three categories at 1% level of significant which indicates difference of adoption of potato technology packages among sampled households (Table 2).

As indicated in Table 2, non-adopter accounts for 46.4% with the mean adoption index of 0.0000. This indicated that non-adopters were not practicing any of the recommended package and the technologies in the production year of 2010. Next to non-adopters, low and high adopters constituted about 27% each. Low adopters

Table 2. Distribution of respondents by adoption category of improved potato technologies

Adoption category	N	Percent	Adoption index score	Mean	SD	F-value	P-value
Non-adopter	52	46.4	0.00	0.000	0.00000		
Low	30	26.8	0.01 - 0.60	0.5233	0.09353		
High	30	26.8	0.61 - 1	0.7467	0.05713		
Total	112	100	0.00 - 1	0.3482	0.32604	50.79***	0.000

Source: Own survey data (2011). *** = significant at 1% level.

Table 3. Distribution of adopter households by the area coverage under improved potato variety.

Adoption category	N	Mean	SD	t	P-value
Low	30	0.3047	0.24237		
High	30	0.6858	0.41552		
Total	60	0.4952	0.38817	30.85***	0.0000

Source: Own survey data (2011); *** = the mean difference is significant at 1% level.

Table 4. Average seeding rate applied by sample adopter households in quintal/ha.

Adoption category	N	Mean	SD	t-value	P-value
Low	30	18.2667	2.93		
High	30	19.83	2.46		
Total	60	19.05	2.8008	2.575**	0.029

Source: Own survey data (2011); ** = the mean difference is significant at 95 % level.

have mean adoption index of 0.5233, while high adopters mean adoption index were 0.74.

Current practices of improved potato production package

Farmers' current practice of technologies components consists of varieties, seed rate, seed size, methods of planting, spacing, fertilizer rate, chemical spraying, weeding/hoeing and hilling frequency, number of seed in a hole, and seed storage type are discussed as follows:

Improved potato varieties

Gudena and Jalene improved potato varieties are widely grown in the study area. These varieties were released from Holetta Research Center located near by the study area, and so most farmers have been growing these varieties. The extent of varietal adoption was measured by the area covered by improved potato varieties by each farmer.

As shown in Table 3, the total household's average area coverage under improved potato varieties was 0.49 ha. The minimum and maximum area covered by improved potato varieties were 0.13 and 1.50 ha. The statistical analysis of t-test revealed that the existence of significant difference between the two groups at 1% significant level (Table 3).

Seed rate

Using disease free seed and appropriate seed rate are the most critical component of improved potato production. Using over and lower seed rate will cause excessive plant population or less plant population both leading to lower production. Research recommends a specified level of seed rate, however depending on availability of quality seed and farmers interest the recommended seed rate may or not applied as recommended. In the study area farmers were used a minimum of 12 quintal and maximum of 28 quintal seed rate per hectare, whereas the recommended rate of potato production is 20 quintal/ha.

According to Table 4, low and high adopters were using on average 18.29 and 19.86 qt/ha, respectively. However, both categories have used the seed rate below the recommended rate but the seed rate used by high adopter category is close to the recommended rate. From group discussion farmers were explained that increasing

Table 5. Sources of potato seed for sample households in 2010/2011 production year.

Source of seed	Frequency	Percent
Local market	5	8.3
Own /home	6	10.0
MOA	3	5.0
Research center	30	50.0
Cooperative	5	8.3
NGO	2	3.3
Individual producer	9	15.0
Total	60	100

MOA, Ministry of Agriculture Source: own survey data (2011).

Table 6. Fertilizer rate applied by sample households in 2010/2011 production year.

Adoption category	N	Mean	SD	t	P-value
Low	30	95.9333	37.74591		
High	30	125.0667	37.75322		
Total	60	110.50	40.2077	2.328***	0.004

Source: Own survey (2011), ***the mean difference at 1% significant level.

the space between rows was minimized the quantity of the average seed rates. In 2010/2011 production year potato grower's farmers average seed cost incurred in one production season was 4560 birr. In the same year the seed purchasing price was ranged between 600 to 800 birr per quintal. Most recently, most farmers became seed producers and mostly used their own seed source. On the other hand, Holetta Agricultural Research Center delivered good quality seed for scaling up the technologies.

As shown in Table 5, half of adopters obtained seed from Holetta Research Center. From group discussion farmers said that "Holetta Research Center gave them clear seed without cash and they will repay next year in kind (potato itself)". On the other hand, the current seed price is expensive so it minimized the amount of credit farmers took for seed. Few numbers of sample households purchased improved seeds from cooperatives and local markets (Table 5).

Fertilizer application

In the study area farmers have good knowledge on fertilizer application. However, some farmers did not apply the recommended rate of fertilizer of 120 kg (urea and DAP) per hectare. The maximum and minimum fertilizers applied by respondent farmers were 50 and 200 kg/ha.

According to Table 6, the average fertilizer rates applied by the respondent farmers were 110.50 kg/ha. However, the rate of fertilizer application across adopter categories has showed high variation. This was also revealed by statistical analysis of independent t-test (t = 2.328, P = 0.004); it shows that there is significant mean difference between the two groups. In the group discussion some farmers said that a "few years back fertilizer price became expensive because of that they could not apply the recommended fertilizer rate as needed."

Weeding/hoeing and hilling

The recommended number of weeding operations per cropping season is two times. The first weeding and hilling operation should be after 40 and 60 days of planting (HARC, 2008).

The sampled respondent farmers practice weeding and hilling on average 2.27 times (Table 7). And independent t-test also revealed that there is no weeding and hilling practice difference between the two groups (t = 1.161, P = 0.250). On the other hand, all sampled farmers practiced more than the recommended rates. From group discussion, farmers were said that they have good knowledge on the advantage of timely weeding and hilling the plant. Depending on the incidence of the weed and the amount of the rain could increase frequency of weeding and hilling practice. Moreover, the area is highland and the farmers are enforced to hill the plant more than the recommended number of weeding and hilling frequency.

Number of seed in a hole

After the rows are prepared farmers put the seed with an

Table 7. Weeding frequency by sample households in 2010/2011 production year.

Adoption category	N	Mean	SD	t	P-value
Low	30	2.20	0.407		
High	30	2.33	0.479		
Total	60	2.27	0.446	5.458	0.250

Source: Own survey (2011). NS = Not significant.

Table 8. Number of seed put in a hole by sample households in 2010/2011 production year.

Adoption category	N	Mean	SD	t	P-value
Low	30	1.17	0.379		
High	30	1.03	0.183		
Total	60	1.10	0.303	1.736	0.880

Source: Own survey (2011). NS = Not significant.

Table 9. Planting method by sample households in 2010/2011 production year.

Adopter category	Planting method		Total	χ^2
	Broadcasting	Row		
Low	2	28	30	
High	0	30	30	
Total	2	58	60	2.069

Source: own survey, (2011), (χ^2 = 2.069, p = 0.150) = Not significant

appropriate spacing. The recommended number of seed in each hole is one. The respondent farmers put on average 1.10 seeds in each hole. As shown in Table 8, the low and high adopters were putting on average 1.17 and 1.03 seeds in each hole, respectively. This implies that there are no statistical differences between the two groups. This was also revealed by independent t-test (t = 1.736, P = 0.880).

Planting method

Before plantation, farmers prepared rows to prevent water logging and to manage the farm very easily. Majority of potato producer farmers use row planting method.

As shown in Table 9, all high adopters used row planting method and all except two of the low adopters also used the same method. There is no difference between the two groups on method of planting. This was also revealed by statistical analysis of Chi-square (χ^2 = 2.069, P = 0.150) which showed insignificant difference between the two groups. On the study area farmers have enough knowledge on row planting methods and almost all respondent farmers prefer row planting method rather than broadcasting method.

Chemical application

"Late blight" and "Bacteria wilt" are the common diseases that occurred during production period. Both affect potato production very seriously and minimize the quantity and quality of the product. To prevent the Late blight disease, two times spray of redomile chemical at after 50 and 70 days of plantation is recommended. The incidence of the disease is common in all farmers farm. So, farmers are enforced to spray the chemical before the incidence of potato blight goes worse. All high adopters and 19 low adopter's farmers used radomile chemical to prevent "Late blight" disease (Table 10). The chi-square test also revealed the existence of strong correlation between the two groups (χ^2 = 13.469, P = 0.000) at 1% significant level.

From group discussion farmers said that radomile chemical is expensive so it is very difficult to buy at the beginning of technology adoption. Even after having few years of potato production experience some famers do not need to buy chemical sprayer because they use one by borrowing or renting.

Seed storage

In the study area, a few years ago most farmers reserved

Table 10. Chemical use by sample households in 2010/2011 production year.

Adoption category	used chemicals/Radomile		Total	χ^2
	"yes"	"No"		
Low	19	11	30	
High	30	0	30	
Total	49	11	60	13.5***

Source: own survey (2011), *** (χ^2 =13.469, p = 0.000) significant 1% level.

Table 11. Number of DLS constructer by sample households in 2010/2011 production year.

Category		Storage type		Total	χ^2
		Floor	DLS		
Adopter and non-adopter	low	12	18	30	
	high	5	25	30	
Total		17	43	60	9.17***

Source: Own survey (2011), (χ^2 = 9.170, p = 0.010) = significant at 1% significant level.

Table 12. Seed size selection by sample households in 2010/2011 production year.

Category	Seed size		Total	χ^2
	Large	Medium		
Low	1	29	30	
High	0	30	30	3.018
Total	1	59	60	

Source: Own Survey (2011), (χ^2 = 3.018, p = 0.221) = Not significant.

their seed in the hole. However, at present time most potato producer farmers use the diffused light store (DLS) for seed reservation. It helps the seed tuber to have enough sun light, ventilation and is easy for management.

Farmers who used DLS for seed reservation can have disease free and good quality seeds for the next season. Twenty-five (25) and 18 high and low adopters, respectively used DLS for seed reservation (Table 11). The statistical analysis of Chi-square also shows positive correlation between the two groups (χ^2 = 9.170, p = 0.010) at 1% significant level.

Seed size

The major seed tuber selection criteria employed by farmers were by healthiness and size of tuber (Gebermedihn, 2005). Medium sized tubers are recommended for seed purpose. Before the product is supplied to market, farmers commonly sort the potato tuber by three types of size. They use most of the time small sized tuber for home consumption, whereas medium and big sized tubers for seed and marketing purpose. Almost all respondents used medium sized potato tubers for seed purpose. As shown in Table 12, there is no significant difference between the two groups by statistical analysis of Chi-square (χ^2 = 3.018, p = 0.221).

Spacing

Spacing is one of the management practices that help to minimize mineral competition and have sufficient spacing between plants. The recommended spacing for potato production is 30 cm between plant and 75 cm between rows. From sampled households, high adopters used more than the recommended spacing (30 × 80), and more than low adopters. As shown in Table 13 and 19 of low and high adopters used (30 × 80) spacing practices. However, statistical analysis of Chi-square (χ^2 = 4.519, P = 0.477) showed that there was no significant difference between the two groups.

From group discussion farmers said that increasing the space between rows would help for management like hoeing and hilling, and chemical spraying. Beyond that the tuber grows bigger in size, and is highly preferred for market.

Table 13. Spacing by sample households in 2010/2011 production year.

Adoption category	spacing		Total	χ^2
	30 × 75	30 × 80		
Low	17	13	30	
High	11	19	30	4.519
Total	28	32	60	

Source: Own Survey (2011), (χ^2 = 4.519, p = 0.477) = Not significant.

CONCLUSION AND RECOMMENDATIONS

1) Non adopters, low adopter and high adopters account 46.4, 27 and 27% with the mean adoption index of 0.0000, 0.5233 and 0.74. Moreover, all recommended potato production packages were not implemented by all adopter farmers. Low adopters used seed rate, fertilizer rate, DLS and chemical application below the recommended packages. On the other hand, high adopters used seed rate and DLS below recommended quality and spacing and fertilizer application above recommended rate. The overall finding of the study underlined the high importance of institutional support in improving seed rate, DLS construction, spacing fertilizer rate and chemical application. Therefore, the research center, development agents and other stakeholders should provide on farm extension training to fill knowledge gaps in adoption of improved potato production packages. Because, horticultural crops like potatoes demand skill in the production process.

2) From the study, half of adopters obtained potato seeds from the Holetta Research Center. Few numbers of sample households purchased improved seeds from cooperatives and local markets. So, the district administrative and related offices should increase number of cooperatives and improve the services provided by the existing cooperatives.

3) It is reported that the potato seed quality has gradually decreased in terms of yield performance and resistance to diseases. To keep the seed quality to the regained standard, there should be monitoring mechanisms and training facilities to seed producers. So, the district agricultural office and Holetta Research Center should strengthen the training and seed monitoring system.

4) Producers and extension agents need adequate skills in production management practices starting from seed selection to post harvest technology suitable at their level. Marketing principles, bargaining skills, business planning, quality management and post harvest handling of horticultural products are some of the interventions needed in study area. Thus, the district agricultural office and Holetta Agricultural Research Center should take the responsibilities to meet the above requirements.

Conflict of Interest

The authors have not declared any conflict of interest.

REFERENCES

CSA Report of Federal Democratic Republic of Ethiopia (2008/2009). Statistical Report on Socio-Economic Characteristics of the Population in Agricultural Households, Land Use, Area and Production of Crops. Addis Ababa: CSA, pp.193-211.

Cornell Chronicle (2014). Potato may help feed Ethiopia in era of climate change 1. Available at http://www.news.cornell.edu/stories/2013/05/potato-may-help-feed-ethiopia-era-climate-change: Verified on 27 May 2014. Cornell Chronicle.

FAOSTAT (Food and Agriculture Organization)(2008). FAOSTAT Agricultural Data. Agricultural production, crops, primary. Available at http://faostat.fao.org/faostat/collections subset agriculture Accessed on 15 September; verified on 17 September 2011. United Nations Food and Agriculture Organization.

HARC (Holetta Agricultural Research Center) (2004). Holetta Research Center Annual Report 2008-2009. EARO/HARC, Addis Ababa, Ethiopia. P. 14.

Lumpkin T, Weinberger A, Mooree S (2005). Increasing Income through Fruits and Vegetable Production, Opportunities and Challenges. Marrakech,Morocco. PMCid:PMC2387001.

NKonya E, Ted S, David N (1997). Factors affecting adoption of improved maize seed and fertilizer in northern Tanzania.' J. Agric. Econ. 48(1):1-12. http://dx.doi.org/10.1111/j.1477-9552.1997.tb01126.x

Stevenson W, Rosemary R, Franc G, Weingartnur D (2001). Compendium of potato diseases. The American Phyto Pathological society. Second edition, USA.

Round potato (*Solanum tuberosum*) production in southern highlands of Tanzania: Are smallholder farmers becoming commercial?

Hosea Mpogole[1]*, Melchior E. Mlambiti[2] and Reuben M. J. Kadigi[1]

[1]Department of Agricultural Economics, Sokoine University of Agriculture, Tanzania.
[2]St. Augustine University of Tanzania, Mwanza, Tanzania.

The subsistence production is chosen by farmers because it is subjectively the best option. However, subsistence farming on small and fragmented plots risks not reaching even farmers' own food requirements, let alone producing a surplus. Therefore, it would have been expected that farmers would be commercialising production of profitable food crops. One of the more profitable food crops than many others as studies show is round potato (*Solanum tuberosum*). Nonetheless, the extent to which farmers have commercialised round potato production was not known. This study was then undertaken to fill this knowledge gap by analysing the proportion of land allotted to round potato production and the extent to which the crop was oriented towards the market by using the commercialisation index. A sample of 510 farmers was visited from three districts of the southern highlands of Tanzania. Results showed that round potato production was highly commercialised. This was evidenced by the proportion of land that was allotted for round potato cultivation and the commercialisation index. About 20 to 67% of the total land under cultivation was allotted to round potato production and 88% of the produce was sold. However, both the cultivated land and output per capita were small.

Key words: Commercialisation, commercialisation index, Irish potato, market based production, subsistence, subsistence farming.

INTRODUCTION

Over the past years, subsistence farming has neither improved the livelihoods of the rural population nor guaranteed food security. As a result, smallholder agricultural production has become a policy priority for the United Republic of Tanzania, which seeks to commercialise it (URT, 1997, 2008). Commercialisation of agricultural production in situations where the majority of farmers are engaged in food crops means making these crops profitable (Wolter, 2008). Food crops on which the government has focused include the major staples namely, maize and rice, for which there are restrictions on sales and export (Gabagambi, 2009). While such restrictions lower the prospect of maize and rice as commercial engagement, opportunities do exist in other sub-staples, such as round potato (*Solanum tuberosum*) in areas of favourable climatic conditions (Kelly, 2006). Studies on round potato have shown that production and consumption of the crop is increasingly becoming popular (Anderson, 1996, 2008; Kabungo, 2008; Koizumi, 2007; Namwata et al., 2010). The potato has great potential in both national and regional markets, due to growing demand for chips and snacks/crisps (Anderson, 2008). This growth in demand can be traced to many factors, including increasing economic activities, urbanisation, tourism, and changing lifestyles, all of which are shifting consumer food preferences towards easy to cook and processed foods (Anderson, 2008; CIP, 2008; FAOSTAT, 2008).

Apart from the fact that round potato is easy to cook and to process, as studies have shown, it produces

*Corresponding author. E-mail: hmpogole@yahoo.com.

remarkable quantities of calories comparable to cereal crops (Blanken et al., 1994; Scott et al., 2000). This means that round potato can address both food security as well as profit. Indeed, as existing data show, this round potato is more profitable than traditional staples, as it has higher yield per unit of land, matures earlier, and provides a larger income (Blanken et al., 1994; CIP, 2008; Goossens, 2002; Namwata et al., 2010; UARC, 1990). The maturity period of round potato is about three months as compared to maize (the major staple in Tanzania), which takes about eight to ten months to be ready for harvest (UARC, 1990). Also, one acre of round potato produces up to 120 (100 kg) bags versus about 20 bags of maize. Also, prices per unit are comparable although round potato sales are often higher than maize (BOT, 2010; UARC, 1990). For the round potato, these data show that it is a good income earner, and because of its potentiality, the crop is considered to be a hidden treasure for smallholder farmers (Blanken et al., 1994; CIP, 2008).

Given the prospect and from economic point of view, smallholder farmers would have been expected to commercialise the crop by allotting more acreage to it and sell a bigger part of the produce than they keep for home consumption (Nyikai, 2003; Pingali et al., 2005; Rudra, 1983; Sokoni, 2008; von Braun and Kennedy, 1994). However, the extent to which farmers in the southern highlands of Tanzania (SHT) have commercialised their round potato production was not known. This study was then undertaken to fill this knowledge gap by analysing the proportion of land allotted to round potato production and the extent to which round potato production was oriented towards the market by using the commercialisation index (CI).

The study has both descriptive and normative aims. Firstly, it is aimed at increasing knowledge on the commercial orientation of smallholder farmers. The prospect here is often overlooked because subsistence farming on small and fragmented plots generally risks not reaching even farmers' own food requirements, let alone producing a surplus. Secondly, this study is aimed to show policymakers and stakeholders in agriculture who can develop policies and strategies to help smallholder farmers with the commercialisation process. Commercialisation of smallholder production is essential in improving the income and livelihoods of these farmers and assuring their food security (Ahmed, 1994, Blanken et al., 1994; Nyikai, 2003; OECD, 2008; Sokoni, 2008; von Braun and Kennedy, 1994). It is not the aim of this study to provide a normative formula to help subsistence farmers commercialise, but rather, to develop a deepest understanding of their commercial orientation.

Study area

This study was conducted in Njombe, Mbeya Rural and Nkasi districts in Iringa, Mbeya, and Rukwa regions, respectively. These three districts are all found in the southern highlands of Tanzania. Njombe and Mbeya Rural districts were purposively selected because they were the leading producers of round potato in their respective regions (URT, 2007), and the characteristics of their farmers and farming practices differ. Njombe produces predominantly one variety of round potato, which is *Kikondo*. Mbeya Rural district produces a number of varieties including *Kikondo*, *Arka*, *Kidinya*, *Kagiri*, and *Tigoni*. Also, Mbeya Rural district is wet almost throughout the year, making it possible to cultivate round potato more than once, while this is not the case in Njombe.

Njombe and Mbeya Rural districts are within or close to better transport networks than Nkasi district. They have more access to input and output markets as well as to extension services than Nkasi district. Although farmers in Nkasi produce only small quantities of round potato, the district was taken for comparison purposes because this makes it possible to compare levels of commercialisation in areas with high potential (in terms of access to input and output markets) with low potential areas.

METHODOLOGY

Sampling and data collection

A pilot survey to test data collection instruments and to gain familiarisation with the study areas was conducted in two villages, one in Njombe district and another in Mbeya Rural district. This main survey was conducted in two seasons, from March 2010 to June 2011. Data was collected from 15 villages, which were purposively selected based on the volume of production of round potato. In those villages, respondents were randomly selected from farmers' meetings called by village executive officers (VEOs). The VEOs were informed at least a day prior to the visit of our request that they call for round potato farmers' meeting on the day of the visit. In total, 510 farmers were included in this study. However, the proportion of women who showed up for the meetings was relatively small. This phenomenon is not uncommon for it has been well documented that the gender division of labour which allocates all childcare, household activities, and water and wood carrying to women, constraints their capacity to participate in the market based production irrespective of opportunities (Ellis, 1988; Kaaria et al., 2007; World Bank, 2009).

Data was then collected on general characteristics of round potato farmers, varieties of round potato produced, land ownership, relative acreage allotted to round potato production, reasons for acreage allocation to various crops, input use, output per acre, volume sold, volume consumed and/or stored, and average selling price.

Analytical technique

Commercialisation of subsistence agriculture can take place on the output side as well as on the input side (von Braun and Kennedy, 1994). On the output side of production, commercialisation is manifested by the increased marketed surplus while on the input side, it is shown by the increased use of purchased inputs. This study assessed the commercialisation of round potato production from the output side. According to von Braun and Kennedy (1994), commercialisation on the output side is defined as in Equation (1):

$$Commercialisation\,of\,agriculture = \frac{Value\,of\,agricultural\,sales\,in\,markets}{Agricultural\,production\,value} \quad (1)$$

The commercialisation index (CI) was used to determine the extent of commercialisation of round potato production. Deriving from Bekele et al. (2011), Strasberg et al. (1999), and von Braun and Kennedy (1994), the commercialisation index (CI) for round potato production can be defined as:

$$CI = \frac{Gross\,value\,of\,all\,round\,potato\,sales}{Gross\,value\,of\,all\,round\,potato\,production} \times 100\% \quad (2)$$

Commercialisation index measures the extent to which round potato production is oriented toward the market, so, a value of zero would signify a totally subsistence-oriented farmer while the closer the index is to 100%, the higher the degree of market orientation (Strasberg et al., 1999). Since CI depends on the output *Y*, and assuming that farmers consume a fixed amount, *c*, of round potato, then:

$$CI = \frac{Y - c}{Y} \times 100\% \quad (3)$$

This assumption is realistic since farmers' consumption of a particular food crop cannot increase indefinitely with increasing production, for instance, if a farmer or rather a household consumes an amount equal to *c*, then any excess above *c* should be sold. The relationship in Equation (3) is desirable since the higher the production (*Y*) the higher the CI, that is:

$$\lim_{Y\to\infty}(CI) = 100\% \lim_{Y\to\infty}\left(\frac{Y-c}{Y}\right) = 100\% \lim_{Y\to\infty}\left(1 - c/Y\right) = 100\% \quad (4)$$

Equation (4) means that as *Y* becomes very large, relative to *c*, CI approaches 100%.

It is noteworthy that this study was descriptive in nature aiming at obtaining a general picture of the market orientation of round potato farmers by analysing the proportion of land allotted to round potato production and the commercialisation index.

RESULTS AND DISCUSSION

Description of round potato farmers visited

Results showed that the percentage of male respondents was 62, 79, and 67% in Njombe, Mbeya Rural, and Nkasi respectively (Table 1). On average, 69.6% of the all respondents were male while 30.4% were females. As aforementioned, more men showed up to the round potato farmers' meetings than women. However, statistics show that in Tanzania, female constitute about 51% of the total population. Thus, it can be inferred that the higher percentage of men in this study might be a reflection of the commercial nature of round potato production in the study areas. In this situation, it is likely that more men would be involved in it, leaving women with other food crops for home consumption and other household activities. It has widely been shown that in Africa, when a crop is perceived as commercial, men are more likely to take over from women (Kaaria et al., 2007; World Bank, 2009).

Majority of the respondents were in the 30 to 44 years age group (Table 1). This age group accounted for about 58% in Njombe and Nkasi districts and about 54% in Mbeya Rural district. Also, the 45 to 64 years age group accounted for about 31, 26, and 22% in Njombe, Mbeya Rural, and Nkasis districts respectively. The proportion of respondents in the 14 to 29 years group and the 65 years and above was relatively small. This result indicates that few youths, for example, primary and secondary school leavers were involved in round potato production. Quite often, age is used as an indicator of farming experience. This experience makes certain informational and search costs easier (Luh, 1995).

The survey results on the educational level of respondents indicated that about 82, 85, and 92% of respondents from Njombe, Mbeya Rural, and Nkasi districts respectively had primary education (Table 1). Also, about 14, 5, and 4% of respondents from Njombe, Mbeya Rural, and Nkasi respectively had secondary education. The proportion of respondents with secondary education was nearly 3 times those of Mbeya Rural or Nkasi, but the trend is similar. Educational level is said to affect market orientation and productivity. Other studies such as Hawassi (2006) and Nkumba (2007) find that educational level influences productivity and market access. It also influences the cost of information seeking and negotiating, and hence, market orientation (Asrat et al., 2009; Pingali et al., 2005; von Braun and Kennedy, 1994).

Results on marital status showed that about 86% of respondents from Njombe district, 87% from Mbeya Rural district, and 96% from Nkasi district were married while the rest were living single, widowed or separated (Table 1). Also, marital status is said to influence farm practices (World Bank, 2009). For instance, Kilima et al. (2010) indicated that married household heads had significantly less maize area under improved technologies than single household heads.

Cultivated land and the proportion for round potato production

The average land sizes under cultivation in the three districts were small (Table 2). In itself, this is unsurprising, given that it has often been reported that smallholder farmers in Tanzania farm on small and fragmented plots (Wolter, 2008; Sokoni, 2008). For instance, Wolter (2008) shows that land sizes for smallholder farmers in Tanzania range from about 2.0 to

Table 1. Characteristics of round potato farmers visited.

Variable	Njombe		Mbeya Rural		Nkasi	
	Frequency	Percent	Frequency	Percent	Frequency	Percent
Sex						
Female	64	37.6	35	20.6	56	32.9
Male	106	62.4	135	79.4	114	67.1
Total	170	100.0	170	100.0	170	100.0
Age						
14-29 years	14	8.2	32	18.8	33	19.4
30-44 years	99	58.2	91	53.5	98	57.6
45-64 years	52	30.6	44	25.9	38	22.4
65 years and above	5	2.9	3	1.8	1	.6
Total	170	100.0	170	100.0	170	100.0
Education level						
No formal education	6	3.5	13	7.6	6	3.5
Primary education	139	81.8	145	85.3	157	92.4
O-level secondary education	24	14.1	9	5.3	7	4.1
A-level secondary education/certificate	1	.6	3	1.8	0	0.0
Total	170	100.0	170	100.0	170	100.0
Marital status						
Married	146	85.9	148	87.1	157	92.4
Single	7	4.1	9	5.3	6	3.5
Separated/widowed	17	10.0	13	7.6	7	4.1
Total	170	100.0	170	100.0	170	100.0

Table 2. Acreage allocation for round potato and maize.

District	Land description	Minimum	Maximum	Mean
Njombe	Total land under cultivation (acres)	1.00	13.00	4.71
	Proportion of land under cultivation (ratio of total)	0.17	1.00	0.82
	Land size under maize production (acres)	0.50	6.00	2.17
	Land under round potato production (acres)	0.25	7.00	1.99
	Land under round potato per total under cultivation	0.08	0.67	0.42
Mbeya Rural	Total land under cultivation (acres)	1.00	40.00	3.74
	Proportion of land under cultivation (ratio of total)	0.13	1.00	0.90
	Land size under maize production (acres)	0.50	10.00	1.69
	Land under round potato production (acres)	0.50	40.00	2.63
	Land under round potato per total under cultivation	0.10	1.00	0.71
Nkasi	Total land under cultivation (acres)	3.00	21.00	7.08
	Proportion of land under cultivation (ratio of total)	0.20	1.00	0.71
	Land size under maize production (acres)	1.00	14.00	3.87
	Land under round potato production (acres)	0.50	3.00	1.14
	Land under round potato per total under cultivation	0.03	0.75	0.20

Table 3. Commercialisation index (CI).

District	Total output (100 kg bags)		Amount consumed (100 kg bags)		Amount stored (100 kg bags)		Amount sold (100 kg bags)		CI	
	Mean	Std. dev.	Mean	Std. dev.	Mean	Std. dev.	Mean	Std. dev.	Mean	Std. dev.
Njombe	91.4	84.0	0.5	1.2	10.6	10.7	80.6	79.1	0.864	0.135
Mbeya rural	67.6	80.1	0.4	1.2	5.3	9.5	62.3	74.2	0.913	0.147
Nkasi	12.9	13.4	0.5	0.8	0.9	2.1	11.6	12.4	0.875	0.168
Total	57.1	74.8	0.5	1.1	5.5	9.2	51.3	69.2	0.884	0.152

7.5 acres (or about 0.9 to 3 ha). What is surprising, however, is the proportion of land that is allotted to round potato production. This was 20% of the total land under cultivation in Nkasi, 42% in Njombe and 67% in Mbeya Rural.

Mbeya Rural had the highest proportion of land allotted to round potato production but its per capita land size was small. This was due to the fact that land in Mbeya Rural is characterised by mountain slopes and valleys, making producers farm on small and fragmented plots. Also, in this area, maize for instance, takes too long to be ready for harvest compared to Njombe and Nkasi districts. This makes round potato a potential crop around this area.

The proportion of land under round potato production at Nkasi was relatively small. This could be due to the remoteness of the district, as it is very far from potential market centres and has poor transport infrastructure. This gives farmers in this area impetus to produce more of maize than round potato because maize does not require heavy inputs as compared to round potato. Also, maize can be stored for a longer time compared to round potato, which is highly perishable.

From the commercial point of view, the average acreage under production was too small to provide any meaningful output for business purposes. However, in the theory of peasant economics, there appears to be an inverse relationship between farm size and productivity (Carter, 1984; Ellis, 1988). The gross output or income per unit of land for a small farm has been shown to be higher than that of a bigger farm. The farm size used in this context is, as defined by Ellis (1988), the strict area size of the farm (as opposed to farm scale which refers to the overall economic size of farms). The proposition that Ellis (1988) develops to explain this is that small farms generally make more efficient use of resources than large farms, an argument which is similar to Schumacher (1989), who also advocates for small farms.

Although, the total land sizes under cultivation were comparable to previous studies, the proportion of land devoted to round potato production in the current study was high especially in Njombe and Mbeya Rural districts. This is so because the study locations are considered to be the major cereal producers in Tanzania (URT, 2006). Hence, having between 20 and 67% of the land allotted for round potato production is by any means high. This high proportion of land allotted to round potato is consistent with our expectation that since the crop is more profitable than (for example) maize, which is popular in SHT, farmers would increasingly allocate more land to it. Therefore, this indicates a movement towards commercial round potato production, a conclusion that does not disregard the inverse relationship between farm size and productivity as discussed by Carter (1984) and Ellis (1988). However, our argument is that commercially-oriented farmers would allocate more acreage to a crop that, subject to probabilities, promises to yield higher profits.

Commercialisation index (CI)

The findings showed that CI for round potato was 86% in Njombe, 88% in Nkasi and 91% in Mbeya Rural (Table 3). On average, 88% of all round potato produce was being sold. The remaining output was either consumed or stored as seed tubers for the following season. The CI of 88% found in this current study is surprisingly high. According to Bekele et al. (2011) and Strasberg et al. (1999), a crop commercialisation index greater than 50% signifies a commercial oriented farmer for a crop under consideration. Since the commercialisation index for this study is about 88%, then round potato production is highly commercialised in the study areas. Since farmers in these areas did not grow a single crop, it is likely that they produced other crops for home consumption and the round potato for the market.

Although no studies have assessed the commercialisation index of farmers in Tanzania, Nyikai (2003) argues that the majority of smallholder farmers in sub-Saharan Africa are neither purely subsistence nor purely commercial. They are either semi-commercial or semi-subsistence. Some farmers usually produce certain crops for home consumption and some specific crops for sale (Bekele et al., 2011). In this case, the commercial orientation of farmers should be measured with reference to a specific crop rather than the farmer in general.

Despite the fact that the commercialisation index was very high, the per capita output of round potato was very small. This is due to two main reasons. Firstly, the small and fragmented plots that farmers cultivated, and secondly, due to low productivity. Low productivity is a result of little or non-use of agricultural inputs. Therefore, given that only about 20% of all arable land is cultivated in Tanzania (URT, 2009), then opportunities exist in terms of land expansion by mechanisation and improved productivity. The current land holdings among famers are too small to provide meaningful commercial outputs.

Conclusion

This study found that round potato production in Southern highlands of Tanzania is highly commercialised. This was evidenced by the proportion of land that was allotted for round potato cultivation and the commercialisation index. About 20 to 67% of the total land under cultivation was allotted to round potato production and about 88% of the produce was sold. This means that farmers produced round potato primarily for the market. However, both the cultivated land and the output per capita are very small. This calls for both land expansion and improved productivity. Since commercialisation of agriculture directly generates income, creates employment and increased agricultural labour productivity, recommendation for extension services that are directed towards training of rural farmers on the use of appropriate farm inputs, agronomic practices and access to markets is envisaged.

ACKNOWLEDGEMENTS

The authors would like to thank the Research on Poverty Alleviation (REPOA) and Tumaini University, Iringa University College for funding this study. We thank Dr. Mishili and Dr. Hella of the Department of Agricultural Economics and Agribusiness at Sokoine University of Agriculture for their initial ideas about the possibility of conducting this study. Finally, authors would like to thank Mr. Fidelis Mawona, Mr. Geofrey Matata, and other research assistants for their help in data collection.

REFERENCES

Ahmed R (1994). Investment in Rural Infrastructure: Critical for Commercialisation in Bangladesh. In: von Braun J and Kennedy E (eds) Agricultural Commercialisation, Economic Development, and Nutrition. Hopkins University Press, Baltimore and London pp. 141-152.

Anderson JA (1996). Potato cultivation in the Uporoto mountains, Tanzania: An analysis of social nature of agro-technological change. Afr. Affairs 95:85-106.

Anderson PK (2008). A Global Perspective of Potato Production in Emerging Markets. International Potato Centre, SCRI, Dundee, Scotland. http://www.scri.ac.uk/scri/files/IIPP%20presentations/PamelaAnderson.pdf site visited on 23/11/2008.

Asrat S, Yesuf M, Carlsson F, Wale E (2009). Farmers' Preferences for Crop Variety Traits: Lessons for on-farm conservation and technology adoption. Environment for Development, Discussion paper series, EfD DP 09-15.

Bank of Tanzania (BOT) (2010). Monthly Economic Review. September 2010.

Bekele A, Belay K, Legesse B, Lemma T (2011). Effects of crop commercial orientation on productivity of smallholder farmers in drought-prone areas of the central Rift Valley of Ethiopia. J. Rural Dev. 33(4):105-128.

Blanken J, von Braun J, De Haen H (1994). The triple role of potatoes as a source of cash, food, and employment: effects on nutritional improvement in Rwanda. In: von Braun J and Kennedy (eds) Agricultural Commercialization, Economic Development, and Nutrition. The Johns Hopkins University Press, Baltimore and London, pp. 276-294.

Carter MR (1984). Identification of the inverse relationship between farm size and productivity: an empirical analysis of peasant agricultural production. Oxford Econ. 36(1):131-145.

Centro Internacional de la Papa (CIP) (2008). 2008 – The International Year of Potato. http://www.cipotato.org/pressroom/facts_figures/2008_international_year_of_the_potato.asp site visited on 1/5/2009.

Ellis F (1988). Peasant Economics: Farm households and agrarian development. Cambridge, Cambridge University Press, U. K. pp. 196-201.

FAOSTAT (2008). Potato World: Production and Consumption. http://faostat.fao.org/default.aspx site visited on 1/5/2009.

Gabagambi DM (2009). Towards a paradigm shift in revolutionalising Tanzania's agriculture. In: Batamuzi EK, Gimbi DM, Max RA, Nombo CI, Msoffe GE (eds) Research initiatives for improvement of livelihoods: Contribution of SUA and her partners. Proceedings of the Fourth Annual PANTIL Scientific Conference, Morogoro, Tanzania, October 19-21. pp. 25-38.

Goossens F (2002). Potato marketing in Rwanda: Agricultural Policy Development Project. Research Report No. 12. Abt Associates Inc., USAID/Rwanda.

Hawassi FGH (2006). Analysis of Processing, Marketing and Demand for Processed Fruits and Vegetables in Tanzania. Unpublished Thesis for Award of PhD Degree at Sokoine University of Agriculture, Morogoro, Tanzania.

Kaaria S, Sanginga P, Njuki J, Delve R, Chitsike C, Best R (2007). Enabling rural innovation in Africa: An approach for empowering smallholder farmers to access market opportunities for improved livelihoods, [http://www.future-agricultures.org/farmerfirst] site visited on 12/06/2009.

Kabungo CD (2008). Evaluation of Irish Potato Production and Marketing Performance: A Case Study of Mbeya Rural District. MSc Dissertation, Sokoine University of Agriculture.

Kelly T (2006). Tanzania Potato production. http://research.cip.cgiar.org/confluence/display/wpa/Tanzania site visited on 15/02/2009.

Kilima FTM, Mbiha ER, Erbaugh JM, Larson DW (2010). Adoption of improved agricultural technologies by smallholder maize and sorghum farmers in Central Tanzania. Eastern and Southern Afr. J. Agric. Econ. Dev. 7:24-54

Koizumi M (2007). Comparative study of farming systems in South-western Tanzania: Agrarian adaptation in a socio-historical perspective. Afr. Stud. Monogr. 34:3-20.

Luh Y (1995). Are Farmers Learning by Doing? Experience In Taiwan. Rev. Agric. Econ. 17(2):213-227.

Namwata BML, Lwelamira J, Mzirai OB (2010). Adoption of improved agricultural technologies for Irish potatoes (*Solanum tuberosum*) among farmers in Mbeya Rural district, Tanzania: A case of Ilungu ward. J. Anim. Plant Sci. 8(1):927-935.

Nkumba JM (2007). Assessing adoption and economic impacts of new banana varieties on livelihoods of farmers in Kagera region in Tanzania. Unpublished Thesis for Award of PhD Degree at Sokoine University of Agriculture, Morogoro, Tanzania.

Nyikai RA (2003). Commercial and Subsistence Farming: What is the future for smallholder Kenyan agriculture? Afr. Crop Sci. Soc. 6:591-596.

Organisation for Economic Cooperation and Development (OECD) (2008). Turning African Agriculture into a Business: Lessons Learnt.

www.oecd.org/dataoecd/58/56/42987772.pdf site visited on 12/08/2010.
Pingali P, Khwaja Y, Meijer M (2005).Commercialising Small Farms: Reducing Transaction Costs. In: The Future of Small Farms. Workshop Proceedings, IFPRI and Overseas Development Institute (ODI), Wye, UK, June 26-29, pp. 61-79.
Rudra A (1983). Non-Maximising Behaviour of Farmers: Crop Selection. Econ. Polit. Weekl. 18(40):1717-1722.
Schumacher EF (1989). Small is beautiful: Economics as if people mattered. Harper Perennial, London pp. 155-169.
Scott GJ, Rosegrant MW, Ringler C (2000). Roots and Tubers for the 21st Century: Trends, Projections, and Policy Options. Food, Agriculture, and the Environment Discussion Paper 31, International Food Policy Research Institute (IFPRI).
Sokoni CH (2008). Commercialisation of Smallholder Production in Tanzania: implications to sustainable resource management. Geogr. J. 174(2):158-161.
Strasberg PJ, Jayne TS, Yamano T, Nyoro J, Karanja D, Strauss J (1999). Effects of Agricultural Commercialisation on Food Crop Input Use and Productivity in Kenya. Office of Sustainable Development; Policy Synthesis for USAID-Africa Bureau p. 41.
United Republic of Tanzania (URT) (1997). Agricultural and Livestock Policy of 1997. Ministry of Agriculture and Cooperative Development.
United Republic of Tanzania (URT) (2006). National Sample Census of Agriculture 2002/03; Statistics Unit – Ministry of Agriculture, Food Security and Cooperatives.
United Republic of Tanzania (URT) (2007). Iringa Region Socio-Economic Profile. National Bureau of Statistics (NBS) and Iringa Regional Commissioner's Office.
United Republic of Tanzania (URT) (2008). Agricultural Marketing Policy. Ministry of Industry, Trade and Marketing.
United Republic of Tanzania (URT) (2009). Investment Potential and Opportunities in Agriculture (Crop Sub-Sector). Ministry of Agriculture, Food Security and Cooperatives.
Uyole Agricultural Research Centre (UARC) (1990). Kilimo Bora cha Viazi Mviringo. Extension Leaflet No. 49. Shirika la Kilimo Uyole, Mbeya, Tanzania.
von Braun J, Kennedy E (1994). Agricultural Commercialization, Economic Development, and Nutrition. The Johns Hopkins University Press, Baltimore and London pp. 11-33.
Wolter D (2008). Tanzania – Why a Potential Food Exporter is still Importing Food. OECD Development Centre, Paris. www.oecd.org site visited on 29/07/2010.
World Bank (2009). Gender in agriculture sourcebook. The International Bank for Reconstruction and Development/The World Bank, Washington DC, USA.

Determinants of adoption of improved maize varieties in Osun State, Nigeria

Wasiu Olalekan Akinbode and Adebayo Simeon Bamire

Department of Agricultural Economics, Faculty of Agriculture, Obafemi Awolowo University, Ile-Ife, Osun State, Nigeria.

The paper examines the level of awareness and identified the various improved maize varieties (IMVs) cultivated in Osun state, Nigeria. It also analyzed the socioeconomic factors influencing the adoption and intensity of use of improved maize varieties. A multistage sampling design was used to select 360 farming households that were interviewed using structured questionnaire. Data collected included demographic and socioeconomic characteristics of respondents such as age, household size, gender, farm size and other improved maize production related activities. Descriptive statistics and double hurdle model were used as analytical tools. Results showed that the level of awareness of improved maize varieties was 97.8%. About 91% of these estimates were adopters while 8.8% were non adopters. The double hurdle model estimates showed that age ($t=4.50$, $p<0.05$), level of education ($t=3.33$, $p<0.05$), farming experience ($t=4.33$, $p<0.05$), household size ($t=2.18$, $p<0.05$), farm size ($t=4.02$, $p<0.05$), and household's distance to market ($t=2.26$, $p<0.05$) were significant determinants of adoption of IMVs while age of respondents ($t=2.31$, $p<0.05$), level of education ($t=2.27$, $p<0.05$), household size ($t=2.79$, $p<0.05$), farm size ($t=2.51$, $p<0.05$), frequency of contact with extension agent ($t=10.46$, $p<0.05$), off farm income ($t=2.19$, $p<0.05$), and membership in association ($t=2.46$, $p<0.05$) determined use intensity of improved maize varieties. The study concluded that policies that increase farmers' level of education and effectiveness of extension services contact will facilitate adoption and use intensity of improved maize varieties.

Key words: Adoption, use intensity, improved maize varieties; and double hurdle model.

INTRODUCTION

For most developing countries, agriculture provides a leading source of employment and contributes large fractions of their national income. In addition, provision of adequate food for an increasing population and supplying of adequate raw materials to and providing markets for the product of a growing industrial sector have been identified as part of the major roles of agriculture in the economy of Nigeria. However, for agriculture to perform its roles towards economic development depends largely on agricultural productivity. High yielding seed varieties that are fertilizer responsive, tolerant to drought and resistant to pest is one of the key elements that constitute the pivot on which increased agricultural productivity per unit of land is rested (Idachaba, 1994). As noted by

Duflo et al. (2006), the rapid population growth has made countries in Africa to be no longer viewed as a land-abundant region where food crop supply could be increased by expansion of land used in agriculture. Demographic and environmental pressures have made arable land to become scarce and increasingly marginal for food production in Africa.

Maize is one of the major cereal crops grown and consumed across all agro ecological zones of Nigeria. It currently accounts for approximately 20% of domestic food production in West and Central Africa. It has also achieved the highest growth rate of the major crops since the 1970s (Kamara et al., 2006). Despite the high yield potential of maize; its production is faced with numerous constraints. Studies (Babatunde et al., 2008, Kudi et al., 2011) have shown that maize average yield is still low compared to its potential yield. Thus, enough maize has not been produced in Nigeria to meet both the food and industrial needs of Nigeria. The International Institute of Tropical Agriculture (IITA) has developed extra early maturing, decreasing susceptibility to drought, disease tolerant and high yielding maize varieties that are adapted for growth in West Africa. All these positive attributes of improved maize varieties will reduce the chronic food shortages, stabilizes rural income and lessening the risk of farming.

The importance of farmers' adoption of new agricultural technology has long been of interest to agricultural economists, extensionists and rural sociologists. It is believed that an effective way to increase productivity is broad based adoption of new farming technologies (Minten and Barret, 2008). This hypothesis is supported by the substantial improvement in productivity of cereal crop in mid-1990's following extensive promotion of improved technologies by Sasakawa Global 2000, an international NGO working to improve productivity of smallholder agriculture (Tura et al., 2009). Adoption of agricultural technologies refers to the decision to apply a technology and to continue with its use. The adoption decision is divided into three phases: acceptance, actual adoption, and continued use. It is generally a multistage process undertaken most often sequentially and being influenced by a wide range of economic, social, physical and technical aspects of farming (Paudel and Thapa, 2004; De Graff et al., 2005). The low productivity levels have been attributed to low yield potential of seed cultivars, susceptibility of seeds to biotic and abiotic stress, low adoption rate and other recommended management practices (Asnake et al., 2005).

The objective of this paper is to assess the level of technology (IMVs) adoption, identify IMVs cultivated in the study location, and analyze the socioeconomic determinants of adoption and intensity of use of IMVs. By understanding farming households' adoption pattern of improved maize varieties (IMVs), extension programmes can be better designed. Hence, the outcome of this study will enable agricultural policy makers to design policies that will address factors determining the adoption of IMVs.

STUDY AREA AND SAMPLING TECHNIQUE

The study was conducted in Osun State, southwestern part of Nigeria and lies between latitude 05° 58'N and 08° 07'N and longitude 04° 00'E and 05° 05'E. It covers a total land area of approximately 14,875 km^2 with total population of 3,423,535 with sex distribution of 1,740,619 male and 1,682,916 female and population density of 238.1/km^2.The state has three Agro Ecological Zones (AEZs) namely rain forest (Ife/Ijesa), derived savannah (Osogbo), and savannah (Iwo) zones.

The climate is tropical and characterized by bi-modal rainfall pattern with the annual rainfall ranges from 800 mm in the derived savannah to 1500 mm in the rain forest while the mean annual temperature varies from 21.1 to 31.1°C (OSSG, 2004). The state's soil type is of the highly ferruginous tropical red soil and the vegetation is mostly rain forest.

The people of the State are mostly farmers, traders and artisans with larger percentage being farmers. The farmers cultivate permanent crops such as cocoa (*Theobroma cacao*), kolanut (*Cola nitida and C. acuminata*), plantain and bananas (*Musa spp*), Oil palm (*Elias guinensis*) and citrus (*Citrus Spp*). They also cultivate arable crops especially maize (*Zea mays*) with different varieties widely cultivated. Other arable crops cultivated include yam (*Discorea spp*), cassava (*Manihot esculenta*), rice (*Oryza sativa*) and cocoyam (*Colocasia spp).*

Multi-stage random sampling technique was used to select sample of 360 maize farmers. The first stage involved purposive selection of four Local Government Areas (LGAs) noted for maize production in each of the three agro-ecological zones (AEZs) in Osun State, based on the classification of the state's Agricultural Development Programme (ADP). The second stage also involved purposive selection of three high maize producing villages in each of the LGAs. In the third stage, stratified random sampling was used to categorize maize farmers into adopters and non adopters of improved maize varieties in each of the village. The fourth stage involved simple random selection of five maize farmers in each of the two categories making a total of 360 respondents for the study.

Data collected include the socioeconomic variables such as: Age, sex, farming experience, level of education, frequency of contact with extension service, credit availability, market access and farm size. The data also included the level of awareness and IMVs cultivated in the study area.

Data analysis

Descriptive statistics was used to assess the level of awareness and adoption of IMVs as well as identifying IMVs cultivated while double hurdle model was used to analyze socioeconomic factors determining adoption and intensity of use of improved maize varieties.

Specification of double-hurdle model

The underlying assumption in the double-hurdle approach is that a farmer makes two decisions with regard to his willingness to grow improved maize. The first decision is whether to allocate a positive amount of land to improved maize variety at all while the second decision is about the share of land to allocate, conditional on the first decision.

Originally proposed by Cragg (1971), the double- hurdle model is a parametric generalization of the Tobit model, in which two

Table 1. Awareness and adoption of improved maize varieties.

Variable	Frequency	Percentage
Level of awareness		
Households with awareness	352	97.8
Households without awareness	8	2.2
Total (sample size)	360	100.0
Level of adoption based on awareness		
Adopters	321	91.2
Non adopter	31	8.8
Total (level of awareness)	352	100.0

Source: Field Survey, 2011.

separate stochastic processes determine the decision to adopt and the intensity of adoption of the technology (Greene, 2000). The double-hurdle model has an adoption (D) equation:

$D_i = 1$ *if* $D^*_i > 0$
$D_i = 0$ *Otherwise*

$$D^*_i = \alpha Z'_i + \mu_i \quad (1)$$

Where D* is a latent variable that takes the value 1 if the farmer adopts improved maize varieties and zero otherwise, D_i is the observed variable which represent farmers adoption decision, Z' is a vector of explanatory variables hypothesized to influence adoption, α is a vector of parameters and μ is the error term. The level of adoption (Y) has an equation of the following:

$$Y^*_i = \begin{cases} Y_i = \beta X_i' + v_i & if\ Y^*_i > 0 \ \ and \ \ D^*_i > 0 \\ 0 & \text{otherwise} \end{cases} \quad (2)$$

Where Y_i is the proportion of land area planted with improved maize varieties (signifying the extent or intensity of adoption), Y^*_i is the unobserved or latent variable for the intensity of adoption, X is a vector of explanatory variables hypothesized to influence intensity of use of improved maize varieties, β is a vector of parameters to be estimated and v_i is the error term.
The error terms, μ_i and v_i are distributed as follows.

$$\left. \begin{array}{l} \mu_i \sim N(0, 1) \\ v_i \sim N(0, \sigma^2) \end{array} \right\}$$

The error terms, μ_i and v_i are assumed to be independent of each other and normally distributed with zero mean and constant variance. The log-likelihood function for the double-hurdle model following Greene (2000) is:

$$\text{Log} L = \sum_0 \ln\left[1 - \Phi(\alpha z_i')\left(\frac{\beta x_i'}{\sigma}\right)\right] + \sum_+ \ln\left[\Phi(\alpha z_i')\frac{1}{\sigma}\phi\left(\frac{y_i - \beta x_i'}{\sigma}\right)\right] \quad (3)$$

In this case, the model relating to adoption was specified as:

$$Y_1 = \beta_0 + \beta_1X_1 + \beta_2X_2 + \ldots\ldots\ldots\ldots + \beta_{12}X_{12} \quad (4)$$

Where the explanatory variables are defined as: X_1 = Education of the household head (years), X_2 = Age of household head (years), X_3 = Farming experience (years), X_4 = Household size (number), X_5= Farm size (ha), X_6 = Off-farm income (₦), X_7 = Access to credit (1 if yes, 0 otherwise), X_8 = Frequency of extension service contacts (frequency), X_9 = Distance to market (km), X_{10} = Membership in association (1 if yes, 0 otherwise), X_{11} = Seed availability (1 if adequate, 0 otherwise), X_{12} = Land security (1 if secured, 0 otherwise) while the dependent variable, Y_1 = adoption of improved maize varieties (1 if adopted, 0 otherwise) and βs are coefficients of parameters to be estimated.

RESULTS AND DISCUSSION

Level of awareness and adoption of IMVs

Descriptive statistics results showed that majority (97.8%) of sampled households were aware of improved maize varieties. Of the aware households, majority (91.2%) were adopters while 8.8% were non adopters (Table 1).

Improved maize varieties cultivated in Osun state

Five improved maize varieties were commonly cultivated in the study area. DMR-ESR-Y was most widely adopted (61.9%) while 3.1% TZMSR-W was least adopted (3.1%) as shown in Table 2.

Socioeconomic characteristics of respondents

The mean age of the total respondents was 50.6±13.3 years which was an indication that the respondents were fairly in their active years. The mean age of the adopters and non adopters were 52.1±9.4 years and 54.2±10.8 years respectively. The mean age difference between adopters and non adopters was observed to be significant at 5% level. This indicated that age influences adoption of improved maize varieties.

Respondents composed of both male and female household heads. Majority (83.6%) were male while 16.4% were female headed. The male headed households' proportion for adopters and non adopters

Table 2. Improved maize varieties cultivated.

Variety	Frequency	Percentage
DMR-ESR-W	47	14.6
DMR-ESR-Y	199	61.9
DMR-LSR-W	29	9.0
DMR-LSR-Y	36	11.2
TZMSR-W	10	3.1
Total	321	100

Source: Field survey (2011).

Table 3. Socioeconomic characteristics of respondents.

Variable	Pooled (n=360)	Adopter (n=321)	Non-adopters (n=39)	t-value
Mean age (year)	58.6(13.3)	52.1(9.4)	54.2(10.8)	2.8824**
Sex (%)				
Male	83.6	84.7	74.4	
Female	16.4	15.3	25.6	
Education (%)				
No education	36.7	35.2	48.7	
Primary	21.9	23.7	7.7	
Secondary	28.6	26.8	43.6	
Tertiary	12.8	14.3	0.0	
Mean year of education (year)	7.1(6.3)	6.3(2.1)	5.6(2.2)	4.5768**
Mean farmsize (ha)	3.2(2.1)	2.8(2.0)	2.6(2.0)	3.5540**
Mean farming experience (year)	31.9(13.8)	30.5(13.4)	26.9(12.3)	2.6911**
Income sources (%)				
Farming only	54.7	57.1	35.9	
Farming/petty trading	19.4	16.5	43.6	
Farming/artisan	14.7	14.6	15.4	
Farming/civil service	11.2	11.8	5.1	
Membership in Association (%)				
Yes	52.8	54.2	35.9	
No	47.2	45.8	64.1	

ha = hectare, % = percentage, **=5% level of significance, standard deviation in parenthesis.
Source: Field survey (2011).

were 84.7 and 74.4% respectively. This shows that male headed were higher than female headed households for each category of respondents. This could be attributed to various reasons such as economic and social position of female headed households, including labour shortages and limited access to required information and inputs.

Education is an important determinant of adoption decisions. It helps to understand and interpret the information coming from any direction to farmers (Bekele and Mekonnen, 2010). To a greater degree, education determines the ability to read and/or write by farmers. As shown in Table 3, 36.7% of the total respondents had no formal education while 21.9, 28.6 and 12.8% completed primary, secondary and tertiary education respectively. If completion of primary school is taken to measure ability to read and/or write, the finding revealed that 63.3% of the total respondents could read and/or write, while 36.7% of them could not. This indicated that the level of literacy is high; this may be the reason for high rate of adoption as shown in Table 1. The mean year spent in acquiring formal education for the total respondents was 7.1±6.3 years while it was about 6.3±2.5 years and 5.6±2.2 years for adopters and non adopters respectively. The mean difference of the year spent in

Table 4. Probit estimates of socioeconomic factors determining adoption.

Variable	Coefficient	Standard error	P-value	Marginal effect	t-value
Age	-0.0702***	0.0156	0.000	-0.0099	4.50
Education level	0.0789***	0.0237	0.001	0.0113	3.33
Farming experience	0.0705***	0.0163	0.000	0.0101	4.33
Household size	0.1241**	0.0569	0.029	0.0167	2.18
Farm size	0.0949**	0.0236	0.000	0.0128	4.02
Access to credit	0.3327	0.0234	0.155	-0.0467	2.26
Seed availability	0.8346	0.5609	0.137	0.1518	
Distance to market	-0.0782**	0.0346	0.024	-0.0978	
Extension contact	-0.1483	0.2393	0.536	0.0012	
Land security	0.0797	0.2629	0.762	0.0061	
Off farm Income	0.0001	0.0015	0.967	0.0000	
Membership	0.2946	0.2273	0.195	0.0456	
Constant	0.2887	0.9300	0.756		
LR chi2(13)	112.06				
Prob > chi2	0.0000				
Log likelihood	-104.5536				

*** = 1% significance, ** = 5% significance.
Source: Field survey (2011).

school between adopters and non adopters was significant at 5% level. This indicated that there is a relationship between education and adoption of improved maize varieties in the study area.

Membership of respondents in different farmers' association is assumed to have influence on adoption decision of farm households. It makes farmers to have more access to input, information and better interpretation of available information related to new technology. More than half of the respondents (52.8%) were members in one farmers association or the other while, 47.2% were not. Further, within each category, greater percentages of adopters (54.2%) were members in farmers' association while greater percentages of non adopters (64.1%) were not.

The mean farm size of the total respondents was 3.2±2.2 ha. The mean farm size for adopters category was the highest (2.8±2.1 ha) whereas it was 2.6±2.0 ha for non adopters. The mean difference of the total farm size between the adopter and non adopters' categories was found to be significant at 5% level.

This shows that farm size has a relationship with adoption of improved maize varieties in the study area. Farming experience is likely to have a range of Influences on adoption. A more experienced farmer appears to be more knowledgeable and may have a lower level of uncertainty about new technologies. Table 3 shows that the mean years of farming experience was 31.9±13.8 years for the total respondents. The adopters and non adopters category of the respondent had mean years of 30.5±13.4 and 26.9±12.3 respectively. The mean difference of farming experience between adopters and non adopters was found to be significant at 5% level. This indicates that farming experience influences adoption of improved maize varieties in the study area. Some respondents had other sources of income besides farming which implied that they engaged in off farm activities. Of the total respondents, 54.7% got their income solely from farming. Others combined petty trading (19.4%), artisans (14.7%) and civil service (11.2%) with farming. Further, within each category; majority of adopters (57.1%) got their income solely from farming. For the non adopters, majority (43.6%) earned their income from farming and petty trading.

Determinants of adoption of improved maize varieties

The double hurdle model shows the result of probit model for adoption and the truncated regression model for use intensity of improved maize varieties in the study area. The estimated coefficients of probit model and truncated regression model are presented in Tables 3 and 4 respectively.

Factors determining adoption of improved maize varieties

Age of the household's head was found to be a statistically significant variable at 1% level with negative relationship. The negative relationship implies that age reduces adoption probability. A unit increase in the age of the respondent reduces probability of adoption by 0.9%. This implies that the older the respondent, the lower the probability of adoption. This finding agrees with previous studies on technology adoption such as Bamire et al. (2002) and Akinola et al. (2008).

The coefficient of level of formal education of the household head was positive and statistically significant at 1% level. Education which is the ability of respondents to read and/or write increased adoption probability in the study area. Educated farmers are more analytical and observe easily the obvious advantages of new technologies. The positively significant influence implies that the higher the level of formal education, the higher the probability of adoption of improved maize varieties. This agreed with previous studies on technology adoption such as Lemchi et al. (2005) and Nnadi and Akwiwu (2008).

The coefficient of farming experience was positive and statistically significant at 1%. This implies that the more the years of experience in farming, the higher the likelihood of adopting improved maize varieties. Increased years of farming experience furnished farmers with more knowledge that increased their rationality in the use of innovations. This is in consonance with Nnadi and Amaechi (2007) that explained increased years of farming experience as a valuable asset in adoption decision making. However, this contradicted Bamire et al. (2002) that indicated that the older the farmer, the less likely he is to adopt new ideas as he gains more confidence in his accustomed ways and method because experience affects individual mental attitude to new ideas differently and influences adoption in several ways.

Household size was statistically significant and positively related to the probability of adoption at 5%. The direct relationship implies that large household size predisposes adoption of improved maize varieties. This may be due to the fact that large household size is assumed to be an indicator of labour availability and that such a household would like to improve its food security. This is in agreement with the study conducted by Nnadi and Akwiwu (2006).

Total farm size of the respondent was positive and had statistically significant influence at 1% level on the adoption of improved maize varieties. Nowak (1987) argues that larger farm owners have more flexibility in their decision making, greater access to discretionary resources, and more opportunities to use new practices on a trial basis with more ability to deal with risk. This could be explained by the fact that large farm size presupposes large farm asset. Thus, farmers who had more assets had more dispositions to adopt new technologies than those who had less. A similar result was reported by Nkonya et al. (1997) and Aklilu and De Graaf (2007).

Distance of the farmers' village to market center was found to be statistically significant with negative relationship at 5% level. The negative relationship implies that the farther the distance between farmers' village and the market center, the lower the probability of adopting improved maize varieties. This may be due to the fact that relative proximity to market reduces marketing cost. Therefore, longer distance to market, better yield associated with improved maize and huge associated marketing costs might be responsible for the lower probability of adoption decision. This result is consistent with other studies such as Tesfaye and Alemu. (2001) and Kebede (2006).

With respect to other variables, none was statistically significant. Off farm income, membership in association, land security and improved seed availability were positive as expected. However, access to credit and frequency of extension services paradoxically had negative influence on adoption of improved maize varieties in the study area.

Factors determining use intensity of improved maize varieties

The result of the truncated regression model for the use intensity of improved maize varieties upon adoption showed that age of the household head was statistically significant at 5% level but had a negative influence on the hectares of land cultivated (Table 5). This implies that the older the respondents, the smaller the land area planted with improved maize varieties. An increase in the age of the respondent reduced the use intensity of improved maize varieties. This agreed with previous studies on technology adoption, such as Bamire et al. (2002) and Akinola et al. (2007).

The level of formal education was positive and statistically significant at 5% level. This outcome was expected and conforms to the study conducted by Bekele and Mekonnen (2010). An increase in the level of education of the respondents increased the intensity of use of improved maize varieties. The more educated a farmer, the more he is to diagnose and observe the benefits of new technologies, hence, more hectares of land was put into cultivation of improved maize varieties.

Household size which is an indication of labour availability had a positive influence on the intensity of use of improved maize varieties and was significant at 1% level. As the household size increased, the size of land in hectares planted with improved maize varieties increased. An increase in household size of the respondents increased the use intensity of improved maize varieties. This outcome was in line with the expectation as the sign could be either positive or negative (Zeller et al., 1998). The larger the household, the more the pressure to ensure food security, hence, the cultivation of more hectares of land with improved maize varieties.

The coefficient of farm size of the respondent was positive and statistically significant at 5%. As expected, the larger the farm size, the more the areas planted with improved maize varieties. An increase in the farm size of the respondent increased the areas planted with

Table 5. Truncated regression estimates for use intensity of improved maize varieties.

Variable	Coefficient	Standard error	p-value	t-value
Age	-0.0615**	0.0266	0.021	2.31
Level of education	0.0833**	0.0367	0.024	2.27
Farming experience	0.0223	0.2625	0.395	
Household size	0.2446***	0.0877	0.000	2.79
Farm size	0.0675**	0.0269	0.013	2.51
Access to credit	0.0675	0.3336	0.013	
Seed availability	-1.1708	1.0051	0.245	
Distance to market	-0.0610	0.0542	0.262	
Extension contact	3.9693***	0.3793	0.000	10.46
Land security	0.2037	0.3837	0.0596	
Off farm income	0.0046**	0.0021	0.030	2.19
Membership in association	0.4849**	0.1969	0.014	2.46
Constant	-2.0506	1.6209	0.207	
Sigma	2.3317	0.1433		
LR chi2 (1)	238.82			
Prob > chi2	0.0000			
Log likelihood	-418.4937			

*** = 1% significance, ** = significance at 5%. Source: Data Analysis, 2011.

improved maize varieties. This agreed with Gebremedhin and Swinton (2003) and Kabubo-Mariaura et al. (2010).

Frequency of extension service contact was a positive and statistically significant variable in determining intensity of use at 1% level. Households that had regular contacts with extension agents are more enlightened through advisory services and therefore appreciate the more, benefits of a new technology. An increase in frequency of contact with extension agent increased the intensity of use of improved varieties. This finding agrees with Knowler and Bradshaw (2007).

Coefficient of off farm income was statistically significant positively in determining the use intensity of improved maize varieties at 1% level. This implies that the more the income realized from off farm engagements, the more the hectares of land cultivated with improved maize varieties. Increase in the income from off farm engagements increased the intensity of use of improved maize. This may be due to the fact that more money is available to acquire more hectares of land, more seeds and associated inputs. The positive sign was in line with "a priori" expectation and agreed with Lapar and Pandey (1999).

Membership in association was statistically significant and had a positive relationship at 5% level. The more the respondents join farmer's association, the more the hectares of land cultivated with improved maize varieties. This implies that information related to procurement and benefits of improved maize varieties were discussed and disseminated at farmers' meetings. An increase in the number of respondents in farmers' association increased the use intensity of improved maize varieties. This finding is consistent with the expectation and agrees with Akinola et al. (2008).

However, farming experience, access to credit and land security variables had positive influence on intensity of use of improved maize varieties as anticipated but statistically insignificant. On the other hand, improved seeds availability and distance of farmers' village to market were also statistically insignificant with negative relationship.

Conclusion

The study revealed that the level of awareness and adoption of improved maize varieties were high in the study area and that IMV adoption was profitable. The study concluded that the adoption decision of improved maize varieties was driven by a host of socioeconomic factors such as age, level of education, farm size, farming experience, household size and distance to the nearest market center while insight into key socioeconomic factors influencing use intensity was provided which included age, level of education, farm size, household size, frequency of contact with extension agent, off farm income and membership in associations. Policies should target strengthening maize farmers to have access to improved education and frequent extension service contact for aiding acceptance and dissemination of agricultural technology information which has the potential to increase the rate of adoption and intensity of use of improved maize varieties.

Conflict of Interest

The authors have not declared any conflict of interest.

REFERENCES

Akinola AA, Arega DA, Adeyemo R, Sanogo D, Olanrewaju AS, Nwoke C, Nziguheba G, Diels J (2008). Determinants of adoption and intensity of use of balanced nutrient management systems technologies in the Northern guinea savanna of Nigeria. Afr. Assoc. Agric. Econ. Confer. Proceed. pp.111-118.

Aklilu A De Graaff J (2007). Determinants of adoption and continued use of stone terraces for soil and water conservation in an Ethiopia highland watershed. Ecol. Econ. 61: 294-302.

Asnake F, Gowda CL, Demisie L, Legesse M, Mcs Batiln D, Gaur PM, Senait RM (2005). Adoption of Improved chick pea varieties in Ethiopia. Ethiopian Agricultural Research Organization, Addis Ababa, Ethiopia.

Babatunde RO, Fayode SB, O AA (2008). Fadama maize production in Nigeria: Case study from Kwara state. Res. J. Agric. Biol. Sci. 4(5):340-345.

Bamire AS, Fabiyi YL, Manyong VM (2002). Adoption pattern of fertilizer technology among farmers in the ecological zones of Southwestern Nigeria: A Tobit analysis. Austr. J. Agric. Res. 53:901–910.

Bekele G, Mekonnen A (2010). Investment in land conservation in Ethiopia highlands: A household plot-Level analysis of the roles of poverty, tenure security, and market incentives. Environ. Dev. 10(9):1-41.

Cragg J (1971). Some Statistical models for limited dependent variable with application to the demand for durable goods. Econometrica, 39:829-844.

De Graff J, Amsalu A, Bodnar F, Kessler A, Posthumus H, Tenge A (2005). Adoption of soil and water conservation measures. Paper presented at EFARD Conference on Agricultural Research for Development: European Responses to Changing Global Needs. Zurich, Switzerland.

Gebremedhin B, Swinton SM (2003). Investment in soil conservation in Northern Ethiopia: The role of land tenure security and public programmes. Agric. Econ. 9:69-84.

Greene WH (2000). Econometric Analysis. Fifth Edition. Prentice-Hall International, Inc, USA.

Idachaba FS (1994). The Dilemma of fertilizer Subsidies in African Agriculture. Paper Delivered at the International Fertilizer Industry Association (IFA), Regional Conference for Africa, Dakar, Senegal.

Kabubo-Mariaura J, Linderhof V, Kruseman G, Atieno R, Mwabu G (2010). Household welfare, investment in soil and water conservation and tenure security: Evidence from Kenya. Environ. Dev. Econ. 15(2):1-83.

Kamara AY, Kureh I, Menkir A, Kartung P, Tarfa B, Amaza P (2006). Participating on-farm evaluation of the performance of the Drought-Tolerant maize varieties in the Guinea Savannahs of Nigeria. J. Food Agric. Environ. 4(1):192-196.

Kebede B (2006). Farmers' perception and determinants of land management practices on Ofla Woreda, Southern Tigray, Ethiopia. Unpublished M.Sc. Thesis presented to School of Graduate Studies of Haramaya University.

Knowler D, Bradshaw B (2007). Farmers' adoption of conservation agriculture: A Review and Synthesis of Recent Res. Food Policy 32:25-48.

Kudi TM, Bolaji M, Akinola MO, Nasa'l DH (2011). Analysis of adoption of improved maize varieties among farmers in Kwara State, Nigeria. Int. J. Peace Dev. Stud. 1(13):8-12.

Lapar MLA, Pandey S (1999). Adoption of soil conservation: The case of Philippine Uplands. Agric. Econ. 21:241-256.

Lemchi JI, Tshiunsa M, Onyeka U, Tenkouano A (2005). Factors driving the adoption of cooking banana processing and utilization methods in Nigeria. Afr. J. Biotechnol. 4(11):1335-1347.

Minten B, Barrett BC (2008). Agricultural Technology, productivity, and Poverty in Madagascar. World Dev. 36(35):797-822.

Nkonya E, Schroeder T, Norman D (1997). Factors affecting adoption of improved maize seed and fertilizer in Northern Tanzania. J. Agric. Econ. 4:1-12.

Nnadi FN, Akwiwu CD (2006). Strategies for overcoming economic marginalization among rural women in Imo State, Nigeria. Int. J. Natural Appl. Sci. 2:1-6.

Nnadi FN, Amaechi CO (2007). Rural sociology for Development Studies, Owerri. Custodab Investments.

Nnadi FN, Akwiwu CD (2008). Determinants of youths participation in rural agriculture in Imo state, Nigeria. J. Appl. Sci. 8(2):328-333.

Nowak PJ (1987). The adoption of conservation technologies: Economics and Diffusion explanations. Rural Sociol. 42:208-220.

Osun State Government. (2004). Resources of Osun State. Available at: http//www.osunstate.gov.ng/resources2.htm

Paudel GS, Thapa GP (2004). Impact of social, institutional and ecological factors on land management in mountain watershed of Nepal. Appl. Geogr. 24:35-55.

Tesfaye Z, Alemu H (2001). Adoption of improved maize technologies and inorganic fertilizer in north western Ethiopia. Ethiopia Agricultural Research Organisation. Research report No. 40 Addis Ababa, Ethiopia. P 51.

Tura M, Dejene A, Tsegaye W, La Rovere, R, Mwangi W, Mwabu, G. (2009). The Disadoption of Agricultural Technologies: The case of hybrid maize seeds in Central-western, Ethiopia. CIMMYT/SG2000 Nippon IA Research Report. P. 9.

Zeller M, Diagne A, Mataya C (1998). Market access by smallholder farmers in Malawi: implication for technology adoption, agricultural productivity and crop income. Agric. Econ. 19:219-229.

Utilisation of sunflower crop among smallholder farmers in sub-Saharan Africa: Evidence from Nigeria and Botswana

Torimiro D. O.[1], Yusuf O. J.[2], Subair S. K.[2], Amujoyegbe B. J.[3], Tselaesele N.[1] and Ayinde J. O.[3]

[1]Department of Agricultural Economics, Education and Extension, Botswana College of Agriculture, Gaborone.
[2]Department of Agricultural Economics and Extension Services, Kwara State University, Malete, Nigeria.
[3]Department of Crop Production and Protection, Obafemi Awolowo University, Ile-Ife, Nigeria.

Sunflower is a crop with high oil and economic values which portends its great potential for enhancing agricultural productivity and poverty alleviation among smallholder farmers. These notwithstanding, optimum benefit of the sunflower value chain have not been adequately harnessed. Conceived with the backdrop of lack of awareness about the commercial, nutritional and medicinal potentials of sunflower among the rural farm families, the study assessed the utilisation of sunflower among smallholder farmers in two southwestern states (Ogun and Ekiti) of Nigeria, and two districts (southeast and Kgatleng) of Botswana in sub-Saharan Africa. A total of two hundred smallholder farmers were randomly chosen from four purposively selected farming communities in southwestern Nigeria and Botswana districts. Results showed that 49 and 84% of the respondents were aware about the sunflower crop in Nigeria and Botswana, respectively, while only 10 and 25% cultivated the crop, respectively. The results further showed that all those (25%) who cultivated the crop in Botswana utilised it as animal feed only, while their Nigerian counterpart utilised it for variety of purposes. While 6 and 7% utilised it as animal feed and manure/fertiliser production, respectively, in Nigeria, 2, 5 and 8%, respectively, used it for cake/snack production, and traditional and ornamental purposes. Yet, another 6% cultivated the sunflower crop for seed oil extraction. The study concluded that, although, higher proportion of smallholder farmers from Botswana knew about and cultivated the sunflower crop, yet, their Nigeria counterpart utilised the crop more. The need for continuous popularization of the production and utilisation of sunflower crop among smallholder farmers was therefore recommended in both Bostwana and Nigeria.

Key words: Sunflower utilisation, smallholder farmers, value chain.

INTRODUCTION

Smallholder farmers are vital for agriculture and rural economy of every developing nation especially in the sub-Saharan Africa and other parts of Africa continent where they largely rely on rain-fed agriculture. The smallholder farmers are generally marginal and sub-marginal farm households that own or/and cultivate less than 2.0 ha of land and are characterized by a narrow spectrum of education, mass illiteracy with high level of poverty and poor standard of living (Adedoyin et al., 1996; Ekong, 2003). They usually produce crops in mixed intercropping, which often result in persistent and continuous poor yield in agricultural production leading to poor financial return thereby exacerbating their poverty condition (Amujoyegbe et al., 2011).

The sunflower, *Helianthus annuus* L., is one of major crops of global importance native to the United States. Archeological evidence suggests that Native Americans began cultivating and improving the sunflower as early as 2300 B.C. (Rindels, 1996). Although, alien to the tropical and subtropical Africa, it could be suitable for planting in most areas of Nigeria and Botswana, considering its tolerance to drought and great variety of soil (CAADP, 2008; Amujoyegbe et al., 2012).

The sunflower crop has a lot of potentials and has been utilised for a wide variety of purposes from time immemorial. For instance, 'native Americans in the U.S. have been using wild sunflower for food and medicine for at least 8,000 years, the seeds were usually roasted and ground into a fine meal for baking or used to thicken soups and stews. "Seed-balls", similar to peanut butter, made from sunflower butter made a convenient carry-along food for traveling. Roasted sunflower hulls were steeped in boiling water to make a coffee-like beverage. Dye was extracted from hulls and petals. Face paint was made from dried petals and pollen. Oil, extracted from the ground seeds by boiling, provided many tribes with cooking oil and hair treatment. Medicinal uses included everything from wart removal to snake bite treatment to sunstroke treatment' (Rindels, 1996). Also, by adding sunflower to an existing crop rotation, pest problems such as corn borer or soybean cyst nematode can be reduced (Myers, 2002). A field of sunflowers in bloom is a striking sight, and many farmers remark about the pleasure they, and people passing by, get from seeing the flowers (Myers, 2002).

According to Johnson et al. (2009) cited in McClure et al. (2013), there are basically two types of sunflower hybrids: the oilseed type that is grown for vegetable oil and the confection or non-oilseed type. The oilseed type has a higher oil composition in the seeds than the non-oilseed type. Oilseed types produce smaller black seeds and the oil is primarily used for human consumption. The oilseed types are also marketed as a sole ingredient for birdseed or in birdseed blends. The non-oilseed type produces the large, striped seeds that are used for human food snacks in the shell or as kernels, in baking ingredients, and in birdseed mixes (Johnson et al., 2009). Currently, production of edible oil in sub Saharan Africa has been largely from oil palm and soybean, which remain inadequate for the ever growing teeming population (FAO, 2003). Sunflower, which is the fourth leading oil consumed (7.8%) behind (oil) palm (31.8%),soybean (30%) and rapeseed (14%) (ASA, 2008) has not been adequately exploited for its commercial edible oil potential in most African countries, especially Nigeria and Botswana. Even in South Africa, for instance, where approximately 95% of all sunflower seed produced is processed for sunflower oil production, the dilemma of the sunflower market is that not enough sunflower seed is produced locally for the oil industry. With the total demand for sunflower seed, resulting from the total demand of sunflower oil, increased to over 1 million tonnes, South Africa has become a huge importer of sunflower crude oil in the last decade (FPMC, 2003).

Enhancing the value chain can improve the livelihood of smallholder farmers, ensure competitiveness in the global market, and ultimately contribute to economic growth. However, imperfections along the chain continue to widen the disparity between farm gate and retail prices, leaving poor farmers with the least value (IFPRI, 2008). Despite the promising potentials of the sunflower which portends great potential for enhancing agricultural productivity and poverty alleviation among smallholder farmers, optimum benefit of the sunflower value chain have not been adequately harnessed. Conceived with the backdrop of lack of awareness about the commercial, nutritional and medicinal potentials of sunflower among the rural farm families, the study assessed the utilisation of sunflower among smallholder farmers in two southwestern states (Ogun and Ekiti) of Nigeria, and two districts (southeast and Kgatleng) of Botswana in sub-Saharan Africa. Specifically, socioeconomic characteristics of smallholder farmers were described, and their awareness about and utilisation of the sunflower crop were assessed.

METHODOLOGY

The study was conducted in two Southwestern states (Ogun and Ekiti) of Nigeria and two districts (Southwest and Kgatleng) of Botswana. Ogun state is an important state where the principal partner of the project is located (Crawford University, Igbesa) while Ekiti state is the site of the Faculty of Agriculture of the Crawford University where the Sunflower Extension Centre is located. Ogun state which is an agrarian state with extensive industrial encroachment has a population of about 2, 338,570 (NPC, 2006) and occupies a land area of 16, 762 km^2. The coordinates of the state is 7°00'N 3°35'E/ 7°N 3.583°E with transitional savanna vegetation. It experiences approximately eight months (March – October) of bimodal rainfall and five months (November – March) of dry season each year with slightly irregularity in the rainfall distribution annually. On the other hand, Ekiti state is an inland state with a total land area of 9, 251 km^2 and a population of 1,628,762 (NPC, 2006). The state falls within 7°30'N 4°30'E/ 7.5°N 4.5°E, and the vegetation is rainforest with some patches of guinea savanna. It experiences approximately eight months (March – October) of bimodal rainfall and four months (November – February) of dry season each year with slightly irregularity in the rainfall distribution yearly.

Botswana, where the second institution (Botswana College of Agriculture) is located is predominantly flat, tending toward gently rolling tableland. Botswana is dominated by the Kalahari Desert, which covers up to 70% of its land surface. It covers about 600,370 km^2 (231,804 sq mi) and lies between latitudes 17° and 27°S, and longitudes 20° and 30°E. According to 2011 population Census, Botswana has a population of about 2,024,787 people (BPC, 2012)

Two farming communities were purposively selected from each of the countries because of their proximity to the participating institutions. These communities were Adie-Owe Igbesa in Ogun State and Oye-Ekiti, Ekiti State of Nigeria; and Oodi in Kgatleng District and Mogobane in South-East District of Botswana. Fifty farmers were also purposefully selected at each of the locations making a total of 200 participants for the study within a time frame of twelve months. Structured interview schedule, which was developed by the researchers using literature and professional experience, was used for data collection. Reliability test of the instruments was determined using a pre-test technique. With correlation coefficient (r = 0.71) obtained from the analysis, the research instrument was adjudged reliable since the correlation value of 0.7 and above are considered as satisfactory or good for a test-retest reliability (Statistics.com, 2014). The data collected were processed using SPSS version 14. Descriptive statistics such as percentages, means, line graph and charts were used to analyze the data.

RESULTS AND DISCUSSION

Socio-economic characteristics of the respondents

Results in Table 1 showed that 82% of the respondents from Nigeria were male, while 18% were female. Conversely, 56% were female in Botswana, while 44% were male. The results indicated that more female were engaged in smallholder farming in Botswana compared to Nigeria, where more male were involved. Majority (72.0%) of the respondents from both countries were aged between 31 and 60 years old. Mean age of respondent from Nigeria was 46.4 years, with standard deviation of 12.45 while mean age of respondents from Botswana was 56.13 with standard deviation of 10.85. The results thus indicated that respondents from Botswana were relatively older than their Nigerian counterparts. Similarly, majority (83% Nigerian and 86% Botswana) were married, while few others were either single or separated from their spouses. Majority (89% Nigerian and 95% Botswana) were Christians while very few were of the Islamic faith.

Results in Table 1 further showed that in Nigeria, 89% of the respondents could read and write, while 52% of the respondents from Botswana could read and write. Also, while 9% of the Nigerian could neither read nor write, 38% Botswana could neither read nor write. The results indicated that respondents from Nigeria had higher literacy level than their Botswana counterpart. These results have implications for educational techniques and approaches to adopt in continuous popularization of sunflower crop in both countries, in order not side line the unlettered ones amongst the smallholder farmers.

Similarly, in Nigeria, 72, 58 and 53% of the respondents belonged to religious association, community development association and farmers' association, respectively. While half (50%) of the respondents belonged to cooperative societies, 41 and 30% belonged to social organization and political group, respectively. Conversely, while 40% of the respondents from Botswana belonged to religious association, very few (less than 10%) were members of other associations. The results indicated that level of association membership was higher amongst Nigerian smallholder farmers compared to their Botswana counterpart.

Results in Table 2 showed that majority of the respondents (75.0% Nigerian and 87.0% Botswana) had farm size below 20 acres, while few cultivated above 20 acres in both countries. Mean farm size was 14.45 acres in Nigeria and 13.1 acres in Botswana. In addition, 59% of respondents from Nigeria acquired their farm land through inheritance, while 26% cultivated rented land area. Conversely, in Botswana, 56% of the respondents acquired their farm land through gift. The results indicated that respondents from both countries own sizeable portion of cultivable land area, from which they might dedicate some portion to sunflower crop, if they are convinced about its economic importance and become interested in its cultivation.

Furthermore, results in Table 2 showed that 61 and 28% of the respondents from Nigeria engaged in mixed cropping and mixed farming, respectively, while 11% engaged in mono cropping. In Botswana, however, 55% engaged in mixed farming, while 40 and 5%, respectively, engaged in mixed cropping and mono cropping. The results indicated that respondents from both countries already engaged in 'multi-farming' system, and therefore the integration of sunflower cultivation into the existing system could significantly enhance smallholder farmers' productivity.

Results in Table 3 showed that 49% of the respondents from Nigerian indicated they know about sunflower crop, while 84% of their Botswana counterpart knew about the crop. Results further showed that 24 and 26% of the respondents indicated parent and co-farmer as their source of awareness about sunflower in Nigeria, while 49 and 39% of their Botswana indicated these same sources, respectively. 17 and 8% indicated extension agent as source of awareness in Nigeria and Botswana, respectively, while very few (7% and below) indicated other sources friends, mass media and market. These results implied that parents and co-famers were most prominent ways through which respondents knew about the sunflower crop. The results underscored the need for popularization of sunflower amongst smallholder famers through use of mass media, radio, for example, which will gain wider coverage and audience.

Sunflower cultivation and utilization

Results in Figure 1 showed that, despite the proportion of respondents that knew about the sunflower crop as earlier discussed, only 10 and 25% of the respondents cultivated the crop from Nigeria and Botswana, respectively.

Table 1. Respondents' socio-economic characteristics.

Variable	Nigeria (n = 100)		Botswana (n = 100)	
	Frequency	Percentage	Frequency	Percentage
Sex				
Male	82	82.0	44	44.0
Female	18	18.0	56	56.0
Age (Years)				
Below 30	13	13.0	0	0.0
31 – 60	72	72.0	72	72.0
Above 60	15	15.0	28	28.0
Mean	46.40		56.13	
Standard deviation	12.45		10.85	
Marital status				
Single	13	13.0	7	7.0
Married	83	83.0	86	86.0
Divorced/widowed/Separated	4	4.0	7	7.0
Religion				
Islam	8	8.0	5	5.0
Christianity	92	92.0	95	95.0
Literacy level				
Can read and write	89	89.0	52	52.0
Can read but can't write/Can write but can't read	2	2.0	10	10.0
Can neither read nor write	9	9.0	38	38.0
Association membership				
Religious association	72	72.0	40	40.0
Village council	48	48.0	0	0.0
Cooperative society	50	50.0	9	9.0
Community development				
Association	58	58.0	3	3.0
Social organization	41	41.0	4	4.0
Political group	30	30.0	0	0.0
Farmers' association	53	53.0	9	9.0
Farming experience (Years)				
Below 20	31	31.0	12	12.0
21 – 40	23	23.0	19	19.0
41 – 60	4	4.0	2	2.0
Above 60	0	0.0	2	2.0
No response	42	42.0	65	65.0
Mean	21.19		27.07	

Source: Computed from field survey, 2013.

This indicated that low cultivation of sunflower amongst smallholder farmers in both countries. Results in Figure 2 showed that all those (25%) who cultivated the crop in Botswana utilised it as animal feed only, while their Nigerian counterpart utilised it for a variety of purposes. While 6 and 7% utilized it as animal feed and manure/fertiliser production, respectively, in Nigeria, 2, 5 and 8%, respectively, used it for cake/snack production, and traditional and ornamental purposes. Yet, another 6% cultivated the sunflower crop for seed oil extraction. The results indicated that the potentials for enhancing agricultural productivity and reducing poverty amongst

Table 2. Respondents' farm size, land acquisition and type of farming.

Variable	Nigeria (n = 100)		Botswana (n = 100)	
	Frequency	Percentage	Frequency	Percentage
Farm size (Acres)				
Below 20	75	75.0	87	87.0
21 – 40	6	6.0	13	13.0
41 – 60	4	4.0	0	0.0
Above 60	1	1.0	0	0.0
No response	14	14.0	-	-
Mean	**14.45**		**13.1**	
Land acquisition type				
Inheritance	59	59.0	39	39.0
Rent/lease	26	26.0	3	3.0
Outright purchase	9	9.0	2	2.0
Gift	8	8.0	56	56.0
Type of farming				
Mono cropping	11	11.0	5	5.0
Mixed cropping	61	61.0	40	40.0
Mixed farming	28	28.0	55	55.0

Source: Computed from field survey, 2013.

Table 3. Awareness about the sunflower crop.

Variable	Nigeria (n = 100)		Botswana (n = 100)	
	Frequency	Percentage	Frequency	Percentage
Do you know about the crop?				
Yes	49	49.0	84	84.0
No	51	51.0	16	16.0
Source of awareness*				
Parent	24	24.0	49	49.0
Co-farmer	26	26.0	30	30.0
Friends	7	7.0	5	5.0
Extension agent	17	17.0	8	8.0
Mass media	7	7.0	3	3.0
Market	1	1.0	3	3.0

*Multiple responses applicable, Source: Computed from field survey, 2013.

smallholder farmers through sunflower utilisation have not been well harnessed in both countries.

CONCLUSION AND RECOMMENDATION

The study concluded that higher proportion of smallholder farmers from Botswana were aware about and cultivated the sunflower crop. However, their Nigeria counterpart utilised the crop more. Notwithstanding, optimum benefit that could be derived from utilisation of sunflower have not been adequately harnessed in both countries. The need for continuous popularization of the production and utilisation of sunflower crop among smallholder farmers was therefore recommended in both Bostwana and Nigeria.

Conflict of Interest

The authors have not declared any conflict of interest.

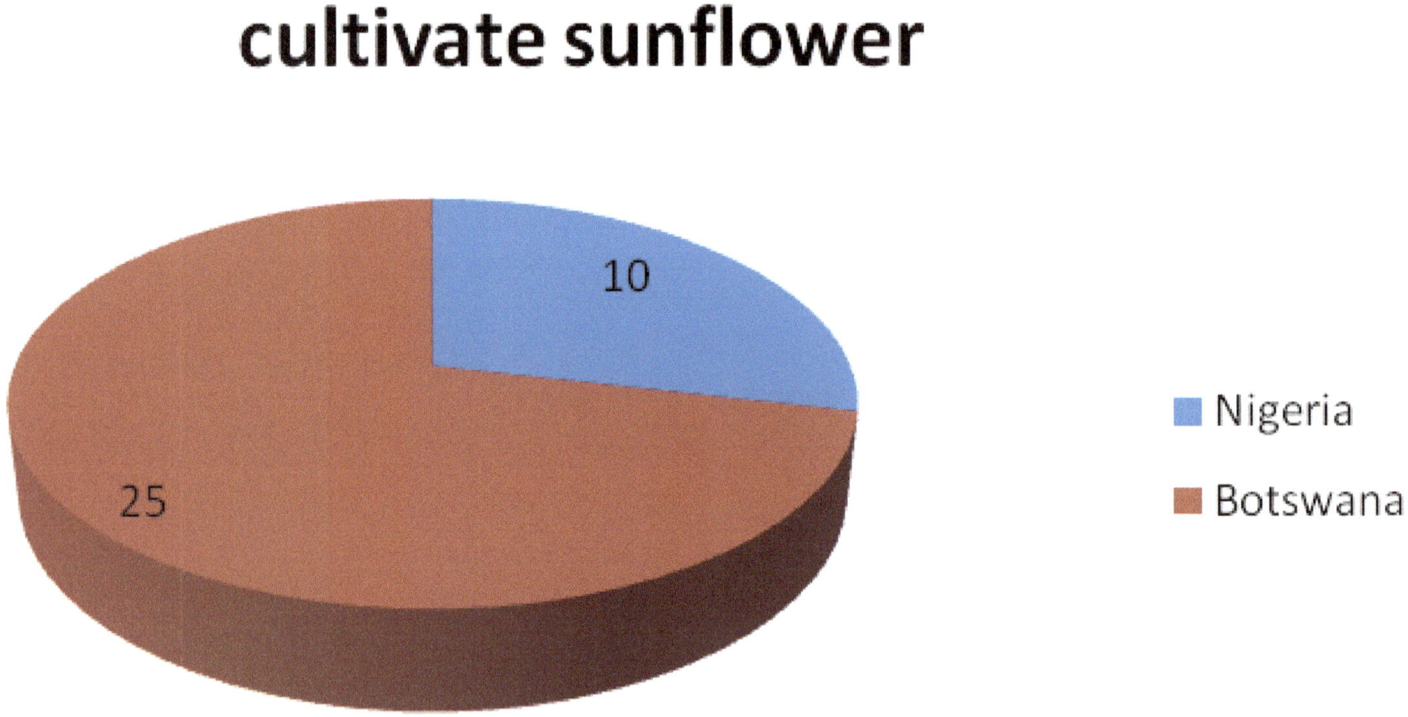

Figure 1. Percentage distribution of respondents that cultivated sunflower crop. Source: Computed from field survey, 2013.

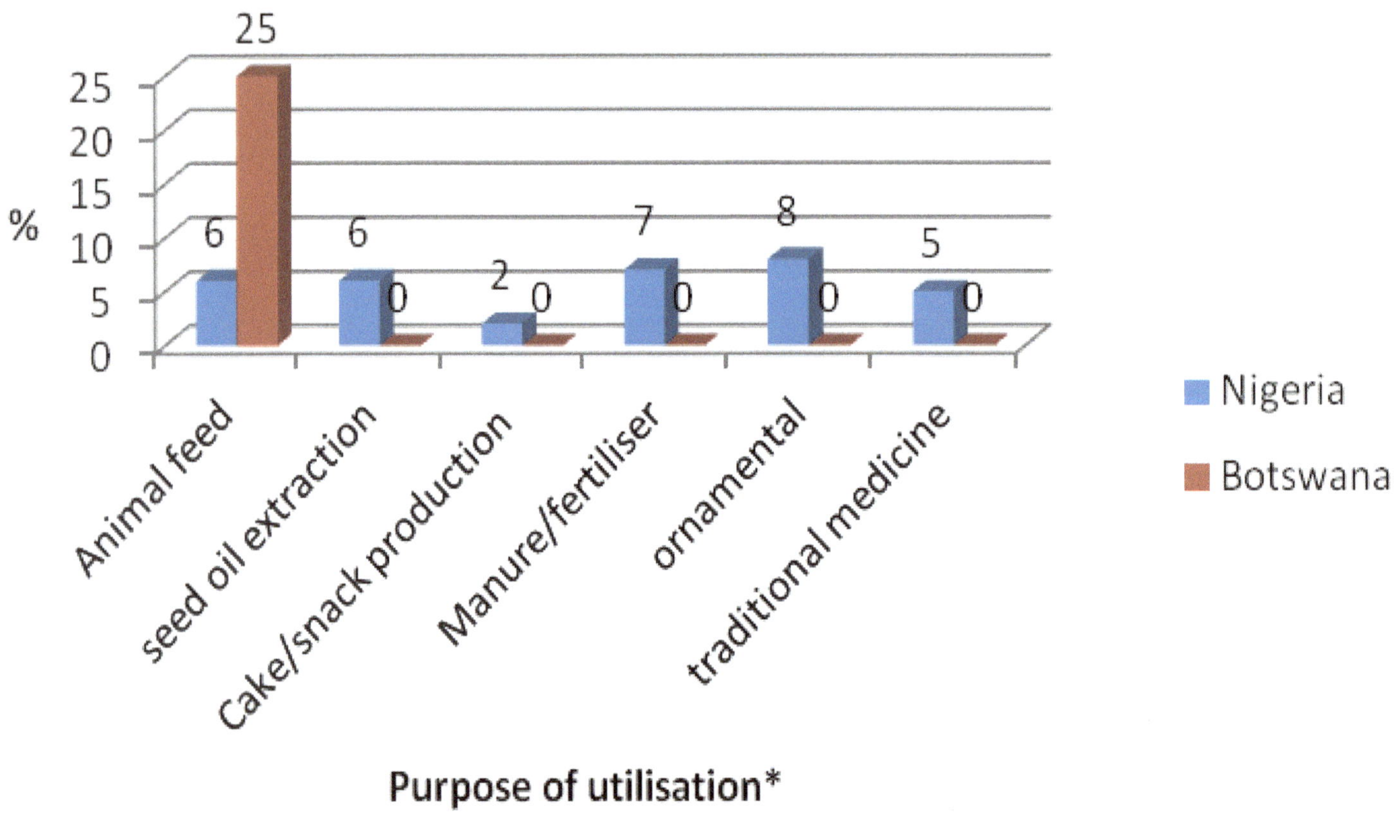

Figure 2. Percentage distribution of respondents that utilised sunflower crop. *Multiple responses applicable. Source: Computed from field survey, 2013.

REFERENCES

Adedoyin SF Torimiro DO Ogunbanwo AS (1996). Technologies for Leafy Vegetable Production: A Case Study of Peri-urban Farmers in Ojo Area of Lagos State. Proceedings of the 14th Conference of HORTSON, Ago-Iwoye.

American Soybean Association (ASA). (2008). Soy Stats: World vegetable oil consumption (2007). American Soybean Association., Online material available at *http://www.soystats.com/2008/page 35.htm* (Accessed July 24, 2013)

Amujoyegbe BJ, Torimiro DO, Ige MT, Subair SK, Tselaesele N, Balole TV, Batlang U (2012). ICT Support for Popularization of Sunflower Intercropping with Arable Crops: A Case Study of Nigerian and Botswana. Agric. J. 7:230-235.

Amujoyegbe BJ, Torimiro DO, Ige MT, Subair SK, Tselaesele N, Balole TV, Batlang U (2011). Popularization of Sunflower Intercropping With Arable Crops for Food Security and Poverty Alleviation: A Case Study of Nigerian and Botswana Higher Educational Institutions. Proceedings of the international stakeholders' meeting on e-agriculture for farm productivity and poverty alleviation in Africa held in the centre for in-service and continue education, Botswana College of Agriculture, Gaborone, Botswana from Tuesday 27 to Thursday 29, September, 2011

BPC (2012). Botswana Population Census Project Document: Population and Housing Census. Gaborone, Central Statistics Office.

CAADP (2008). Pillar 3. Food supply and hunger, Comprehensive Africa Agricultural Development Programme (CAADP), Online document available at http://www.nepad-caadp.net/pillar-3.php (Accessed February, 5 2010 and used July 24, 2013).

Ekong EE (2003). Introduction to rural sociology. Dove publishers, Akwa-Ibom, Nigeria.

Food and Agriculture Organization (FAO) (2003). The state of food insecurity in the world.

Food Pricing Monitoring Committee (FPMC) (2003). The Sunflower cooking-oil seed supply chain. In: Value chain of dairy products. FPMC final report, Department of Agriculture, Republic of South Africa.

International Food Policy Research (IFPRI) (2008). Enhancing Commodity Value Chains in Nigeria. Nigeria Strategy Support Program (NSSP) Workshop Report No. NSSP 004. Online document available at www.ifpri.org (Accessed June 26, 2013).

Johnson JJ, Meyer RF, Krall JM, Shroyer JP, Schlegel AJ, Falk JS, Lee CD (2009). Agronomic Practices. In: High Plains Sunflower Production Handbook. MF-2384. Kansas State Univ. Manhattan, KS. pp. 1-4. http://www.ksre.ksu.edu/library/crpsl2/mf2384.pdf

McClure MA, Allen FL, Johnson RD, Heatherly LG (2013). Sunflower: An Alternative Crop for Tennessee Producers Production Guidelines and Tennessee Hybrid Trials. Online document available at *http://varietytrials.tennessee.edu* and *www.UTCrops.com* (Accessed July 1, 2013).

Myers RL (2002). Sunflower: A native oilseed with growing markets. Online document available at *www.jeffersoninstitute.org* (Accessed July 1, 2013).

Nigeria Population Commision (NPC) (2006). Population census of National Republic of Nigeria: Analytical report at the National level, pp. 22-54.

Rindels S (1996). Sunflower horticulture and home pest use Department of Horticulture Iowa State University of Science and Technology. Online document available at (Accessed July 1, 2013).

Statistics.com (2014). Glossary of statistical terms: test-retest reliability. Accessed online on 24th June, 2014 from http://www.statistics.com/glossary&term_id=867

Cropping system analysis of two agro ecological zones of Southwestern Nigeria

B. J. Amujoyegbe[1]* and O. S. Alabi[2]

[1]Department of Crop Production and Protection, Faculty of Agriculture, Obafemi Awolowo University, Ile-Ife, Osun State, Nigeria.
[2]Department of Agricultural Extension and Rural Development, Faculty of Agriculture, Obafemi Awolowo University, Ile-Ife, Osun State, Nigeria.

Proper understanding of cropping system within agro ecological zone is needed if adequate agricultural development programme will be planned. Therefore, the study looked into the cropping systems practiced by farmers in Southwestern Nigeria, described the cropping system and crop combination of farmers and also described and compared the socio-economic characteristics of farmers in the agro ecological zones studied. Ogun (derived savannah zone) and Osun States (forest zone) were purposively sampled while multi-stage sampling procedure was used to select 298 farmers in both States. Structured interview schedule consisting of open and close ended questionnaire in addition to focus group discussion was used to quantify the socio-economic status of the farmers. The average age of farmers in the two agro ecological zones were 45.00 ± 3.24 and 48.00 ± 2.98 for forest and derived savannah zones respectively. The average year of formal education of the farmers were 6.00 ± 2.74 years and 7.00 ± 2.11 years for the forest and derived savannah zones respectively. Also, the average farm size of the respondents was 0.52 ± 0.17 ha and 0.47 ± 0.13 ha for forest and derived savannah zones, respectively. There was a significant difference between the farm sizes of farmers in the two agro ecological zones ($P \leq 0.01$). About 43.00 and 34.00% apiece of respondents in forest and derived savannah zones practiced arable plus permanent crop farming system. Mixed cropping was the most preferred cropping system in the forest zone while mixed cropping and crop rotation were the dominant cropping systems in the derived savannah zone. The study therefore, concluded that cereal/tuber crop combination is the most practiced among farmers in the two agro ecological zones in Southwestern Nigeria which may affect the final productivity of the available land as the two crops combination depletes soil nutrient.

Key words: Farming system, cropping system, crop combination.

INTRODUCTION

Traditional cropping systems vary, since they have evolved in response to prevailing soil and climatic conditions and social and ethnological preferences (Okigbo, 1980; Kang, 1986). Traditional farmers often plant more than one crop species in a small patch of cleared and burnt land after several years of bush fallow.

*Corresponding author. E-mail: bamujo2002@yahoo.com, bamujo@oauife.edu.ng.

Intercropping, the practice of growing two or more crops simultaneously in the same field is common throughout the tropics. Major food crops in humid tropical regions include; plantain / banana (*Musa* spp), rice (*Oryza sativa* L.), and root crops (such as cassava (*Manihot esculenta*), yam (*Dioscorea* spp), sweet potato (*Ipomea batata*) and cocoyam in the humid zone; sorghum (*Sorghum bicolor* L.) Moench), maize (*Zea mays* L.) and cowpea (*Vigna unguiculata* (L.) Walp) in the sub-humid zone; millet (*Pennisetum glaucum* (L.) R.Br and cowpea in the semi-arid zone (Okigbo, 1980; Mudahar, 1986) with little external input and minimum livestock integration. Consequently, farm productivity is low and the cycle of low input, low yield and low income perpetuates poverty (IITA, 1998). As a result of the traditional cropping system, crop production expands into less fertile land. The poor productivity of this system and the emergence of unsustainable farming pose the greatest threat to food security in the sub-region (Woomer et al., 2001).

Crop production systems in West Africa and Nigeria in particular, involve several different cropping patterns. Among these systems is intercropping which, according to Andrews and Kassan (1976), involves growing two or more crops simultaneously on the same field and the ways in which the crops are managed. It was noted that this form of crop production is common among subsistence farmers who practice low-input agriculture (Ntare, 1990) and those who lack land and / or capital in combination with plentiful labour (Hildebrand, 1976). Apart from ensuring greater yield stability, intercropping has been found to be beneficial in reducing damage caused by pests and diseases (Andrews, 1974).

A farming system could be defined as a population of individual and homogeneous farm systems that have broadly similar resource bases, enterprise patterns, household livelihoods and constraints, and for which similar development strategies and interventions would be appropriate (FAO, 2001; Doppler, 2002). The main emphasis in farming systems analysis is a holistic approach, whereby household structure, gender, social networks, local institutions, information, policies, markets and all other factors as they affect the livelihood of the various homogeneous groups are brought into play. The analysis could be static, dynamic and comparative depending on the objectives of the study. The dynamic analysis is the analysis of the developments over time in each of the identified homogeneous farming groups; static is the comparison of the farming systems at a point in time while comparative analysis of the farming systems is the analysis of similarities and differences among the identified farming systems. All relies on parameters and interpretation that are supported by statistical calculations (Doppler, 2002). However, cropping system is a subset of farming system in which the available resources of the farmer is evaluated in the production of crops within a unit area of land. The resources combination in addition to the socio-economic condition of the farmer affects the final output of the farmer.

Given the wide range of climates and types of soil that exist around the globe, there are substantially different constraints to productivity from countries or site to another. Francis (1993) indicated that there are so many constraints of soil and climatological factors, and interaction among these factors make it difficult to establish the precise physical or climatical constraints for a given area without "on the ground" experience or survey. According to Francis (1993), the participation of farmers in this process can be crucial. There may be economic, political or social reasons why productivity is limited in a given situation and this reason may not be apparent to a researcher or extension specialist who is familiar only to crop and soil.

Many farmers in developing countries are struggling to produce in poor environmental conditions with few tools for coping with drought, pests, and disease. Agricultural research is needed to help these farmers reduce their risk, improve their productivity, and protect their natural resources. Therefore, a sound knowledge of the prevailing cropping systems in any agro ecological zone is pertinent to the generation of agricultural and food security policies for farming families. This ensures that innovations to be developed are suitable to meet the economic needs of the various homogeneous groups and have a high probability of being accepted by the farm families. The analysis of the cropping systems of two main agro ecological zones in Southwestern Nigeria is the focus of this study. The specific objectives of the study were to:

1. Describe and compare the socio-economic characteristics of farmers in the two agro ecological zones in the study areas and
2. Describe cropping system and crop combination in the two agro ecological zones in the study areas.

STUDY SETTING

The study was conducted in two agro-ecological regions of rain forest and derived savanna States of Southwestern Nigeria. Ogun and Osun States (Figure 1) are both agrarian States. Ogun State has extensive industrial encroachment with a population of about 2, 338,570 (NPC, 1991) and covers a land area of 16, 762 km^2. The coordinates of the state is 7°00'N 3°35'E/ 7°N 3.583°E with transitional savanna vegetation. It experiences approximately eight months (March to October) of bimodal rainfall and five months (November to March) of dry season each year with slightly irregularity in the rainfall distribution annually. Osun is an inland state with a total land area of 9, 251 km^2 and a population of 2,203,016 (NPC, 1991). The state falls within 7°30'N 4°30'E/ 7.5°N 4.5°E, and the vegetation is rainforest with some patches of Guinea savanna. It experiences approximately eight months (March to October) of bimodal rainfall and four months (November to February) of dry season each year with slightly irregularity in the rainfall distribution yearly. Weather data for Osun State was obtained from the meteorological stations of the Obafemi Awolowo University, Ile-Ife, located in the centre of Osun State while that of Ogun State was obtained from the meteorological station of the Agricultural Development Authority (ADP), a parastatal of Ogun State Government which is located within the State capital city of Abeokuta. The soil of Osun State is classified as ultisol (low base status forest soils), well drained, grayish brown to brownish red while, that of Ogun state is an underlain basement complex rock with quartz schist, coarse grained and fine grained granite and gneiss as parent material (Aiboni, 2001).

Sampling process

Ogun (derived savannah zone) and Osun States (forest zone) were purposively sampled while multi-stage sampling procedure was

Figure 1. Map of Nigeria showing the 36 States of Nigeria. (www.theodora.com/maps 2011).

used to select 298 farmers in both states. At the first stage, five Local Government Areas were purposively selected in each state for their agro-ecological representation. At the second stage, three communities were randomly selected from each of the selected LGAs in the states respectively to give 15 communities both for the rainforest and the derived savannah agro ecological zones. At the last stage, 10 farmers from the list of the extension agent covering each of the selected communities were randomly selected to give 150 farmers for each of the zones respectively. In all, 298 (148 from the rainforest zone and 150 from the derived savannah zone) respondents were interviewed for the study within a time frame of eight weeks. A structured interview schedule consisting of open and close ended questionnaire in addition to focus group discussion was used to quantify the socio-economic statues of the farmers. Simple description statistical techniques such as frequency counts, percentages, means and standard deviation were used to analyze and summarized the data collected.

Measurement of variables

Selected socio-economic characteristics of farmers like age, years of formal education, household size, farm size in hectare and distance of farm to home/farmstead were measured by their absolute values. Furthermore, farmers were asked to supply information on the type of farming system they practice and the type of cropping system they practiced. Farmers were also asked to supply information on the crop combination planted if they were not planting by monocropping. Responses to these questions were summarized and analyzed. Data were summarized using frequency, percentage and standard deviation and a two-sample t-test was used to compare the selected socio-economic characteristics of farmers across the two agro ecological zones of the study areas.

RESULTS AND DISCUSSION

Data in Table 1 revealed the characteristics of the respondents in the two agro ecological zones of the study areas. The data revealed that the mean age of the respondents was 45.00 ± 3.24 years in the forest zone and 48.00 ± 2.98 years in the derived savannah zone. The T-test analysis also established that there is no significant difference between the ages of respondents in the two agro ecological zones in the study areas. On

Table 1. Selected socio-economic characteristics of respondents in the study area.

Characteristics	Forest zone (n = 148)		Derived savannah zone (n = 150)		
	Percentage	Mean ±SD	Percentage	Mean ±SD	t_c
Age (Years)		45.00 ±3.24		48.00 ±2.98	1.85
Literacy levels (Years)		6.00 ±2.74		7.00 ±2.11	1.12
Primary occupation (%)	60.00		78.00		
Household size (number of persons)		5.00 ±2.45		6.00 ±2.30	0.65
Farm size (ha)		0.52 ±0.17		0.47 ±0.13	2.54**
Farm distance from home/farmstead (km)		4.32 ±1.42		1.50 ±0.80	1.99**

SD = standard deviation; t_c = T-test values between respondents in the forest zone and the derived savannah agroecosystem of South-western Nigeria; ** = Significant at 0.01 level, n = number of respondents.

literacy, the average years of formal education was 6.00 ± 2.74 years for respondents in the forest zone and 7.00 ± 2.11 years for respondents in the derived savannah zone. Further analysis also showed that there exists no significant difference among the years of formal education of the respondents in the two zones as seen in the result of the T-test conducted. About 60.00% of the respondents in the forest zone had agriculture as their primary occupation while 78.00% of the respondents in the derived savannah zone had agriculture as their primary occupation. The average household size of the respondents in the forest zone is 5.00 (± 2.45) people and 6.00 (± 2.30) in the derived savannah zone. The T-test conducted showed no significant difference in the average household size of the respondents in the two agro ecological zones in the study areas. The average household size in the two zones were in tandem to the stand of Ekong (2011) who established that average household size in Southwestern Nigeria is 6 people.

Significant differences were found in farm size between the two agro ecological zones ($p < 0.01$). The average farm size (ha) of the respondents in the forest zone is 0.52 ± 0.17 and 0.47 ± 0.13 in the derived savannah zone. Farm distance from home/farmstead also showed a significant different ($P < 0.01$) across the two agro ecological zones. The average farm distance from home/farmstead of respondents in the forest zone was 4.32 ± 1.42 km while that of derived savanna was 4.32 (± 0.80). It could be deduced that farmers in the derived savannah zone are closer to their farms compared to their counterparts in the forest zone in the study areas. There may be likelihood of the concept of urbanization which is more prominent in the forest zone. The settlements in the area are more of urban set up compare to the savanna that are scattered and dispersed (Morgan and Moss, 1965). Blench and Dendo (1994) also found that the expansion and invasion of the fulbe pastorals to the sub-humid and humid conditions of the Southwestern Nigeria might have contributed to the type of cluster and small settlements which existed in the zone.

The farm sizes of the respondents in the two agro ecological zones (Table 2) revealed that about 39.00% of the respondents in the forest zone had 0.51 to 1.00 ha of farmland while 43.00% of the respondents in the derived savannah zone had 0.51 to 1.00 ha of farmland. Also from the table, 21.00% (forest zone) and 22.00% (derived savannah zone) of the respondents had about 0.01 to 0.50 ha of farmland in the study area. About 18.00% of the respondents in the forest zone had 1.01 to 1.50 ha of farmland while 19.00% had 1.01 to 1.50 ha in the derived savannah zone study area. Only 10.00 and 5.00% of the respondents in forest zone and derived savannah zones, respectively, had above 2.01 ha of farmland in the study area. The average of the farmland in hectare of the respondents in the two agro-ecological zones was given as 0.52 and 0.47 for forest and derived savannah zones, respectively. This difference is statistically significant ($p < 0.01$) as seen from the Table 1.

Data in Table 3 showed the different farming systems in the two agro-ecological zones of the study areas. The table showed that the farming systems showed a similar pattern with little variations in the two agro-ecological zones. From the results, 43.00% of the respondents in the forest zone practice arable plus permanent cropping while 34.00% in the derived savannah zone practice arable plus permanent cropping. Also, 22.00 and 25.00% of the respondents in forest and derived savannah zones, respectively, practice arable cropping only while 12.00 and 13.00% of the respondents in forest and derived savannah zones, respectively, practice livestock plus arable plus permanent cropping in the study area.

Livestock plus arable cropping is another major farming system adopted in the derived savannah zone with about 18.00% of the respondents practicing it while the other major cropping system in the forest zone is permanent cropping only with about 10.00% of the respondents involved in it. The study therefore, further confirmed the earlier findings of Blench and Dendo (1994) concerning the intrusion of the fulbre pastorals in the zone hence, the pockets of variations in farming systems in the two agro ecological zones studied although, it could not establish whether these changes is due to the variation in ecology alone or whether it is also been influenced by other factors.

Data in Table 4 revealed the farm size and the cropping system of the respondents in the two agro-ecological

Table 2. Respondents farm size (ha) in the agro ecological zones in the study area.

Farm size (ha)	Forest zone		Derived savannah zone	
	Frequency	Percentage	Frequency	Percentage
0.01 - 0.50	32.0	21.00	34.0	22.00
0.51 – 1.00	57.0	39.00	65.0	43.00
1.01 – 1.50	27.0	18.00	28.0	19.00
1.51 – 2.00	18.0	12.00	15.0	10.00
Above 2.01	14.0	10.00	8.0	5.00
Total	148	100	150	100

Field survey.

Table 3. Respondents farming system in the agro ecological zones in the study area.

Farm size (ha)	Forest zone		Derived savannah zone	
	Frequency	Percentage	Frequency	Percentage
Arable cropping only	33.0	22.0	38.0	25.0
Permanent cropping only	15.0	10.0	5.0	3.0
Arable + Permanent cropping	64.0	43.0	50.0	34.0
Livestock only	5.0	4.0	10.0	7.0
Livestock + Arable	14.0	9.0	27.0	18.0
Livestock + Arable + Permanent	17.0	12.0	20.0	13.0
Total	148	100	150	100

Field survey.

Table 4. Average arable farm size and the cropping system of the respondents*.

Farm size (ha)	Forest zone (%)				Derived savannah Zone (%)			
	Sole cropping	Mixed cropping	Crop rotation	Shifting cultivation	Sole cropping	Mixed cropping	Crop rotation	Shifting cultivation
0.01 - 0.50	9	56	20	15	10	50	22	18
0.51 – 1.00	8	57	25	9	15	48	33	4
1.01 – 1.50	20	45	24	11	24	36	38	2
1.51 – 2.00	19	50	28	3	30	25	36	9
Above 2.01	22	40	31	7	32	24	35	8

Respondents in percentage.

zones. A pattern was discovered in the forest zone irrespective of the farm size, mixed cropping was mostly practiced by the respondents. The mixed cropping and crop rotation was the prominent cropping systems among respondents in the derived savannah zone. It was found that the respondents in the derived savannah zone practiced crop rotation cropping system. This revealed that while a single cropping system (mixed cropping) dominated the landscape in the forest zone, two (crop rotation and mixed cropping) cropping systems held sway in the derived savannah zone.

Data in Table 5 revealed the arable farm size and the proportion crop combination of respondents in the two agro ecological zones in the study area. Across the different farm sizes in hectare in the two zones, cereals/tuber crops combination is the most practiced by the farmers. This could be seen as 63.00 and 65.00% (0.01 to 0.50 ha), 67.00 and 62.00% (0.51 to 1.00 ha), 70.00 and 63.00% (1.01 to 1.50 ha), 68.00 and 70.00% (1.51 to 2.00 ha) and 71.00 and 70.00% (above 2.01 ha) in the two zones respectively plant cereal/tuber crops combination. This could be explained to mean that irrespective of the agro-ecological zone and the farm size cultivated in the study area, cereal/tuber crops combination is the most important option among the arable crop farmers. This finding from the crop combination of the arable farmers in both agro-ecological zones also revealed the state of things in Southwestern Nigeria about the cultivation of legumes which could actually reduce the rate of nutrient loss in the soil by fixing nitrogen in the soil. It then means that serious effort should be made in encouraging farmers in the two agro

Table 5. Average arable farm size and proportion crop combination of percentage respondents.

Farm size (ha)	Forest zone (%)					Derived savannah zone (%)		
	Cereal/cereal crops	Cereal/legumes crops	Cereal/tubers crops	Legume/tuber crops	Cereal/cereal crops	Cereal/legumes crops	Cereal/tubers crops	Legume/tuber crops
0.01- 0.50	6	18	63	13	10	10	65	15
0.51– 1.00	8	13	67	12	12	12	62	14
1.01– 1.50	5	15	70	10	12	19	63	6
1.51– 2.00	7	15	68	10	10	14	70	6
Above 2.01	9	12	71	8	5	15	70	5

Field survey.

ecological zones to plant legumes both for economic reason and that of soi improvement.

CONCLUSION AND RECOMMENDATION

The study concluded that the socio-economic characteristics of the farmers in the two agro ecological zones were invariably similar except for significant difference in their farm sizes and the distance from their farm to home or farmsteac. Also, the study revealed that 39 00 and 43.00% cf the respondents in forest and derived savannah agro ecological zones respectively had 0.5 to 1.00 ha farmland. This established that small land holding is still prevalent in the rural areas cf Southwestern Nigeria irrespective of the agro ecological zone. Lastly, cereal/tuber crop combination is the preferred crop combination by farmers in the two agro-ecological zones. This showed that despite the changes in ecology of the areas, the same cultural background shared actually influence the choice of crop combination of farmers. The study therefore, recommends that agricultural development programme compatible with farmers farming system, cropping system and crop combination should be planned to improve the standard of living of the rural people. Alsc, information needed to correct the wrong practices or crop combination should be given to the farmers so as to maximize agricultural productivity.

REFERENCES

Aiboni VU (2001). Characteristics and classification of soils of a representative topographyic locations in the University of Agriculture, Abeokuta. ASSET Series A. 1(1): 51-62.

Andrews DJ (1974). Response of sorghum varieties to intercropping. Exp. Agric. 10:57-63

Andrews DJ, Kassam AH (1976). Importance of multiple cropping in increasing world food supply In ed. R.I.Papendick, A. Sanchez and G.B. Triplett ASA Special Publication 27, Madison W.I. USA pp. 1-10

Blench R, Dendo M (1994). The expansion and adaptation of fulbe pastorialism to sub-humid and humid conditions of Nigeria. Cambridge, U.K. p. 230.

Doppler W (2002). Farming and Rural Systems Approaches. Published Lecturing Material.Hohenheim, Stuttgart, Germany.

Ekong EE (2011). An introduction to rural sociology. 3rd edition, Dove education publishers, Uyo. University Press Nig. Ltd. p. 352.

FAO (2001). Food and Agriculture Organization of the United Nations. Farming Systems andPoverty: Improving farmers' livelihood in a changing world. John Dixon and Aidan Gulliver with David Gibbon. Principal Editor: Malcolm Hall. FAO and World Bank Rome and Washington D.C.

Francis CA (1993). Designing future tropical agricultural system: challenges for research and extension. In: Ragland J, Rattan L eds., Technologies for Sustainable Agriculture in the Tropics, ASA Special Publication 56:184-209.

Hildebrand PE (1976). Generating technology for traditional farmers. A Multidisciplinary technology.

International Institute of Tropical Agriculture (1998). Farmers acclaim Crop livestock system. In IITA Annual Report pp.36-39.

Kang BT (1986). Cropping systems and soil fertility management in the humid and subhumid tropics with special reference to West Africa. In: Management of Nitrogen and Phosphorus Fertilizers in Sub-Saharan Africa, Mokwunye AU, Vlek PLG (Eds.), Proceedings of a Symposium, Lome, Togo, March 25-28, 1985. Martinus Nijhoff Publishers, Boston, USA pp. 83-94.

Mudahar MS (1986). Fertilizer problems and policies in sub-Sahran Africa. In: Mokwunye AU, Vlek PLG (eds.), Managemetn of nitrogen and phosphorus fertilizers in sub-Saharan Africa. Proceedings of a Symposium, Lome, Togo, March 25-28, 1985. Martinus Nijhoff Publishers p. 1-32.

WB Morgan, RP Moss (1965). Savanna and forest in Western Nigeria. Africa: J. Int. Afr. Inst. 35(3): 286-300.

National Population Commission (NPC) (1991). Nigeria demographic figure. Nigeria exchange 2011 http://www.ngex.com/nigeria/places/ cited 21 December. p. 50.

Ntare BR (1990). Intercropping morphological different cowpea with pearl millet in a short season environment in the Sahel. Exp. Agric. 26:41-47.

Okigbo BN (1980). A review of cropping systems in relation to residue management in the humid tropics of Africa. In: Organic Recycling in Africa. FAO Soil Bulletin 53, FAO, Rome pp. 13-37.

Woomer PL, Langat M, Mukhwana E (2001). Stratified intercropping of fertilized maize and legumes in Western Kenya. In 5th Biennial Conference of the African Crop Science Society.

Farmer Field Fora and adoption of yam integrated pest and disease management technologies in Northern Ghana

Daniel Y. Opare-Atakora[1], Samuel A. Donkoh[2] and Amin Alhassan[3]

[1]CSIR-Savanna Agricultural Research Institute, Tamale, Ghana.
[2]Department of Agricultural and Resource Economics, University for Development Studies, Tamale, Ghana.
[3]Department of Communication Sciences, University for Development Studies, Tamale, Ghana.

Justification for the use of public funds on programmes and the determination of their effectiveness among other factors call for their evaluation. This study was therefore conducted to investigate the effectiveness of Farmer Field Fora (FFF) and adoption of yam Integrated Pest and Disease Management (IPDM) Technologies in the Nanumba North and Kpandai Districts of the Northern region of Ghana. A multi-stage sampling technique was carried out to select 240 participants and non-participants from the study area. Primary data collection was done in 2012 through individual questionnaire administration and focus group discussions. Data were analyzed using descriptive statistics, regression analysis and budgetary techniques. From the findings, the farmers' perceptions of the effectiveness of the FFF were favorable and the FFF led to a close in knowledge gap and the adoption of IPDM technologies. Factors that positively affected adoption were training such as the FFF, farm size, and research contacts. Variables that had negative effects on adoption were age and house hold size. Also, while the benefit cost ratio for project participants was 2.5 that of non-participants in project community and non-participants outside project community were 1.9 and 2.1 respectively. FFF is therefore an appropriate mechanism to transfer IPDM technologies and the process could be adopted for extension activities. However, for a rapid adoption of the technologies, farmers should be supported with credit and more contacts with researchers. Also, younger farmers and large-scale farmers should be targeted for extra support without neglecting older farmers and small-scale farmers.

Key words: Farmer Field Fora, benefit cost ratio, integrated pest and disease management, participation, technology adoption, yam.

INTRODUCTION

A pre-requisite for the adoption of agricultural technologies is the acquisition of knowledge in the form of awareness, skills and principles (Rogers, 2003). The willingness of farmers to adopt new technologies will, among other factors, include an effective mechanism through which technologies are made available to the end users. In Ghana, various approaches of agricultural extension have been used in this regard (MoFA, 2011). They range from the top-down transmission of information approaches such as the Training and Visit (T&V) to the more recent participatory approaches such as the Farmer Field School (FFS) and Farmer Field Fora (FFF) in addition to innovative ICT based approaches which provide advice to farmers on-line and other approaches such as the promotion of mobile phones and community radio stations. The strengths and weaknesses of the various extension approaches that have been adopted in Ghana have been discussed in MoFA (2011).

The Ghana government and International Fund for Agricultural Development (IFAD)-financed programme aimed at developing the Root and Tuber sub sector from 1999 to 2005 implemented the Root and Tuber Improvement Programme (RTIP), which focused essentially on developing crop production through research and extension (RTIMP, 2010). Due to challenges in the implementation of RTIP, the Root and Tuber Marketing Programme (RTIMP) is being implemented to address them. The programme became effective on the 8th November, 2006 and will end in December 2014. The goal of the RTIMP is to enhance income and food security in order to improve livelihood of the rural poor.

The FFF is under component B, namely, Support to Root and Tuber Production. The main objective of Component B is to enhance the productivity of root and tuber production by facilitating access to new, relevant and adoptable technologies. The thematic area of concern for this study is Integrated Pest and Disease Management (IPDM) which was arrived at through a Participatory Rural Appraisal (PRA) undertaken in the participating project communities. The FFF is a participatory approach to train farmers in relevant yam technologies. It is field based, season long, and based on the principle of learning by doing. The fields are essentially made of a Farmers' Practice (FP) plot where farmers implement what they do in real life and a Participatory Action Research (PAR), where farmers together with researchers and extensionists implement proven technologies. In the course of the training and finally at the end, farmers compare the two plots and make their own informed decisions. Interspersed with these field activities are special topics to boost participants' knowledge of the technologies taught.

The main technologies taught in this study were chemical pest control and botanical pest control using neem. To enhance the use of these pesticides, scouting for pests and safe use of pesticides were included. Participants were also taught how to space the yam mounds for optimum plant population. This training was therefore carried out to enhance farmers' skills with respect to the adoption of these technologies so as to increase their food and cash security.

However, since its inception no evaluation has been done, at least to the best of our knowledge, to find out whether these technologies have been adopted. The aim of this study therefore was to evaluate the effectiveness of the FFF and the adoption of IPDM technologies as a means to increasing the output and incomes of farmers.The specific objectives were as follows:

(i) Investigate farmers' perceptions about the effectiveness of FFF;
(ii) Assess the knowledge change of the respondents (with respect to the IPDM technologies) as a result of the FFF;
(iii) Investigate farmers' perceptions about the IPDM technologies;
(iv) Measure the effects of FFF participation on adoption;
(v) Determine the effects of FFF on the viability of IPDM adoption.

MATERIALS AND METHODS

Literature review-Theoretical framework

Technology adoption

According to Rogers (2003), adoption is a stage in the innovation decision process when an individual decides to use an innovation or reject it while diffusion means the spread of the innovation through certain channels among a community or social system.

Diffusion of Innovation Theories (Rogers, 2003) and Knowles' Adult Learning Theory of Andragogy (Knowles et al., 2011) explain how new ideas and technologies are spread and adopted in a social system over time. The theory of Individual Innovativeness states that in most social systems there are innovators, early adopters, early majority adopters, late majority adopters and laggards. These categories have certain socio-economic characteristics such as age, marital status, educational level, income, farm size, and years of experience, among others, which affect adoption. The Theory of Perceived Attributes also focuses on how programme participants view the characteristics of the practice under investigation. These have been typically categorized as those that relate to the complexity, compatibility, trialability, relative advantage, and observability of a practice or technology. They provide explanations as to why an innovation is accepted or rejected. The innovation decision process theory has five stages as follows: (1) Knowledge stage, where the individual is exposed to the innovation's existence and gains an understanding of how it functions; (2) Persuasion stage, where the individual forms a favorable or unfavorable attitude towards the innovation; (3) Decision stage, where the individual engages in activities that lead to a choice to adopt or reject the innovation; (4) Implementation stage, where the individual puts the innovation into use; and (5) Confirmation stage, where the individual seeks reinforcement for an innovation-decision already made but may reverse the decision if he/she is exposed to conflicting messages about the innovation.

Communication channels such as educational programme type and its characteristics are based on the principles and process of andragogy or how adults learn as is the case of the FFF.

Evaluation of agricultural training programmes

Evaluations of agricultural training have normally been on knowledge gained and adoption of practices (Erbaugh et al., 2010; Hubbard and Sandman, 2007; Amujal et al., 2005). The inclusion of the evaluation of the educational processes gives it a holistic approach since it is important to understand the absolute and relative importance of the information presented and the programme experience. Effective implementation evaluation requires extension educators to clearly understand programme objectives and processes of implementation. This knowledge will empower educators to be able to better understand their current programme offerings, improve future services, and ultimately serve their target audiences better (Duerden and Witt, 2012; Hubbard and Sandman, 2007).

Effectiveness, according to Rogers (2003), is the degree to which an intervention programme fulfills its goals and objectives. The goal of an educational programme is a change in knowledge and behaviour (Hubbard and Sandman, 2007). Its scope of evaluation is the extent of adoption of disseminated technologies and determinants of adoption for a policy design and effective management of educational or extension programmes (Musaba, 2010). Equally important are the contents and conduct of training, participants' perceptions of procedures, techniques, materials, and products used, participants' background, the extent of instructors' credibility and effectiveness in identifying and addressing their needs (Strong et al., 2010; Rezvanfar et al., 2009; Coley and Scheinberg, 2000). Other qualities of a training programme are summarized in Tesfaye et al. (2009) and Wise and Ezell (2003).

Agricultural programme participation

Participation is a concept that has been gaining increasing recognition and prestige in the development discourse and its practices, requiring a shift in the way individuals are considered, from passive recipients to active agents of development efforts (Mefalopulos, 2008). Empirical studies on the effect of participation in training on project participants in the area of knowledge gained and adoption of agricultural technologies are presented in Erbaugh et al. (2010), Musaba (2010), Amujal et al. (2005), Nsabimana and Masabo (2005) and Mauceri (2004).

Study area

The survey was carried out in the Nanumba North and Kpandai Districts in the Northern Region of Northern Ghana in 2012. These districts were purposively chosen for the study because they are important yam production areas and also the Savanna Research Institute (SARI) carried out FFF training in yam Integrated Pest and Disease Management (IPDM) in the area. The target groups were participants of the FFF in project communities, and a control group of non-participants within and outside the project communities. The project communities were Ganguyille and Taali in the Nanumba North District and Nbowura and Onyumbo in the Kpandai District. The non-project communities were Zibaga and Kpalsogu in the Nanumba North District, Balai and Katijeli Damanko in the Kpandai District. Stratified sampling was used to select 20 project participants (PP) from each project community, 20 non-participants in the project community (NPIPC) and 20 non-participants outside the project community (NPOPC). The individual farmers were then selected randomly by a list provided by the trainers. This brought the sample size to 240 consisting of 80 PP, 80 NPIPC and 80 NPOPC. Note however, that in the estimation of the adoption model the sample sizes were 80, 40 and 40 respectively. This was to ensure that the estimated results were not biased. Maddala (1983) argued that for unbiased estimation results, the number of adopters should be equal to or at least proportional to the number of non-adopters.

Data collection

Data collection was done through individual questionnaire administration and focus group discussion. Secondary data from project documents were sourced to know more about the project and the technologies that were extended to the farmers. Likert type scales were used in collecting data on farmers' perceptions of FFF processes, IPDM technology attributes and knowledge levels of both participants and non-participants before and after the training. Nominal scales were also used to collect some dichotomous demographic variables such as marital status and gender; adoption or non-adoption of yam technologies; source of knowledge of technology; and reasons for non-adoption, while numerical data were collected on the other demographic variables as well as input, output and income variables. To determine the reliability of the instruments a pilot survey was done with a group of people similar to those in the study communities.

Data analyses

Data were analyzed using descriptive statistics, regression analysis and budgetary techniques. Specifically, the Chi square was used to test the statistical significance of the differences with respect to the effectiveness of the FFF as well as the knowledge and adoption levels of both participants and non-participants of the FFF before and after the training. Also, a Poisson regression model was estimated to determine the factors that influenced the adoption of IPDM technologies. Lastly, partial budget analysis was done to determine the differences between the Benefit cost ratios of participants and non-participants of the FFF.

The Poisson regression model

According to Greene (1997), the Poisson regression is represented by the basic equation:

$$\Pr(Y = y) = \frac{e^{-\lambda}\lambda^{y}}{y!}, \qquad y = 0, 1, 2 \ldots\ldots\ldots \tag{1}$$

The parameter λ is assumed to be log-linearly related to regressors x_i. Therefore,

$$\ln(\lambda) = \beta' x_i \tag{2}$$

The log-likelihood function is given by the equation:

$$\ln L = \sum_{i=1,2,\ldots,n} [-\lambda_i + y_i \beta' x_i - \ln y_i!] \tag{3}$$

The expected number of IPDM technologies per farm is given by the equation:

$$E[y_i|x_i] = Var[y_i|x_i] = \exp(x_i \beta' + \mu_1) \tag{4}$$

where, β is a $1 \times k$ vector of parameters; x is a $k \times 1$ vector.

According to Octavio and Shultz (2000), the equation can also be expressed as:

$$E[Y_i] = \exp^{(\beta_1 x_{1i})} \exp^{(\beta_2 x_{2i})} \ldots\ldots \exp^{(\beta_k x_{ki})}$$
$$= \exp(\beta_j X_{jn}) C_i \qquad (i = 1, 2 \ldots\ldots\ldots\ldots\ldots n) \tag{5}$$

where, j can take any value from 1 to k and identifies a specific explanatory variable and c_i is a constant representing the product of the remaining exponential terms in Equation (5). For dichotomous explanatory variables, if $x_{ji} = 0, E(Y_i) = C_i$, and when

$$x_{ji} = 1, E(Y_i) = \beta_j C_i. \tag{6}$$

Therefore, $100 \times (exp^{\beta_j} - 1)$ calculates the percentage change on $E(Y)$ when x_i goes from zero to one, for all i observations. In general, for independent variables that take several integer values, the percentage change in the expected level of adoption when x_j goes from x_{j1} to x_{j2} can be calculated as:

$$100 \times \frac{exp^{\beta_j x_{j2}} - exp^{\beta_j x_{j1}}}{exp^{\beta_j x_{j1}}} \tag{7}$$

The empirical model

The empirical model estimated to determine the factors influencing the adoption of yam IPDM technologies in northern Ghana is as follows:

$$Adoption = \beta_0 + \beta_1 Age + \beta_2 Experience + \beta_3 HHsize + \beta_4 Farmsize + \beta_5 Training + \beta_6 Research + \beta_7 OfficeDist + \beta_8 MarketDist + u_i$$

The *a priori* expectations are that experience, household size, training and research contacts have positive influence, while age, office and market distances, have a negative influence on adoption. Farm size, however, is inconclusive (i.e. it may be either positive or negative).

The Benefit/Cost ratio

The Benefit/Cost (B/C) ratio is given by:

$$B/C\ ratio = \frac{\sum_{t=1}^{n} \frac{B_t}{(1+r)^t}}{\sum_{t=1}^{n} \frac{C_t}{(1+r)^t}} \tag{8}$$

Where Bt is benefits in each year, Ct is costs in each year, n is number of years and t = 1, 2, 3, …., n, and r, the discount rate. However this study used cross sectional data since data were collected in one year implying n =1 (that is, the *Bt and* Ct components could not be discounted). The formula therefore assumes:

$$B/C\ ratio = \frac{B_t}{C_t} \tag{9}$$

A benefit cost ratio of one means benefits just offset costs while any B/C ratio more than one indicates the business is profitable.

RESULTS AND DISCUSSION

Socioeconomic characteristics of respondents

Almost all the respondents were males (94%), with only 6% being women. The highest percentage of women farmers was recorded by PP (19%), followed by NPOPC (14%) and NPPIPC (1%). In terms of age, the highest percentage of the respondents were within the 21-40 age group (58.3%) followed by the 41-60 age bracket (30.4%). While only 7.5% were under 21 years, the remaining 3.7% were over 61 years. The age distribution of the PP was not different from the others. Also, as high as 42.1% of the respondents had no formal education while 24.2% had non-formal education. This means that only 33.8% had formal education, out of which 16.3%, 15.0% and 2.5% had primary, secondary and tertiary education respectively. Among the groupings, NPPIPC recorded the highest percentage of respondents who had formal education (42.6%), followed by NPOPC (33.8%) and PP (23.8%). The mean household size was 9 ranging from 8 and 10, while the mean farm size was 5 acres. The percentages of respondents who said they had contacts with extension and research staff during the farming season under review were 66 and 45 respectively. Most of them were project participants. Only 2% of respondents indicated they had credit access, and finally 77% of respondents were involved in off-farm activities such as trading, teaching, bee keeping, drug peddling and food processing.

Farmers' perceptions of the FFF process

Project participants were asked to state the extent to which they agreed or disagreed that the preparatory, implementation stages and other characteristics of the FFF were participatory and contributed to the success of the training. From Table 1, 92.6% of them held a very high positive and favourable perception of the effectiveness of the FFF. However, while 6.4% disagreed with the notion, 2.6% were undecided. In terms of the specific characteristics, the highest commendation was for the training needs assessment (98%), the venue for the training (98%) and the training staff (98%). The lowest commendation (80%), however, was for the timing and duration of the training; the respondents felt that the training should have come far earlier before the farming season and also it should have covered a longer period for a better understanding and application of the knowledge gained. In general, the farmers' perceptions of the effectiveness of the FFF were favorable and therefore the FFF and its procedures could be adopted as an appropriate mechanism to transfer IPDM technologies. This process, if adopted in the country's agricultural extension programmes, could lead to adoption of technologies. The findings of this present study are in sync with that of Knowles et al. (2011), Tesfaye et al. (2009), and Wise and Ezell (2003) who found that these variables are key to the success of non-formal adult education programmes like the FFF.

Knowledge change of respondents as a result of FFF

The effectiveness of the FFF had led to a positive knowledge change among 75.7% of the respondents with

Table 1. Respondents' Perception of the Effectiveness of FFF.

Characteristic	S.D/D		U		A/S.A		Chi square value
	F	%	F	%	F	%	
Training needs assessment	1	1.3	1	1.3	78	98.0	83.3***
Participants selection criteria	3	3.8	1	1.3	76	95.0	73.0***
Content of training	2	3.0	1	1.3	77	96.0	74.1***
Relevance of training	2	3.0	1	1.3	77	96.3	69.7***
Timing of the training	16	20.0	0	0	64	80.0	36.1***
Duration of training	13	18.0	2	22.0	64	80.0	74.4***
Methodology of training	2	3.0	1	1.3	77	96.0	79.5***
Place of training	1	1.3	1	1.3	78	98.0	72.2***
Competence of training staff	1	1.3	1	1.3	78	98.0	77.1***
Participation during the training	3	3.8	1	1.3	76	95.0	67.4***
Sharing of experience	2	3.0	1	1.3	76	95.0	108.9***
Indigenous knowledge use	12	15.0	0	0	68	85.0	14.8***
Solution to problems	4	5.0	1	1.3	75	94.0	67.1***
Learning guide development	6	8.0	1	1.3	73	91.3	82.3***
Mean	4.9	6.39	0.93	2.59	74.07	92.68	

S.D, Strongly Disagree; D, Disagree; U, Undecided; A, Agree; S.A, Strongly Agree; F=Frequency; %=Percentage. Source: Field Survey, 2012.

Table 2. Knowledge change of respondents as a result of FFF.

Technique	PP (%)	NPIPC (%)	NPOPC (%)	Pooled (%)
Appropriate mounding	100	71.3	33.8	68.4
Insecticide application	100	81.3	61.3	80.9
Fungicide application	98.8	73.8	48.8	73.8
Weed management	100	91.3	87.5	92.9
Neem application	96.3	71.3	42.5	70.0
Scouting	96.3	52.5	42.5	63.8
Safe pesticide use	98.8	87.5	61.3	82.5
Pooled	98.6	75.6	53.0	-

PP, Project Participants; **NPIPC,** Non Participants in Project Community; **NPOPC,** Non Participants Outside Project Community. Source: Field Survey, 2012.

respect to the IPDM technologies. From Table 2, PP recorded the highest percentage of respondents who indicated there had been an increase in the knowledge gained with respect to the technologies (98.6%), followed by NPPIPC (75.6%) and NPOPC (53.0%). The findings make sense, as the diffusion rate of a technology is likely to be higher in a project community than non-project communities. Across the technologies, the highest percentage of respondents indicated a change in their knowledge gap with respect to weed management (92.6) followed by safe pesticide use (82.5) and insecticide application (80.9). It should be noted that these techniques were relatively new to the respondents. The FFF therefore served as an eye-opener to them, hence the acknowledgement from a high percentage of them. These findings also support that of Erbaugh et al. (2010) in their study of cowpea.

Knowledge sources of IPDM technologies

The study however, found that SARI was the only source of the respondents' knowledge with respect to the IPDM technologies; while the participants gained the knowledge directly from SARI, the non-participants had their knowledge from their fellow farmers. This finding is consistent with Rogers' (2003) observation that an innovation will normally come from an external source, and then diffuse with time in other communities through the farmers themselves or other means. It needs to be mentioned, however, that about a quarter of the project participants were not trained in neem application. Also, 20% of the farmers who were not originally part of the participating group had the opportunity to be trained in fungicide application. According to the project officers, a leeway was given to farmers who were not originally participants in some of the project communities to join if they so wished. Generally, the results indicate a diffusion of the technology within and outside the community in line with Rogers's (2003) concept of homophily and also knowledge being a prerequisite of the innovation decision process. The findings lend support to that of Mirani (2013) and Asiabaka and Owens (2002), who revealed that neighbouring farmers were a good source of agricultural information.

Perceptions of IPDM technology attributes

When asked to what extent they agreed or disagreed with the attributes of the technologies, the participants had favourable responses. Apart from 45% who thought that the technologies were complex, more than 80% of the respondents respectively indicated that the technologies were trialable, observable, compatible with their practices and also had relative advantage (that is, were viable because the benefits far outweighed the costs). This finding is also in sync with that of Maina et al. (2008) and Asiabaka and Owens (2002).

Adoption levels of IPDM technologies

The favourable perceptions of the technologies had led to their adoption by 64.1% of the respondents, especially among the project participants. Across the various groups, the percentages of respondents who had adopted the technologies as at the time of data collection were PP (76.6), NPPIPC (38.8) and NPOPC (28.8). Across the technologies, the adoption of weed management was highest (85.4%). This is an indication of the economic importance of weeds in yam production. However, the adoption levels were relatively low for fungicide application (25%) and neem (28%). According to project participants, they did not use fungicide because it was not available. For neem, some of the project participants said they were not taught. Non project participants did not use neem and fungicides because they had no knowledge of them. The differences in the adoption levels were also significant at the 1% level except that of weed management which was significant at the 5% level. The significant differences in the levels of adoption could be attributed to the effect of participation in the FFF. This finding supports investigations by Erbaugh et al. (2010) who in a study to assess the impact of cowpea Farmer Field School (FFS), found that participation in training was effective in adoption of IPM strategies. Similar findings were reported by Musaba (2010) and Nsabimana and Masabo (2005).

Reasons for non-adoption

When asked why they did not adopt some technologies, a small proportion of project participants said the pesticides were unavailable, expensive and that they were not knowledgeable of their use. About a quarter of them did not use neem because they did not know how to use them. The results lend support to that of Asante et al. (2011), who stated that farmers did not adopt yam technologies because they were expensive. Sani and Bagna (2012), showed that high cost and unavailability of improved seeds and fertilizer constituted the major constraints to improve seeds utilization. Most of the non-participants did not adopt because they had no knowledge of the technologies. This shows that knowledge is a prerequisite to adoption of technologies (Rogers, 2003). Empirical findings similar to this are by Drost et al. (1996) and Ahmad et al. (2007).

Definition of variables and the determinants of IPDM technologies

The second objective of the study was to determine the factors influencing the adoption of IPDM technologies. In this section, we discuss the extent to which demographic, farm-specific and institutional variables influenced the adoption of the technologies. We also compare the Benefit/cost ratio of PP with the non-participating groups. The section however, begins with the definition and descriptive statistics of variables used for the adoption model.

Definition of variables used in the adoption model

Table 3 shows the definition of variables used in the estimation of the adoption model and their expected signs. Foltz (2003) and Feder et al. (1985), have discussed at length the extent to which household characteristics as well as farm-specific and institutional factors influence the adoption of agricultural technologies.

Descriptive statistics of the variables used in the Poisson model

As indicated already, the total number of IPDM technologies in which the farmers were trained was 7, namely appropriate mounding, insecticide application, fungicide application, weed management, neem application, and scouting and safe pesticide use. The average number of technologies adopted was 3.4 (Table

Table 3. Definition of variables.

Variable	Definition	Expected sign
Age	Age of the farmer in years	+/-
Education	Dummy; 1 if farmer had formal education; 0 if otherwise	+
Household size	No. of people in farmer's house eating in the same bowl	+
Farm size	Size of farmer's yam plot in acres	+
Credit	Dummy; 1 if farmer had access to credit during farm season in question; 0 if otherwise	+
Extension	Dummy; 1 if farmer had access to extension service during farm season in question formal education; 0 if otherwise	+
Research	Dummy; 1 if farmer had made contacts with a researcher during farm season in question; 0 if otherwise	+
FFF Distance	Distance form yam farm to FFF center in kilometers	-
Market Distance	Distance form yam farm to market/input store in kilometers	-
Participation	Dummy; 1 if farmer participated in FFF; 0 if otherwise	+

Table 4. Descriptive statistics of variables.

Variable	Mean	Standard deviation	Minimum	Maximum
Adoption	3.36250	2.24397	.000000	7.00000
Age	37.0958	12.9874	15.0000	85.0000
Household size	8.97500	6.43274	1.00000	45.0000
Research	.454167	.498935	.000000	1.00000
Experience	17.4083	12.7227	2.00000	70.0000
Farms size	4.56042	3.38800	1.00000	30.0000
Office Distance	7.88750	4.01455	1.60000	19.3000
Market Distance	4.59375	2.47469	.100000	12.0000

Source: Field Survey, 2012.

4). Also, the mean age of the farmers was 37, which shows that the farmers in the study area are young adults. Yet on average, they had been farming yam for 17 years. However, the average size of their farms was small (4.5), which is slightly below the national average of 5 acres. The percentages of farmers who received training and also contacts with research officers such as the FFF officials from SARI were 41 and 45 respectively. Lastly the average distances from farmers plot to the FFF center as well as to the major market center were 7.9 and 4.6 km respectively.

The determinants of IPDM technology adoption

In Table 5, the estimation results of the Poisson model are presented. We observe that the p-value of the chi square is zero. This means that the probability of all the coefficients, except the constant, is equal to zero and all the variables jointly determine technology adoption.

Age and Household size had significant and negative effects on adoption of IPDM technologies. On the other hand, farm size, training and research contacts had positive and significant effects on the probability of adoption. Farmers' experience and the distance variables were however not significant.

The negative and significant coefficient of age implies that younger farmers had a greater probability of adoption than their older counterparts. Younger farmers have been found to be more innovative and willing to bear risk, as opposed to older farmers (Asiabaka and Owens, 2002). However, Asante et al. (2012), Wiredu et al. (2011), and Maiangwa et al. (2007) found a positive relationship between age and adoption. It is important that projects like the IPDM while targeting both older and young farmers, give priority to younger farmers because of their flexibility and willingness to try new things.

The probability of adoption was also negatively influenced by household size. This has been linked to increased consumption pressure associated with larger families. The argument is that larger households often have a lot of mouths to feed, to the extent that they do not have enough money to invest in technology adoption, as opposed to smaller families. In addition, farmers with

Table 5. Maximum likelihood estimation results of the poisson model.

Variable	Marginal effects	Standard error	t-ratio	P-value	Mean of x
Constant	3.240	0.735	4.404	0.000	
Age	-0.029	0.018	-1.663	0.096	37.096
Experience	-0.003	0.018	-0.167	0.867	17.408
Household size	-0.043	0.026	-1.674	0.094	8.975
Farm size	0.057	0.034	1.672	0.095	4.560
Training	2.467	0.441	5.594	0.000	0.413
Research	0.706	0.319	2.214	0.027	0.454
Office Distance	0.059	0.043	1.369	0.171	7.887
Market Distance	-0.011	0.068	-0.164	0.870	4.594
Chi Squared	165.471	-	-	0.000	-
Degree of Freedom	8	-	-	-	-

Source: Field Survey, 2012.

relatively large family sizes attach greater importance to other non-farm activities than farm activities such as the adoption of IPDM technologies (Maiangwa et al., 2007; Amaza et al., 2008). On the contrary, Baffoe et al. (2013), found a positive association between household size and adoption, arguing that larger families have greater labour force to be able to engage in the adoption of technologies. In general, from a policy point of view, smaller households should be targeted in promoting IPDM technologies while constraints of larger households are attended to with the view of increasing the probability of adoption.

The findings with respect to farm size in this present study also lend support to that of previous studies (Akudugu et al., 2012; Asante et al., 2012; Ayoade, 2012; Katengeza et al., 2012; Feder et al., 1985). The argument is that, normally farming households with bigger landholdings are supposed to have an enhanced ability to afford improved technologies and a greater capacity to cope with the loss, if the technology fails. These large-scale farmers would rather prefer carrying out recommended farming practices than risk ignoring them and incurring huge financial losses. Also, large scale farmers are often eager to try out new recommended practices in order to improve their yield and productivity. In addition, farmers with larger farm sizes are able to access support in terms of inputs, marketing and pricing much better than their counterparts with smaller farm sizes hence will be more willing to adopt a new technology so long as it will positively affect their farming activities. From the policy point of view, any promotional efforts aimed at boosting the use of IPDM technologies could target farmers with large farms. In addition, any policy which will educate, encourage and support small-scale farmers to expand their farms should be carried out so that their adoption may also be enhanced.

Similarly, the results indicate that farmers who interact with researchers and research organisations are more likely to adopt these technologies as compared to their counterparts who do not get the opportunity. It implies that farm activities involving research such as the FFF, on farm research and demonstrations should be encouraged.

Finally, participation in training such as the FFF had a positive significant relationship with the probability of adoption of IPDM technologies. The FFF serves as a platform for knowledge as well as technology transfer. The platform is interactive and provides the opportunity for feedback and further clarification of technological packages associated with farming. Participation on the platform increased farmers' probability of IPDM technology adoption in the study area. This finding corroborates that of Erbaugh et al. (2010) in a study to assess the impact of cowpea Farmer Field School (FFS) on IPM technology adoption in Uganda. Similar results were found by Mauceri (2004), Musaba (2010), Nsabimana and Masabo (2005) and Amujal et al., (2005). Furthermore, our *a priori* expectations were that distances from the farmers plot to the FFF office as well as to the market would be negatively correlated with the probability of adoption. However, as indicated earlier the two variables were all not significant, compared with the findings of Bonger (2001). Farmers' experience was also not significant contrary to our *a priori* expectations that the longer the years of yam farming, the greater the probability of IPDM adoption.

The benefit cost ratio

A Partial Budget Analysis (PBA) was carried out to find out the viability of yam production for the various farmer groups. From Table 6, we observe that even though the total variable cost for PP was higher than the non-participants, the average yield, and for that matter the revenue for the former, also far outweighed that of the latter. This translated into, again, a higher gross benefit than that of the non-participants. In general, the high total

Table 6. Results of partial budget analysis/acre.

ITEM	PP GH¢	NPPIPC GH¢	NPPOPC GH¢
Land preparation	30.50	32.50	32.00
Cost of setts	235.62	192.25	233.50
Cost of pesticides	14.90	10.42	8.33
Cost of stakes	54.74	51.94	52.87
Labour cost	253.49	283.78	272.96
Transportation	275.60	213.51	224.63
Total variable costs	864.86	784.39	824.29
Average yield(*gmani*)	10.10	7.47	7.76
Price /*gmani*	87.28	82.96	91.89
Revenue/gross income	2159.76	1519.29	1747.53
Gross benefits	1294.91	734.90	923.28
Benefit cost ratio	2.50	1.94	2.12

PP, Project Participants; NPPIPC, Non Project Participants in Project Community; NPPOPC, Non Project Participants Outside Project Community. Source: Field Survey, 2012.

variable cost came from the costs of more yam setts, pesticides, stakes and transportation. However, labour cost and other land preparation costs were lowest, perhaps as a result of the benefits from mutual assistance from the group to which project participants belonged. The net effect was that the benefit cost ratio for project participants was 2.5 while that of non-participants in and outside project communities were 1.9 and 2.1 respectively. Thus, the return to investment was higher for project participants than the non-participants. This finding lends support to similar studies that found a profitable net change of income in activities of participants of training as compared with their untrained counterparts (NORC, 2012; Davis et al., 2010; and Alam et al., 2008). This means that with time if the technologies are well-diffused yields returns for non-participants would also increase. However, it may be necessary for SARI to organize training for these non-participating communities for rapid results.

Conclusions

From the findings, the farmers' perceptions of the effectiveness of the FFF are favourable. That is to say, the procedures used in the training met their expectations and led to the overall success of the training. The FFF achieved its outcomes by increasing the knowledge levels and the adoption by both participants and non-participants. While the FFF was the main information source of the FFF participants, the FFF participants were the main source of knowledge of IPDM technologies of the non-participants. This was an indication that the FFF was a conduit for diffusion of the technologies. Also, the adoption of IPDM technologies led to increased returns to farmers. Since the FFF served as an effective platform for knowledge creation, sharing and acquisition, dissemination, and adoption of technologies leading to higher returns, it may be adopted as an appropriate mechanism for sharing of knowledge in yam IPDM practices and other extension programmes.

FUTURE STUDIES

This study mainly looked at the effectiveness of the FFF educational process and factors likely to affect the probability of adoption of the technologies. It would be necessary to look at the cost effectiveness of the FFF, from the point of view of the organizers.

ACKNOWLEDGEMENT

We wish to thank IFPRI for providing the funds for this study.

ABBREVIATION

B/C, Benefit cost ratio; **FFF,** Farmer Field Fora; **FFS,** Farmer Field School; **FP,** farmers' practice; **ICT,** Information and Communication Technology; **IFAD,** International Fund for Agricultural Development; **IFPRI,** International Food Policy Research Institute; **IPDM,** Integrated Pest and Disease Management; **MoFA,** Ministry of Food and Agriculture; **NPIPC,** non-participants in the project community; **NPOPC,** non-participants outside the project community; **PAR,** Participatory Action Research; **PBA,** Partial Budget Analysis; **PP,** project participants; **PRA,** Participatory Rural Appraisal; **RTIMP,** Root and Tuber Marketing Programme; **RTIP,** Root and Tuber Improvement Programme; **SARI,** Savanna Agricultural Research Institute; **T&V,** Training and Visit.

Conflict of Interests

The author(s) have not declared any conflict of interests.

REFERENCES

Akudugu MA, Guo E, Dadzie N (2012). Adoption of modern agricultural production technologies by farm households in Ghana: What factors influence their decision. J. Biol. Agric. Healthcare 2(3):88-94.

Ahmad M, Akram M, Rauf R, Khan IA (2007). Adoption of and Constraints in use of high yielding varieties: A case study of four VIllages of District Peshawar and Charsada. Sarhad. J. Agric. 23(3).

Alam R, Akteruzzaman M, Hossain MM, Sharmin E (2008). Impact of duck rearing training on technology dissemination and income of women under Microfinance and Technical Support (Mfts) programme of selected area of Bangladesh. J. Animal Sci. 37(1):63-69.

Amaza P, Ayubab K, Kamaraa A (2008). Farmers' perception, profitability and factors influencing the adoption of improved maize varieties in the guinea savannas of Nigeria.Tropentag 2008, University of Hohenheim October 7-9, 2008. Conference on International Research on Food Security, Natural Resource Management and Rural Development.

Amujal M, Semana AR, Erbaugh M, Adipala E (2005). Evaluation of adoption of cowpea technologies among farmer field school farmers in eastern Uganda.African Crop. Sci. Confer. Proc. 7:787 700

Asante BO, Otoo E, Wiredu AN, Acheampong P, Osei-Adu J, Nsiah-Frimpong B (2011). Willingness to adopt the vine multiplication technique in seed yam production in the forest savanna transition agro-ecological zone, Ghana. J. Dev. Agric. Econ. 3(16):710-719

Asiabaka CC, Owens M (2002). Determinants of adoptive behaviors of rural farmers in Nigeria. AIAEE Proceedings of the 18th Annual Conference. Durban, South Africa.

Ayoade AR (2012). Attitude of women farmers towards agricultural extension services in Ifelodun local government area, Osun State. Am. J. Soc. Manage. Sci. 3(3):99-105.

Baffoe-Asare R, Danquah J, Annor-Frempong A (2013). Socioeconomic factors influencing adoption of codapec and cocoa high- tech technologies among small holders in Central Region of Ghana. Am. J. Exper. Agric. 3(2):277-292.

Bonger T (2001). The determinants of the adoption of farm technology by resettled farmers in Chinyika, Zimbabwe. Zambezia, 27(2).

Coley SM, Scheinberg CA (2000). Proposal writing, (2nd edition). California: Sage Publications, Inc. PMCid:PMC1495594

Davis K, Nkonya E, Kato E, Mekonnen DA, Odendo M, Miiro R, Nkuba J (2010). Impact of Farmer Field Schools on agricultural productivity and poverty in East Africa. IFPRI Discussion P. 00992.

Drost D, Long G, Wilson D, Miller B, Campbell W (1996). barriers to adopting sustainable agricultural practices. J. Exten. 34(6).

Duerden MD, Witt PA (2012). Assessing Programme Implementation: What It Is, Why It's Important, and How to Do It. J. Exten. February 50(1).

Erbaugh JM, Donnermeyer J, Amujal M, Kidoido M (2010). Assessing the impact of farmer field school participation on IPM adoption in Uganda. J. Int. Agric. Exten. Edu. 17:(3):5–17. http://dx.doi.org/10.5191/jiaee.2010.17301

Feder G, Just RE, Zilberman D (1985). Adoption of agricultural Innovation in developing countries. Asurvey. Econ. Dev. Cult. Change 33:225-298. http://dx.doi.org/10.1086/451461

Foltz JD (2003). The economics of water-conserving technology adoption in Tunisia: An empirical estimation of farmer technology choice. Econ. Dev. Cult. Change, (January), 51(2):359-373. http://dx.doi.org/10.1086/367627

Greene WH (1997). FIML estimation of sample selection models for count data. Working New York: New York University, Leonard N. Stern School of Business, Department of Economics pp. 97–02.

Hubbard WG, Sandman LR. (2007). Using diffusion of innovation concepts for improved program evaluation. J. Exten. [On-Line] 45(5) available at http://www.joe.org/joe/2007october/a1.shtml

Katengeza SP, Mangisoni JH, Kassie GT, Sutcliffe C, Langyintuo A, Rovere RL, Mwangi W (2012). Drivers of improved maize variety adoption in drought prone area of Malawi. J. Dev. Agric. Econ. 4(14):393-403.

Knowles MS, Holton EF, Swanson RH (2011). The adult learner: the definitive classic in adult education and human resource development. Houston, Texas: Gulf Professional Publishing Co.

Maddala GS (1983). Limited-Dependent and Qualitative Variables in Econometrics. New York: Cambridge University Press. http://dx.doi.org/10.1017/CBO9780511810176

Maiangwa MG, Ogungbile JO, Olukosi JO, Atala TK (2007). Adoption of chemical fertilizer for land management in the North-West Zone of Nigeria. Tropical Agric. Res. Exten. P. 10.

Maina IN, Leonh¨auser IU, Bauer S (2008). Farmers' perceptions of attributes of agricultural technologies and their impact on adoption: A case study of Nakuru District, Kenya. Tropentag, October 7-9, 2008, Hohenheim. "Competition for Resources in a Changing World: New Drive for Rural Development"

Mauceri M (2004). Adoption of Integrated Pest Management technologies: a case study of potato farmers in Carchi, Ecuador. Master of Science Thesis. Virginia Polytechnic Institute and State University.

Mefalopulos P (2008). Development communication source book: broadening the boundaries of communication. Washington D.C. World Bank. http://dx.doi.org/10.1596/978-0-8213-7522-8

Mirani Z (2013). Perception of farmers and extension and research personnel regarding use and effectiveness of sources of agricultural information in Sindh Province of Pakistan. J. Comm. Inform. 9(1).

MOFA (2011). Agricultural extension approaches being implemented in Ghana. Available at http://mofa.gov.gh/site/wp-content/uploads/2011/03/Extension-approaches-in-Ghana-.pdf

Musaba EC (2010). Analysis of factors influencing adoption of cattle management technologies by communal farmers in Northern Namibia. Livestock Res. Rural Develop. 22(6).

NORC (2012). Final report for the impact evaluation of the farmer Training and Development Assistance (FTDA) Project Funded by the Millennium Challenge Corporation. University of Chicago.

Nsabimana JD, Masabo F (2005). Factors influencing adoption of agricultural technologies in Kiruhura District of Rwanda. African Crop Sci. Confer. Proceed. 7:759-760.

Octavio AR, Shultz SD (2000). Poisson count models to adoption of agricultural explain the and natural resource management technologies by small farmers in central. Am. Count. J. Agric. Appl. Econ. 32(1):21–33.

Rezvanfar A, Samiee A, Faham E (2009). Analysis of factors affecting the adoption sustainable soil conservation practices among wheat growers. World Appl. Sci. J. 6(5):644-651.

Rogers EM (2003). Diffusion of innovations (5th Ed.). New York: Free Press.

RTIMP (2010). Root and Tuber Improvement and Marketing Programme: Status Report.

Sani RM, Bagna A (2007). Adoption of improved seeds of millet and cowpea by farmers in Madarounfa District, Niger Republic. Continental J. Agric. Econ. 1:1-6,

Strong AR, Harder A, Carter H (2010). Agricultural extension agents' perceptions of effective teaching strategies for adult learners in the master beef producer program. J. Exten. 48(3). http://www.joe.org/joe/2010june/rb2.php

Tesfayo T, Karippai RS, Tesfayo T (2009). Farmers training programme of Ethiopian Institute of Agricultural Research: An appraisal. Afr. J. Agric. Res. 4(4):409-421. Available online at http://www.academicjournals.org/AJAR

Wiredu AN, Mensah-Bonsu A, Andah EK, Fosu KY (2011). Hybrid cocoa and land productivity of cocoa farmers in Ashanti Region of Ghana. World J. Agric. Sci. 7(2):172-178.

Wise D, Ezell P (2003). Characteristics of effective training: Developing a model to motivate action. J. Exten. [On-line], 41(2). Available at: http://www.joe.org/joe/2003april/a5.shtml

Allelopathic effect by aqueous extracts of different parts of *Croton bonplandianum* Baill. on some crop and weed plants

Swapnal Sisodia* and M. Badruzzaman Siddiqui

Department of Botany, Aligarh Muslim University, Aligarh-20 2002, India.

The present study was conducted to investigate the allelopathic effects of *Croton bonplandianum* weed on seed germination and seedling growth of crop plants (*Triticum aestivum* L., *Brassica oleracea* var. *botrytis* L. and *Brassica rapa* L.) and weed plants (*Melilotus alba* Medik., *Vicia sativa* L. and *Medicago hispida* Gaertn). Root, stem and leaf aqueous extracts of *Croton* at 0.5, 1.0, 2.0 and 4.0% concentrations were applied to determine their effect on seed germination and seedling growth of test plants under laboratory conditions. The aqueous extracts from root, stem and leaf had no effect on seed germination of test plants. The extracts from stem had a stimulatory effect on the shoot length at all concentration levels, as against an inhibitory effect of leaf extracts. Stem extracts at low concentration generally promoted root length but aqueous extracts from leaf and root inhibited root length and dry weight. Root length, shoot length of weed species decreased progressively when plants were exposed to increasing concentration (0.5, 1, 2 and 4%). The pH, osmotic potential and total phenolic contents, the osmotic potential and phenolic content increased with increasing concentrations of aqueous extracts of different parts of *C. bonplandianum* while pH does not have any major change.

Key words: Allelopathy, aqueous leaf extract, aqueous stem extract, aqueous root extract, *Melilotus alba, Medicago hispida, Vicia sativa, Croton bonplandianum.*

INTRODUCTION

The *Croton bonplandianum* is an obnoxious weed of Euphorbiaceae family. It is a native of South America and was reported from India during late 1890 (Kaul, 1967). It now occurs widely along roadsides, railways abandoned fields in wide open ravines and paddy or sugarcane fields and on sandy or sandy clay soils. This species is seldom found in areas enclosed by shrubs and trees where free movement of air sets hindered. For obvious reasons, *C. bonplandianum* is a problematic weed for farmers, ecologists/biologists, horticulturists, environmentalists and common man especially in North India. It competes with other plant species upon invasion resulting in adverse effects on the natural vegetation and standing crops. Its invasion in crop fields hinders the preparation of fields while ploughing. Due to its competitive nature, it reduces the availability of nutrients to crops thereby reducing productivity. DeCandolle (1832) was probably the first person to suggest the possibility that many plants may excrete something from their roots which may be injurious to other plants. Molisch (1937) proposed the term 'allelopathy' for expressing the harmful effects that one plant species may have on another through the mechanism of chemical retardants escaping into the environment. The concept of allelopathy was further supported and further developed by Bonner (1950), Grummer and Beyer (1960), Evenari (1961), Whittaker (1970), Pitman and Duke (1978) and Fischer et al. (1978). According to Lavabre (1991), allelopathic effects are controversial and still poorly understood. Allelo-chemicals (inhibitors) are produced by plants as end products, by-products and metabolites and are contained in the stem, leaves, roots, flowers, inflorescence, fruits and seeds of the plants. Of these plant parts, leaves

*Corresponding author. E-mail: swapnal81@hotmail.com, b.siddiqui@rediffmail.com.

seem to be the most consistent producers of these allelochemicals.

The four ways in which allelochemicals escape from a plant are: (i) volatilization, during which the terpenes are released from the leaves of some plant species; (ii) leaching (which has shown that living or dead leaves of many plants contain growth inhibitors); (iii) exudation in which case roots of several crop and non-crop species release large quantities of organic compounds that inhibit the growth of other plants; and (iv)decomposition, through which allelochemicals are released from the plant residue.

MATERIALS AND METHODS

Preparation of aqueous extract

The *Croton bonplandianus* plants which grew naturally in Aligarh Muslim University campus were uprooted and collected at their mature stage on August 4 and 20, 2007, for the experiment. The plants were brought into the laboratory and were immediately separated into leaf, stem and root parts. Each part of the fresh plant was cut into small pieces, shade dried and then ground separately with help of grinder and made fine powder. 4 g powder of root; stem and leaves of *C. bonplandianum* were steeped in 100 ml of distilled water and filtrate through a muslin cloth followed by filter paper (No. 1. Whatman International, Maidstone, UK). After 24 h of soaking at room temperature 921 - 22°C) and then further diluted so as to get 0.5 to 4% aqueous extract, respectively, as adopted from Singh et al. (1989).

Treatments and experimental design

There were 15 treatments including four concentration levels (0, 0.5, 1, 2 and 4%) of each leaf stem and root extracts. Seeds and filter papers were moistened with 10 ml each of 0.5, 1, 2 and 4% aqueous extracts. 10 ml of distilled water was added to the untreated control (0%). The treatments were arranged in completely randomized design (CRD) with five replicates kept at room temperature on a laboratory bench with 12 h supply of fluorescent light during the night The whole experiment was repeated once.

Determination of pH

The pH of each extract prepared from different parts of *C. bonplandianum* was determined by immersing the electrode of a digital ph meter (EcoScan). The mean of five replicates were taken and presented.

Determination of osmotic potential

The osmotic potential of extracts of different parts of *C. bonplandianum* was determined using the following formula:

Osmotic Potential = 0.36 x Conductivity (in mS)

Determination of total phenolic content

Total water-soluble phenolics in the aqueous residue extracts and residue-amended soils were estimated as per the method of Swain and Hills (1959) using Folin-Ciocalteu reagent. Their amounts were determined spectrophotometrically at 700 nm against the standard of ferulic acid.

Physical parameters

After seven days, the seedling root length (cm), shoot length (cm) and dry weight were determined (mg). The root and shoot length were determined manually while the dry weight with the help of 4 digit digital balance of Scientech, Model ZSA 120, Colorado (USA)

Statistical analysis

After seven days, the seedling root length, shoot length and dry weight were determined. The data were subjected to one way analysis of variance and the mean values were separated at $P < 0.05$ applying 2-sample t-test. The statistical analysis was done using SPSS/PC version 10 software.

RESULTS AND DISCUSSION

As compared to the control (0%), the aqueous extracts of *Croton* from root, stem and leaf parts at 2 and 4% concentration levels exhibited significant ($P < 0.05$) inhibition on seedling growth. The effect was in general, more severe on *Melilotus alba* compared to other weed plants. Among the different parts, leaves were the most allelopathic and stems were least allelopathic. The inhibitory effect was concentration dependent (Figures 1, 2 and 3). The inhibition effect was found to increase with increasing concentrations of different aqueous extracts (Sisodia and Siddiqui, 2008, 2009). The radicle length, plumule length and dry weight of seedlings were reduced significantly in response to all the *C. bonplandianum* extracts. At 4% concentration of the leaves extract, in *Vicia sativa* the radicle length and plumule length were reduced by nearly (41%) and (26%) respectively (Figure 3).

Generally, in studies with aqueous extracts, the observed inhibitory effect are attributed to change in pH and osmotic potential thereby raising concern about allelopathy and its ecological existence and relevance (Harper, 1977; Conway et al., 2002; Sisodia, 2008). In the present study, pH of extracts (that is, root, stem and leaves) ranged from 6.13 to 6.68 (Table 1). Likewise, the osmotic potential range -0.43 bars to -0.96 bars is again unlikely to cause any inhibitory effect on the plant growth (Mersie and Singh, 1987). After making these observations, it could be concluded that extracts might possess growth inhibitors. The amount of phenolics was also determined in extracts from different parts at mature stage. In leaves extract, amount of phenolic was about 828 µg/ml in 4% extract. Among different parts, the content of phenolics was the maximum in leaves and less in stem (Table 1).

The observed different phytotoxicity of *C. bonplandianum* may be attributed to the presence of variable amount of phototoxic substances in different

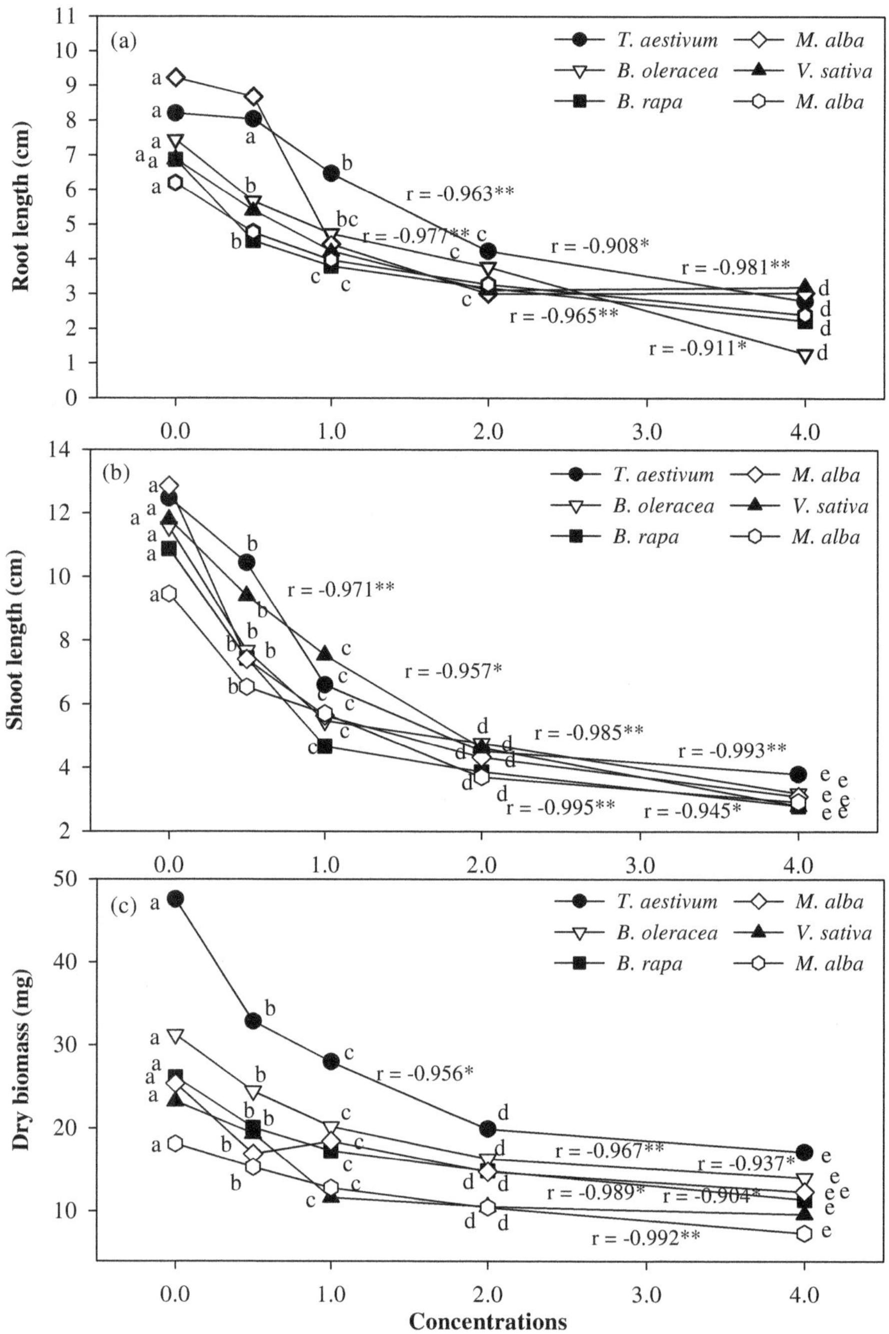

Figure 1. Effect of different concentration of aqueous extracts of root at flowering stage on (a) root length and (b) shoot length and (c) dry biomass of test plants (crops as well as weeds).
Different superscript symbols along a curve respresent signinficant difference among themselves at $P < 0.05$
applying DMRT.
r represent correlation coefficient
* and ** represent significance of correlation at $P < 0.05$ and $P < 0.01$ respectively

parts that leach out under natural conditions. Foliar leachates have been regarded to be most phytotoxic in nature (Xuan et al., 2004) probably owing to their proportionately greater biomass and with greater metabolic activity or production of more metabolites (Xuan et al., 2004).

Some recent studies indicating the phytotoxic/allelopathic effect of aqueous extracts of weeds include *Mikania micrantha* (Ismail and Kumar, 1996), *Vulpia sp.* (An et al., 1999), *Cyperus rotundus* (Quayyaum et al.,

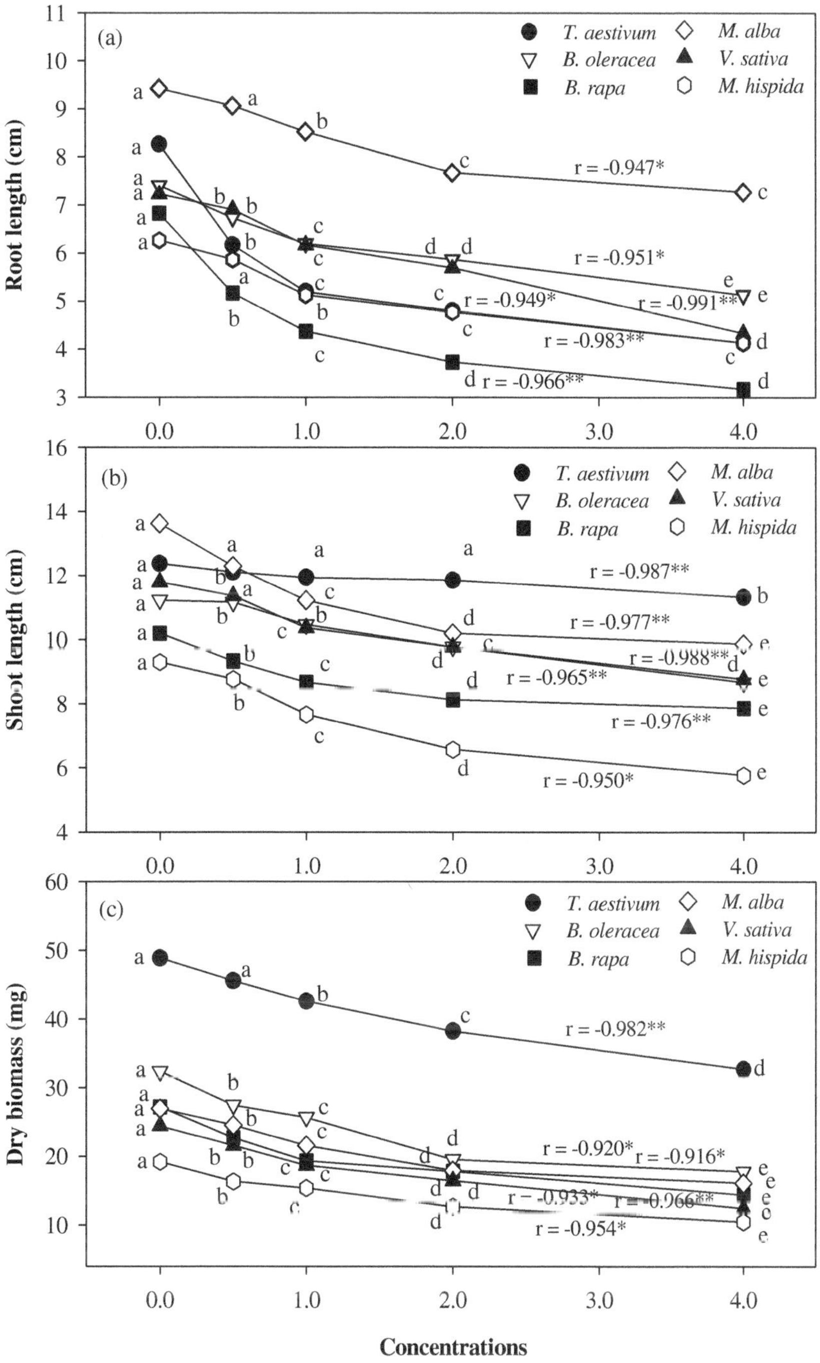

Figure 2. Effect of aqueous extracts of shoot at flowering stage on (a) root length (b) shoot length and (c) dry biomass of test plants (crops as well as weeds).
Different superscript symbols along a curve respresent signinficant difference among themselves at P < 0.05 applying DMRT.
r represent correlation coefficient
* and ** represent significance of correlation at P < 0.05 and P < 0.01 respectively.

a2000), *Cardaria draba* (Kiemnec and McInnis, 2002), *Parthenium hysterophorus* (Batish et al., 2002a; Singh et al., 2003a), *Brassica nigra* (Tawaha and Turk, 2003), *Raphanus raphanistrum* (Norsworthy, 2003) and *Ageratum conyzoides* (Batish et al., 2002b; Singh et al., 2003b,c). All these studies indicate the release of phototoxic chemicals during the preparation of aqueous extracts. Based on this, studies were further extended to explore the impact of *C. bonplandianum* (especially) leaves, as they possessed greater phytotoxicity on the

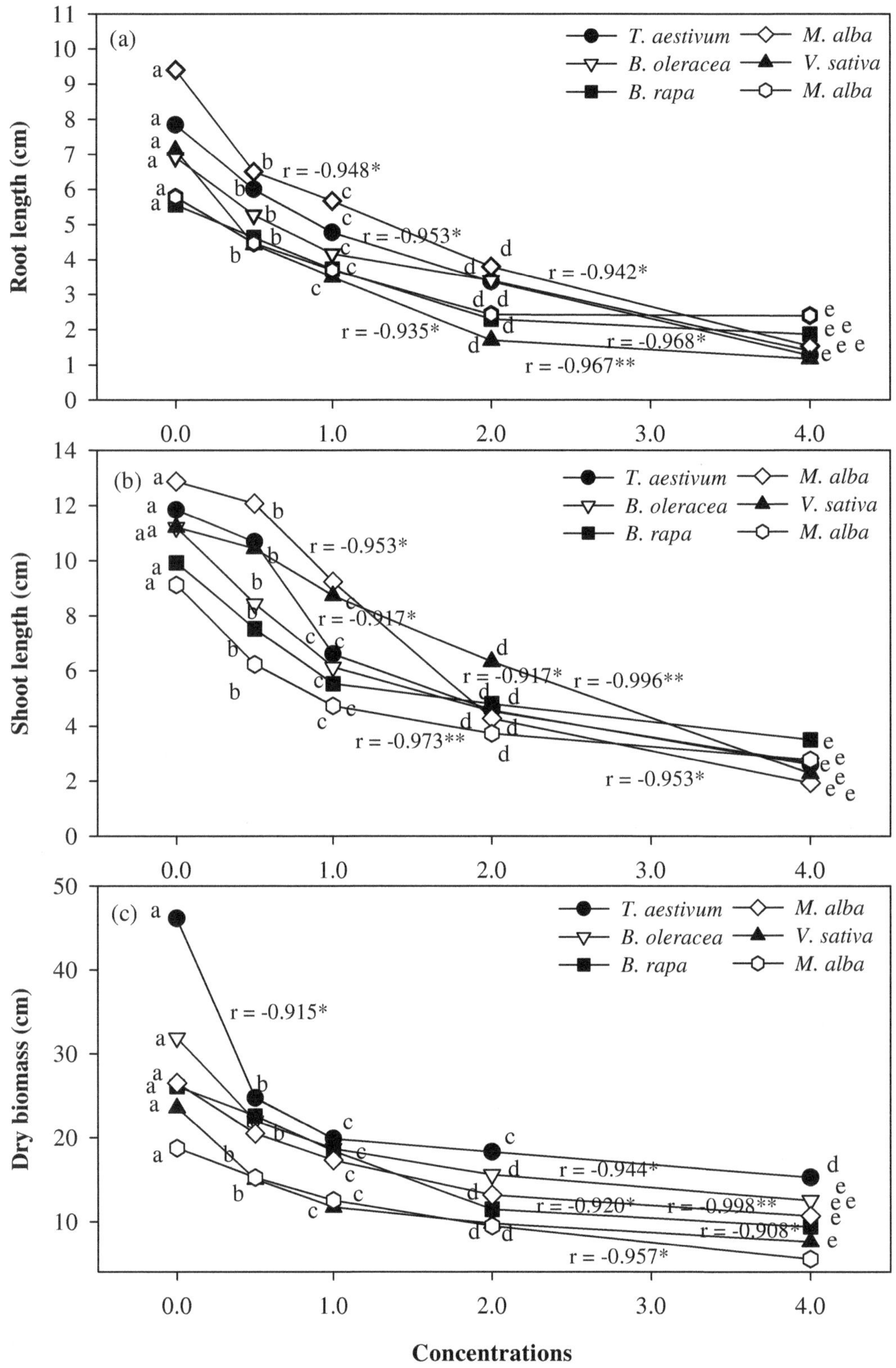

Figure 3. Effect of different concentration of aqueous extracts of leaves at flowering stage on (a) root length (b) shoot length and (c) dry biomass of test plants (crops as well as weeds).

Different superscript symbols along a curve respresent signinficant difference among themselves at $P < 0.05$ applying DMRT.

r represent correlation coefficient

* and ** represent significance of correlation at $P < 0.05$ and $P < 0.01$ respectively

Table 1. Effect on pH, osmotic potential and phenolic content by aqueous extracts root, stem and leaves of *Croton bonplandianum* Baill.

Extracts	pH	Osmotic potential (-bars) mS	Phenolic content (μg/ml)
Root			
0.5	6.59	0.43	113.32
1	6.52	0.59	224.06
2	6.31	0.68	333.03
4	6.12	0.77	543.09
LSD at 5%	0.019	0.036	0.661
Stem			
0.5	6.68	0.59	44.13
1	6.58	0.72	49.73
2	6.41	0.92	58.2
4	6.26	0.99	73.8
LSD at 5%	0.007	0.009	0.318
Leaves			
0.5	6.66	0.46	532.46
1	6.52	0.79	426.13
2	6.41	0.88	512.30
4	6.35	0.96	828.06
LSD at 5%	0.014	0.015	0.445

emergence and growth of weed plants.

ACKNOWLEDGMENT

This research was carried out with financial support from UGC fellowship.

REFERENCES

An M, Pratley JE, Haig T (1999). Allelochemicals dynamics of decaying Vulpia residue and their corresponding biological activity. In Second World Congress on Allelopathy: Critical Analysis and Future Prospects. Abstract no. 50 Lake head University, Thunder Bay, Ontario, Canada. 50: 47-58.

Batish DR, Singh HP, Pandher JK, Arora V, Kohli RK (2002a). Phytotoxic effect of Parthenium residues on the growth of radish and chickpea and selected soil properties. Weed Biol. Manage. 2: 73 78.

Batish DR, Tung P, Singh HP, Kohli RK, (2002b). Phytotoxicity of sunflower residues against some summer season crops. J. Agron. Crop Sci. 188: 19-24.

Bonner J (1950). The role of toxic substances in interaction of higher plants. Bot. Rev. 16: 51-65.

Conway WC, Smith LM, Bergan JF (2002). Potential allelopathic interference by the exotic Chinese tallow tree (*Sapium sebiferum*). Am. Midl. Nat.148: 43-53.

De Candolle MAP (1832). Physiologie Vegetable. Bechet Jeune, Lib., Fac. Med., Paris. Tomme-III pp. 1474-1475.

Evenari, M (1961). Chemical influence of other plants (allelopathy) Handbuch der pflanzen-physiol. 16: 691-736.

Fischer RF, Woods RA, Glavicic MR (1978). Allelopathic effects of goldrod and ashes on young sugar maple. Canadian J. Res. 8: 1-9.

Grummer G, Beyer H (1960). the influence exerted by species of flax by means of toxic substances. In. Harper JL (ed.). The Biology of Weeds. Blackwell, Oxford pp.153-157

Harper LL (1977). Population Biology of Plants. London: Academic Press pp. 273-278.

Ismail BS, Kumar A (1996). Effects of aqueous extracts and residues decomposition of Mikania micrantha H.B.K. on selected crops. Allelopathy. J. 3: 195-206.

Kaul V (1967). Distibution and dispersal of Croton sparsiflorus Morong. J. Ind. Bot. Soc. 46: 154-159

Kiemnec GL, McInnis ML (2002). Hoary cress (Cardaria draba) root extract reduces germination and root growth of five plant species. Weed Technol. 16: 231-234.

Lavabre EM (1991). Weed control. McMillan Education Ltd., London. pp. 169-178.

Mersie W, Singh M (1987). Allelopathic effect of Parthenium (*Parthenium hysterophorus* L.) extract and residue on some agronomic crops and weeds. J. Chem. Ecol. 13: 1739 1749.

Molisch H (1937). Der Einfluss einer pflanze auf die andere. Allelopathic Fischer, Jena pp. 64-67.

Norsworthy JK (2003). Allelopathic potential of wild radish (*Raphanus raphanistrum*). Weed Technol. 17: 07-313.

Pitman AR, WB Duke (1978). Allelopathy in agroecosystems.Annu. Rev. Phytopathology 16: 431-451.

Quayyum HA, Mallik AU, Leach DM, Gottardo C (2000). Growth inhibitory effects of nutgrass (*Cuperus rotundus*) on rice (*Oryza sativa*) seedlings. J. Chem. Ecol. 26: 2221-2231.

Singh SP, Pal UR, Luka K (1989). Allelopathic effect of three serious weeds of Nigerian Savana on germination and seedling vigour of soybean and maize. J. Agron. Crop Sci. 162: 236-240

Singh HP, Batish DR, Pandher JK, Kohli RK (2003a). Assessment of allelopathic properties of *Parthenium hysterophorus* residues. Agric. Ecosyst. Environ. 95: 537-541.

Singh HP, Batish DR, Kaur S, Kohli RK (2003b). Phytotoxic interference of Ageratum conyzoides with wheat (*Triticum aestivum*). J. Agron. Crop Sci. 189: 341-346.

Singh HP, Batish DR, Kaur S, Kohli RK (2003c). Phytotoxic interference of *Ageratum conyzoides*: Role of residue. In Proceedings of Asian Pacific Weed Science Society Conference. Weed Science Society of

Philippines. 19: 739-744.
Sisodia S, Siddiqui MB (2008). Allelopathic effect of Lantana camara on *Bidens pilosa* VEGTOS. 20(1): 29-32.
Sisodia S, Siddiqui MB (2009). Allelopathic potential of rhizosphere soil of Croton bonplandianum on growth and establishment of some crop and weed plants. Afr. J. Agric. Res. 4(5): 461-467.
Sisodia S (2008). "Allelopathic effect of Croton bonplandianum Baill. towards some weed and crop plants" Department of Botany, Aligarh Muslim University, Aligarh pp.181-194.
Tawaha AM, Turk MA (2003). Allelopathic effects of black mustard (Brassica nigra) on germination and growth of wild barley (*Hordeum sponataneum*). J. Agron. Crop Sci. 189: 298-303.
Whittaker RH (1970). The biochemical ecology of higher plants IN: Chemical Ecology. (eds). Sondheimer E, Simeone JB. Academic Press, New York, Pp. 43-70
Xuan TD, Shinjichi T, Hong NH, Khann TD, Min CI (2004). Assessment of phytotoxic action of Ageratum conyzoides L.(Bill goat weed) on weeds. Crop Prot.1:1-8.

Influence of demographic characteristics on adoption of improved potato varieties by smallholder farmers in Mumberes Division, Baringo County, Kenya

Isaac Mwaura Njuguna, Catherine Ngéndo Munyua and Susan Kamuru Makal

Department of Applied Community Development Studies, Faculty of Education and Community Studies, Egerton University, Njoro, Kenya.

Potato (*Solanum tuberosum* L.) is widely consumed as a staple world over with direct consumption by humans as food being 31.3 kg per capita. Ministry of Agriculture through collaborative efforts with other stakeholders has developed improved potato varieties and released them for farmers to plant in Kenya. The main aim of this study was to investigate the influence of selected factors on adoption of improved potato varieties in Mumberes division, Baringo country. The study used a cross sectional survey research design where 128 household heads were sampled and interviewed. Statistical Package for Social Sciences (SPSS) version 17.5 aided data analysis. Descriptive statistics were used to summarize key observations in the study. Ordinal Logit regression was used to test the study hypothesis. It was revealed that perception, access to extension services, family size and access to credit were the key factors influencing adoption of improved potato varieties. The study recommended that there is need to increase access to extension services in the study area. Credit service providers should intensify their provision of these services. The research institutions should endeavor to look for varieties more resistant to blights so as to promote further adoption.

Key words: Adoption, Baringo County, demographic characteristics, improved potato varieties, perception, smallholder farmers.

INTRODUCTION

Irish potato (*Solanum tuberlosum*, L.) is a crop of major economic significance worldwide. Potatoes play a major role in food security due its usefulness as both a cash crop and a food crop. With increasing cereal prices, potatoes have the potential of relieving pressure from the poorest of the poor and promote food security (Hoffler and Ochieng, 2009). Potatoes are ranked fourth after wheat, rice and maize in cultivation on a global scale. Among the tuber crops, potato is the most important and ranks first. It has an estimated annual output of 300 million tonnes cultivated on about 19 million hectares globally (FAO, 2008). Further, FAO asserts that the average production globally is about 17 t/ha whereby the direct consumption by humans as food is 31.3 kg per capita. The importance of potatoes as food in developing countries is because they are grown and eaten locally with little significance in the international market.

In Kenya, potato ranks second after maize in order of importance for utilization purposes (Muthoni and Nyamongo, 2009). About 500, 000 farmers grow this crop on about 120, 000 ha per season where the annual production is about 1 million tons in two growing seasons (MoA, 2008). The average annual production of potato is reported to be 1 million tones and per hectare yield of 7.3 mt/ha (Obare et al., 2010). However, 14.5-20 mt/ha at farm level and 25-35mt/ha under research conditions are attainable.

Food security in Kenya is a measure of available maize in the country. Nonetheless, maize productivity has steadily declined over the years, which has promoted low food security in the country. In addition, the prices of cereals such as maize and rice have been on an upward trend thus increasing the burden on the poorest of the poor (Hoffler and Ochieng, 2009). However, potato may act as a way out of this situation. This is so because potatoes produce more calories compared to cereals and are easy to cook and process. These aspects combined with a short maturity period of about three months imply that potato production is able to address the twin problem of food security and low household incomes that faces smallholder farmers in developing countries (Mpogole et al., 2012).

In the recent years, there has been a steady decline in the output and productivity per unit area of the potato varieties being grown by farmers. This has prompted research institution and organizations to look for newer varieties which are higher in productivity and less resistant to blights and other potato diseases. Improved potato varieties promise increased savings since they are bred with superior characteristics such as pest resistance and high yields (Abong et al., 2009). Through collaborative efforts between Kenya Agricultural Research Institute (KARI), the International Potato Centre (CIP) and the Regional Network for Improvement of Potato and Sweet Potato (PRAPACE), significant developments of several high yielding varieties of potato have been made in Kenya (Kaguongo et al., 2008). The MoA (2007) also indicated that efforts have been focused on development and dissemination of high yielding varieties.

Research on adoption has pointed to factors such as demographic characteristics of famers; for example, level of education, gender, age and experience of the farmers, farmers perception about the technology, family size, technological factors such as complexity, the location of adopter institutions, access to extension services and access to production resources as some of the factors that may influence adoption of agricultural technologies (Ani and Ifah, 2004; Delgado and Johnson, 2005; Idrisa et al., 2010; Mwabu et al., 2006; Namwat et al., 2010; Obare et al., 2010). Given the role that improved potato varieties can play in averting food insecurity, the adoption rate of improved potato varieties is expected to be high for potato farmers world over. To the contrary, this was not the case in the study area. Thus, this study was designed to investigate the influence of demographic factors namely access to extension, access to credit, age, level of education, family size, gender and perception on adoption of improved potato varieties in Mumberes division, Baringo County, Kenya.

Significance of potato sector in food security

Potato (*S. tuberlosum* L.) is a crop of major economic significance worldwide. This crop provides a reliable source of income, employment and food for many populations in the developing countries (FAO, 2008). Globally, potato provides employment and food security to an estimated 800 million people (Hoffler and Ochieng, 2009). As the prices of major staples including wheat, rice and maize continue to rise, millions of people in the developing countries including Kenya are facing food crises. However, the prices of potato have remained relatively stable and thus can be targeted in improving food security in the developing countries (Muthoni and Nyamongo, 2009).

Potatoes have less significance in international trade compared to cereals and therefore they can form a significant source of food in developing countries. Maganga (2012) has pointed out that the potato sector can help improve the incomes of smallholder farmers and subsequently curb food insecurity together with poverty. This is partly because potatoes are grown and consumed locally, mature within a short period (3 to 4 months) and have high yields (about 40 tons/ha). Additionally, potato can be grown in areas with limited land and abundance of labour (FAO, 2008).

According to FAO (2008), Kenya's annual potato output is about 790, 000 tons. Thus, Kenya ranks 8th in Africa and 5th in sub Saharan Africa in potato production. The crop follows maize in order of importance as a staple and plays a vital role in food and nutritional security of the country. Additionally, in high altitudes, it is a major cash crop for smallholder famers and is grown by about 500, 000 farmers on 120, 000 ha. The farmers produce about 1 million tons in two growing seasons (MoA, 2008).

Potato has a superior comparative advantage to major cereals in the high altitude areas. While farmers complete only one planting season for maize, three cropping season are possible with potato in these areas (Muthoni and Nyamongo, 2009). This way, it acts as a more reliable source of income for farmers. In addition, in the high altitude areas, the average revenue for potato is more than twice that of maize (Obare et al., 2010). Thus, in such areas, potato becomes a more realistic venture which should be promoted not only for food security but also to ensure increased revenues for farmers. With the increasing population and low output from agricultural land, use of improved potato varieties may be one way out of the looming food security situation world over.

Adoption of technologies: Significance of demographic factors

For a farmer to decide whether or not to adopt an innovation, there are other aspects and characteristics that come into play. They include farmers' specific demographic characteristics such as age of the household head, education level, family size, marital status and gender of the household head, access to credit services as well as access to extension services. All these factors have been shown to be significant in determining farmer's decision to adopt a technology.

Perceptions act as a filter through which new technologies are taken in and interpreted. Through perception, farmers are able to transform the information they receive into psychological awareness and later make decision of whether or not the information can be utilized. The way a farmer perceives a technology may influence adoption. The technology-specific attributes have been shown in the past to significantly determine farmers' decision to adopt a technology (Idrisa et al., 2010).

Kafle and Shah (2012) noted that farmer's perception on an innovation largely depends on their knowledge and the information they have about the technology. Thus, by evaluating their yield for a short period, farmers are able to prefer a said technology compared to another. More so, base on their perceptions, farmers are able to assess the expected outcomes of any innovation. The actions taken by the farmer many times depend on their evaluation of the outcomes depending on his/her personal perspectives. Since the role of any technological improvement in agriculture is to improve production, adoption of improved potato varieties may be influenced largely by the perceptions that the farmers have (Idrisa et al., 2010). Perception was treated as either positive or negative as this is a subjective aspect emanating from an individual.

Access to extension is a function of whether or not one is visited by extension service providers and the number of times such a farmer is visited. Influence of this factor to adoption is widely acknowledged by various authors (Okoedo-Okojie and Onemolease, 2009; Odoemenem and Obinne, 2010, Namwata et al., 2010). Using knowledge obtained from extension agents, farmers are expected to increase the outputs of their farm enterprises. Extension agents fragment manage the complex agronomic aspects of an agricultural technology by simplifying it for the farmers. This may facilitate adoption of agricultural technologies. Namwata et al. (2010) indicated that access to extension services is a significant factor that influences adoption of improved potato varieties.

Accessing credit helps farmers to invest in buying of raw materials. Therefore, if farmers can access credit within the household, their chances of adopting a said improved crop variety will increase. However in Kenya, there are gender based differences in accessing credit (Mikalista, 2010). Olwade et al. (2010), noted that among the strongest determinants of fertilizer use intensity in Kenya were gender and access to credit. Research has shown that if farmers cannot purchase farm inputs, they are less likely to intensify their adoption of improved agricultural technologies.

Potato production is a relatively labor intensive venture since it increases labor demand during planting, weeding and harvesting (Muthoni and Namgongo, 2009). Thus, it provides employment for family members during peak seasons. Improved potato varieties promise to reduce labor since they are developed with resistance to chemical damage thus facilitating chemical weeding. Subsequently, both families with larger family sizes and those with smaller family sizes would adopt.

Thus adoption may not necessarily be influenced by family size. Gender of the household head is an indicator of marital status. The influence of gender on adoption of agricultural technologies is widely acknowledged. Gender influences division of labor, the type of agricultural commodity to be planted (Nkamleu and Manyong, 2005) and access to extension services (Feder and Umali, 2002; Mikalista, 2010; Namwata et al., 2010; Ofuoko et al., 2009). Delgado and Johnson (2005) reported that gender of the household head would influence the enterprise the family would be involved in. Thus, gender may influence adoption of the improved potato varieties in the study area.

Summary

The past studies reviewed have shown that the said factors are significant in influencing adoption of various agricultural technologies world over. However, specific technologies may vary in their adoption depending on subjective treatment of the technology by the farmers. More so, there are temporal and spatial variations in the adoption patterns of various technologies. In this study, the aim was to investigate if the factors that were believed to influence adoption of other technologies in other parts of the world could have influenced adoption of improved potato varieties in Mumberes division, Baringo County.

Theoretical framework

This study was guided by both the general systems theory as developed by von Bertalanffy (1968) and the diffusion of innovation theory as developed by Rogers (1995). According to the systems theory, components of a system interact with one another to form one whole. The theory's stronghold is its use of the feedback loops to restore equilibrium. The proponents of the theory argue that incase one component of the system changes, the systems equilibrium is compromised and has to be restored using these loops. This is by looking at the two-way flow of technology and information between farmers and the public organizations, through extension, and nature of families and their characteristics that constitute the system. The interdependence of these sub systems within the larger system made it more suitable for a study such as this one.

According to the diffusion of innovation theory, a technology has four main elements that may influence its adoption. They are the innovation itself, the communication channels, time and the social context within which the adopter institutions are located (Rogers, 1995). Viewed from a multidisciplinary perspective, adoption is a multi-dimensional process dependent on many of factors such as perceived profitability, costs of establishment, compatibility with value systems and the ability to com-municate new knowledge and information between developers and among adopters or potential adopters.

Definition of variables

Adoption of an innovation can be viewed from two diverse points. One, it can be viewed as merely the utilization of the technology or innovation which creates a dichotomous dependent variable. Alternatively, it can be viewed as the extent to which the technology is being used or utilized and this creates a continuous dependent variable. This study used dichotomous dependent variable approach where adoption was measured by whether or not famers were growing the improved potato varieties. Lack of adoption was defined by failure to grow the improved potato varieties in the past four years preceding the study.

MATERIALS AND METHODS

Study area

The study was conducted in Mumberes division which is located in Baringo County at an altitude of about 2400 m above sea level. The division has a cool climate with an average annual rainfall of approximately 1425 mm per year and a mean temperature of about 18°C. Night temperatures range between 14 and 18°C. The average slope of the land ranges between 14 and 22% (MoA, 2010). Soils are mainly clay loams with an average pH of between 5.0 and 6.5. These features make this division a high potential area characterized by its ability to promote crop growth. The main agricultural activities in this area are crop and animal agriculture. Various crops are grown which include potatoes, maize, vegetables such as cabbages and kales, and fruits such as plums and pears. However, potato farming is the main agricultural activity for farmers and forms the backbone of agricultural sector in this area. The area has a total population of 21, 891 of which 11,009 are males and 10,882 are females. Also, there are 2660 farming households in Mumberes division (Kenya National Bureau of Statistics, 2010). The division has four locations namely; Mumberes, Chemoswon, Timboroa and Seguton. Their populations are 1253, 482, 581 and 344 households, respectively (KNBS, 2010). All these locations were included in this study.

Population and sample

The study involved a cross sectional survey of all the four locations in Mumberes division. The study was carried out between March and May 2013. A sample size of 128 households was used which was determined using the formula suggested by Gatotoh et al. (2011).

The formula is stated as:

$$n = NC^2/C^2 + (N-1)e^2 .$$

Where n is the sample size, *N* is the population size, *C* is the coefficient of variation, while *e* is the standard error.

Nassiuma (2000) asserts that the coefficient of variation should be less than 30% while the standard error should lie between 2 and 5%. In this study, C was set at 23% while the standard error was set at 2%. Thus,

$$N = 2660(0.23^2)/0.23^2 + (2660-1)0.02^2 = 128$$

The locations were equally represented in the study as the 128 household heads were selected proportionately from all the locations as shown in Table 1.

Data collection

Primary data for this study was collected using pre-tested structured interview guides. The interview guides were preferred because the population is comprised of illiterate and semi literate persons who cannot complete questionnaires (Ministry of Planning and National Development, 2005). Household heads that were purposively selected from the households were interviewed. Heads of households are at the core of decision making in many households of African communities (Idrisa et al., 2010). The instrument contained both close-ended and open-ended questions. Close ended questions provided a basis for quantifying the data obtained while the open ended ones provided useful information that was used in explaining observations in the study.

The instrument had three sections. Section I contained information on demographic characteristics of the farmers, including age, gender, education level, experience and family size. Section II contained information on social aspects of the farmer, including awareness and perceptions about improved potato varieties, and access to extension services. The third section collected data on farmers' economic factors namely access to credit services, other sources of income, land owned and actual land committed to potato production.

Statistical analysis

Descriptive statistics were used to summarize the main findings of the study. Ordinal logit regression model was used to test the study hypothesis. The logistic models have been widely used in adoption studies due to their slightly flatter tails compared to the probit models. The data was analyzed with the aid of SPSS version 17.5. The logit regression equation was stated as follows:

$$Adoption\ (Y_i) = \beta_0 + \beta_1 AGEYR + \beta_2 EDUCLEV + \beta_3 PERCEPT + \beta_4 FAMSIZE + \beta_5 EXTENS + \beta_6 CREDIT + \beta_7 GENDER + \varepsilon$$

Table 1. Summary table for proportionate sampling.

Location	Population (households) (n)	Proportion (p) (n/N)	Number to be sampled (n/N)x 128
Mumberes	1253	0.471	60
Chemoswon	482	0.181	23
Timboroa	581	0.218	28
Seguton	344	0.130	17
Totals (N)	2660	1	128

Adopted from Kenya National Bureau of Statistics (2009).

Table 2. Definition of variables as used in the model.

Variable	Type	Measurement
Dependent variable		
Y_i	Dummy	1 = Adopted; 0 otherwise
Explanatory variables		
Gender (GENDER)	Dummy	1 = Male; 0 otherwise
Age of the household head (AGEHH)	Continuous	Age in years
Education of the household head (EDUCLEV)	Continuous	Total number of years in school
Family size (FAMSIZE)	Continuous	Total number of persons in the household
Access to credit (CREDIT)	Dummy	1 = Household head receives credit; 0 otherwise
Amount of credit received (AMOUCRED)	Continuous	The total amount of credit received for farming per year
Access to extension services (EXTENS)	Dummy	1 = Household head receives extension services; 0 otherwise
Perception (PERCEPT)	Dummy	1 = Farmer perceives that improved potato varieties are tastier, cook faster, increase family incomes, can improve food security, mature faster, have a ready market; 0 otherwise

Measurement of dependent and explanatory variables

Adoption (the dependent variable) was dichotomized such that a value of 1 was for an adopter farmer while 0 was given or a non adopter. The explanatory variables namely age of the household head, education level, family size, access to credit services, access to extension services and marital status and perceptions were defined as shown in Table 2.

RESULTS AND DISCUSSION

The demographic characteristics of the respondents that were studied in this study are gender of the household head, age of the farmer, level of education of the farmer, household size and marital status of the respondents.

Most of the respondents (68%) were males indicating that many of the households in the division are male-headed. Majority of the respondents (more than 70%), were aged between 20 and 40 years with average age being 34.22 years. Additionally, results showed that most of the respondents (about 70%) had at least primary level education with the average years in school being 1.59. Many households (about 58%) had family sizes of between 5 and 9 members with average family size of 4.59 members. The study further showed that only 17% of the respondents were able to access extension services while the rest did not. Furthermore, about 78% of the respondents were able to access credit services while the rest did not. The study also showed that only 26.8% of the farmers had adopted improved potato varieties and were growing them on their farms. Farmers in the study had land sizes ranging between 0.25 and 5 ha allocated to potato production. The average land size allocated for potato production was 1.25 ha.

This study was designed with a key objective of identifying the factors that may be influencing adoption of improved potato varieties in Mumberes division Baringo country. To achieve this, we performed a regression analysis and the results of which are indicated in Table 3. The pseudo R^2 value obtained was 0.645. This indicates that the explanatory variables can accurately predict the dependent variable and thus be able to explain farmer's adoption of improved potato varieties. The overall model thus is significant at 95% confidence interval.

From the results, it can be shown that family size was a significant factor influencing adoption of improved potato varieties. Large families provide readily available and

Table 3. Logit analysis for the adoption of improved potato varieties.

Parameter	B	Std. Error	Z	Significance
(Constant)	.081	.223	.364	.717
Gender of the respondent	-.704	.095	-7.437	.000
Age of the farmer	-.006	.004	-1.379	.171
Family size	.009	.011	.828	.409
Education level	-.074	.064	-1.160	.248
Extension services	.644	.139	4.647	.000
Access to credit	.595	.133	4.461	.000
Farmer perception $R^2 = 0.645$	.276	.095	2.891	.005

cheap labor which is important in adoption decision. Improved potato farming is labor intensive especially during peak seasons, that is, planting, weeding and harvesting. Availability of cheap labor thus will enhance adoption (Muthoni and Namgongo, 2009).

Access to extension services was also another factor that had significant influence on adoption. Although, only 17% of the farmers received extension services, this factor is significant. Low access to extension services has been noted to be a limiting factor to increasing agricultural productivity in Sub-Saharan Africa and in many developing countries (Kabungo, 2008; Okoedo-Okojie and Onemolease, 2009). Access to extension influences farmers' profitability as farmers become equipped with information by the extension agents (Okwoche et al., 2012). This may influence their adoption of agricultural technologies as they know which agricultural investments are more profitable to them. Extension services also provide farmers with the required agronomic knowledge which allows farmers to accept or reject technologies (Namwata et al., 2010). The information required by potato farmers includes availability of seeds, agronomic practices such as fertilizer application and spacing as well as available of markets. The results from this study supported Barungi et al. (2013) who acknowledged that although access to extension services was among the important factors that influenced adoption of agricultural technologies, the level of access to extension is important in the adoption process.

Access to credit services was another significant factor that influenced adoption of improved potato varieties by the farmers. Access to credit was measured by whether or not the farmers were receiving any financial support for their agricultural production. Access to credit allows farmers to purchase farm inputs which are major constraint in agricultural production especially in Sub Saharan Africa (Okwoche et al., 2012). Smallholder farmers face the challenge of lack of collaterals to secure credits thus leading to low access to credit (World Bank, 2008). However, this may lead to a vicious cycle of poverty as the poor are trapped in a cycle of income inequality as they try to repay the credit with other sources of income on the farm or off the farm. Simtowe and Zeller (2006) argued that access to credit allows farmers to adopt even risky technologies since their household's-risk bearing ability is boosted while at the same time relaxing their liquidity constraint. The findings supported Barungi et al. (2013), Namwata et al. (2010) and Simtowe and Zeller (2007) who reviewed various factors that affect adoption in Uganda, Tanzania and Malawi, respectively. The authors independently highlighted credit access as an important factor in adoption of improved agricultural technologies. The authors reported that actual access to credit had a significant influence on adoption in their study areas.

Farmer perception was another key significant factor in influencing adoption in the study area. Farmers had strong positive perceptions about high yields, short time to maturity, good taste, market availability and increased incomes from the improved potato varieties may have influenced the adoption of improved potato varieties in Mumberes division. Farmers agreed that improved potato varieties were tastier, matured early, increased household incomes and yields more compared to the old varieties they had been growing. The attribute of early maturity is especially important in food security as well as helping families to meet their economic obligations such as paying schools fees and loans. Thus, this may have been a key motivating factor to adopting the improved potato varieties. Additionally, it allows farmers to have many cropping seasons in a year thus reducing pressure on land especially where land pressure is dominant.

As noted by Idrisa et al. (2010), early maturity allows a farm produce to be harvested early thus avoiding droughts or heavy rains. The results from this study are in line with Kaguongo et al. (2008), Idrisa et al. (2010) and Chi (2008) who found out that a farmer's perception about a technology actually influences their adoption of such a technology. Attributes related to a technology largely influences how a farmer will perceive its importance. This in turn will influence their decision of whether or not to adopt (Idrisa et al., 2010).

The perception by farmers that improved potato varieties were not resistant to blight may lead to low adoption among the study population. This may be because blights require a lot of capital investment in controlling

controlling since only chemical control is possible. In addition, losses as a result of blights in African countries go as high as 75% on varieties that are susceptible. Kaguongo et al. (2008) attributes this to the high number of times farmers in the tropics have to apply fungicides in order to control late blights. The authors note that farmers can spray up to ten times with an average of 5 sprays per growing season in Kenya.

CONCLUSION AND RECOMMENDATION

The decision of a farmer to adopt a technology or not is influenced by many factors. These factors are either farmer related (farm and nonfarm) or they may be technology related. Technology related factors are mainly judged by famers through their perception. Farmer characteristics including age, gender of the household head, family size, access to extension services and access to credit services are also key factors in in-fluencing adoption. In this study, the key factors that were significant in influencing adoption were perception, family size, access to extension services and access to credit services. Farmer perception was also key in influencing adoption. Short time to maturity, good taste, increased incomes and market availability were the main aspects that motivated farmers to adopt improved potato varieties.

However, susceptibility to blights was also a key issue that farmers felt needs to be addressed in light of the improved potato varieties. The study recommends that there is need to increase farmers' access to extension by increasing the number of extension agents in the study area. Financial institutions also need to increase their presence in the study area and educate farmers on the need to have increased access to credit services. To ensure that there is increased adoption and for it to be main-tained, there is need for research institutions to keep researching on improving the resistance to blight as this may deter further adoption due to the susceptibility of the current grown varieties.

Conflict of Interest

The authors have not declared any conflict of interest.

REFERENCES

Abong GO, Okoth MW, Karuri EG, Kabira JN, Mathooko FM (2009). "Evaluation of selected Kenyan potato cultivar for processing into French fries" J. Anim. Plant Sci. 2(3):141-147

Ani O, Ogunnika O, Ifah SS (2004). "Relationship between socio-economic characteristics of rural women farmers and their adoption of farm technologies in Southern Ebonyi State, Nigeria." Int. J. Agric. Biol. 6(5):802-805

Barungi M, Ng'ong'ola DH, Edriss A, Mugisha, J, Waithaka. M Tukahirwa J (2013). "Factors influencing the adoption of soil erosion control technologies by farmers along the slopes of Mt. Elgon in Eastern Uganda. J. Sustainab. Dev. 6(2):9-25.

Chi TTN (2008). Factors affecting technology adoption among rice farmers in the Mekong Delta through the lens of the local authorial managers: an analysis of qualitative data. Omonrice 16:107-112

Delgado OJ, Johnson JE (2005). Agroforestry adoption in Cape Verde. Small-Scale Forest Econ. Mgt & Policy, 4(2): 205-214.

Food and Agricultural Organization, (FAO) (2008). International year of the potato. Available at: http://www.fao.org/agriculture/crops/thematic-sitemap/theme/hort-indust-crops/international-year-of-the-potato/en/ Retrieved January 20 2012 from

Gatotoh AM, Omulema BEE, Nassiuma D (2011). "Correctional attitudes: an impetus for a paradigm shift in inmate rehabilitation" Int. J. Hum. Soc. Sci. 1(4):263-270

Hoffler H, Ochieng BO (2009). "High Commodity Prices-Who Gets the Money? A Case Study on the Impact of High Food and Factor Prices on Kenyan Farmers". Heinrich Boll Foundation, Berlin, Germany.

Idrisa YL, Ogunbameru BO, Amaza PS (2010). "Influence of farmers' socio-economic and technology characteristics on soybean seeds technology adoption in Borno State, Nigeria" Afr. J. Agric. Res. 5(12):1394-1398.

Junge B, Deji O, Abaidoo R, Chikoye D, Stahr K (2009). "Farmers' Adoption of Soil Conservation Technologies: A Case Study from Osun State, Nigeria" J. Agric. Educ. Ext. 15(3):257- 274

Kabungo CVD (2008). "Evaluation of Irish Potato production and marketing performance: A case study of Mbeya Rural District, Mbeya Region, Tanzania." Msc. Dissertation, Sokoine University of Agriculture, Morogoro, Tanzania.

Kafle B, Shah P (2012). "Adoption of improved potato varieties in Nepal. A case of Bara District" J. Agric. Sci. 7(1):14-22.

Kaguongo W, Gildemacher P, Demo P, Wagolre W, Kinyae P, Andrade J, Forbes G, Fuglle K, Thlele G (2008). "Farmer practices and adoption of improved potato varieties in Kenya and Uganda. International Potato Center (CIP), Lima, Peru". Social. Science Working paper No. 5 http://cipotato.org/wp-content/uploads/2014/08/004365.pdf

Lupilya GS (2007). "Assessment of social support projects for vulnerable groups towards poverty reduction: A case study of TASAF in Bukoba district." M.A Dissertation, Sokoine University of Agriculture, Morogoro, Tanzania.

Maganga AM (2012). "Technical efficiency and its determinants in potato production: evidence from Dedza district, central Malawi" Am. Eur. J. Agric. Environ. Sci. 12(2):192-197.

Meghani Z (2007). Values, Technologies and Epistemology. Agric. Hum. Values. 25:25-34.

Mikalista, S. M., (2010). 'Gender specific constraints affecting technology use and household food security in Western province of Kenya', Afr. J. Food Agric. Nutr. Dev. 10 (4):2324-2343.

Ministry of Agriculture (MoA) (2007). Economic review of agriculture. The Central Planning and Monitoring Unit. Nairobi: Ministry of Agriculture.

Ministry of Agriculture (MoA) (2000). National policy on potato industry. Policy reforms to revitalize the potato industry. Ministry of Agriculture, Nairobi.

Ministry of Agriculture (MoA) (2010). Koibatek district Annual Report. Unpublished

Ministry of Agriculture (MoA) (2011). Food Security Situation As At 30th June, 2011.

Ministry of Planning and National Development (2005). "Koibatek District Strategic Plan 2005-2010 for Implementation of National Population For Sustainable Development". National Coordination Agency for Population and Development, Nairobi

Mpogole H, Mlambiti ME, Kadigi RM (2012). "Round potato (*Solanum tuberosum)* production in south highlands of Tanzania: are smallholder farmers becoming commercial?" J. Agric. Ext. Rural Dev. 4(14):385-391.

Muthoni J, Nyamongo DO (2009). "A review of constraints to ware Irish potatoes production in Kenya" J. Hort. For. 1(17):98-102.

Mwabu G, Mwangi W, Nyangito H (2006). "Does adoption of improved maize varieties reduce poverty? Evidence from Kenya". Poster paper prepared for presentation at the International Association of Agricultural Economists Conference, Gold Coast, Australia.

Namwata BBL, Lwelamira J, Mzirai OB (2010). "Adoption of improved agricultural technologies for potatoes (*Solanum tuberosum*) among farmers in Mbeya Rural District, Tanzania: A case of Ilungu Ward" J. Anim. Plant Sci., 8(1):927-935.

Nkamleu GB, Manyong VM (2005). "Factors affecting the adoption of agroforestry practices by farmers in Cameroon." Small-scale Forest Econ, Mgt. Policy 4(2):135-148.

Obare GA, Nyagaka DO, Nguyo W, Mwakubo SM (2010). "Are Kenyan smallholder allocatively efficient? Evidence from potato producers in Nyandarua North District." J. Dev. Agric. Econ. 2(3):078-085.

Okoedo-Okojie DU, Onemolease EA (2009). "Factors affecting adoption of yam storage technologies in the northern ecological zone of Edo State, Nigeria." J. Hum. Ecol. 27(2):155-160.

Okwoche VA, Asogwa BC, Obinne PC (2012). "Agricultural information utilization among rural sorghum farmers in Benue State of Nigeria." Eur. J. Sci. Res. 76(2):198-207.

Rogers EM (1995). "The diffusion of innovations," 4th Edition. New York: Simon & Schuster.

Simtowe F, Zeller M (2006). "The impact of access to credit on the adoption of hybrid maize in Malawi: an empirical test of agricultural household model under credit market failure." Munich Personal RePec Archive (MPRA) Paper No. 45.

Development and performance of a laboratory-scale passive solar grain dryer in a tropical environment

S. V. Irtwange* and S. Adebayo

Department of Agricultural and Environmental Engineering, University of Agriculture, Makurdi, Nigeria.

A laboratory-scale passive solar grain dryer was developed and its performance evaluated in Makurdi metropolis, Benue State, Nigeria. The dryer consist of a solar collector panel, a thermal storage unit and a drying chamber. The top of the collector is made of one layer of 4 mm thickness of colourless glass sheet as glazing. The absorber material used was corrugated 0.5 mm thick zinc roofing sheet painted black. The thermal storage unit and the drying chamber were built of wood because of its good insulation properties. The dryer was evaluated using 10 kg of freshly harvested maize at 32.8%wb. The performance evaluation results obtained showed that the mean drying rate of the dryer was 0.7 kg/day per every 10 kg of corn whereas sun-drying rate was 0.3125 kg/day comparatively. The solar dryer has considerable advantages over the traditional sun drying method in terms of faster drying rate, less fear of spoilage by micro-organisms when crop is harvested at high moisture content and handling convenience. Savings in time was achieved by using the solar grain dryer as against the traditional sun drying. It took 4-days to dry the corn to moisture content of 13.1%wb using the passive solar dryer while it took 8-days to dry to 13.4%wb under sun drying. Commercial sizes of the solar dryer can be amplified and produced for community level cooperative use and for prospective investors to fast track agricultural development in the rural areas.

Key words: Passive, solar, grain dryer, development, performance, laboratory-scale

INTRODUCTION

The domestic demand for energy now substantially exceeds supply and the trend is ominous. While the gap is not new, its increasing magnitude is alarming. In recent years, the public has become more concerned about the rapid depletion and escalating cost of fossil fuels, which of course are finite in supplies. These concerns have focused worldwide attention to the potential of harnessing the sun's power in new and varied form to meet society's growing energy needs. The importance of solar drying is increasing worldwide, especially in areas where the use of the abundant, renewable and clean solar energy is essentially advantageous. The supply of solar energy is abundant in most locations in Nigeria where solar heat is intense virtually all the year round (Irtwange, 1991). This is because Nigeria lies between latitude 4° to 14° North of equator geographically, which is in the solar belt. On the average, the yearly total solar energy falling on a horizontal surface in Nigeria is about 2,300 kwh/m^2 (Fagbenle, 1991). This energy could be converted into mechanical, electrical or chemical to be used in various fields, such as the production of electricity, irrigation, drying of crops and fishery products, space heating, water heating and air conditioning (Arinze, 1983). Drying crops by solar energy is of great economic importance the world over, especially in Nigeria where most of the crops and grain harvests are lost to fungal and microbial attacks (Itodo et al., 2002). In realisation of these, research and development programmes in solar energy have been carried out in Nigeria. Some methods of utilizing solar energy have reached a stage of development where they can compete economically with methods of using conventional energy source (Arinze, 1983). The methods used are simple and often crude but reasonably effective. Basically, crops are sprayed on the ground or platforms often with no pretreatment and are

*Corresponding author. E -mail: svirtwange@yahoo.com.

are turned regularly until sufficiently dried.

In most advanced segments of the society, artificial drying has in many cases supplemented traditional sun drying in order to achieve better quality control, reduce spoilage and in general cut down on losses. The development of artificial, large scale drying processes has resulted in high cost of labour and fossil fuels (Paul, 2001). Preservation of agricultural products by open air-drying in the sun was presumably one of the first conscious and purposeful technological activities undertaken by humanity (Imre, 1995). Traditionally, open-air solar drying methods are based on long-term experience and continue to be used all over the world to dry agricultural products in order to preserve them. Over the last two decades, open air drying has gradually become more and more limited because of the requirement for a large area, limitation of time, the possibilities of quality degradation, high level of dust and atmospheric pollution from the air, cloudiness and rain, intrusion from animals and man, infestation caused by birds and insects and inherent difficulties in controlling the drying process (Imre, 1995).

Global requirements can be and should be considered from an economic view-point, with the main influencing factors being savings and costs (Sadiku et al., 2001). Solar drying is not simply a method for substituting fossil fuels by solar energy, but is a technology based process for producing dried materials of the required quality (Imre, 1995). It influences the marketing capacity and income generating potential, since a higher price can be obtained for products of improved quality and for these reasons, technologically and scientifically based efforts have been made over the last three decades to development, design, construction and operation of solar dryers. The advent of higher charges for fossil fuels has stimulated the renewed interest in solar agricultural dryers.

Over the past three decades, increasing interest has been paid to the development of solar agricultural dryers which make use of known principles of heliotechnology in order to combat some of the principal disadvantages of classical sun drying. Sato and Irie (1981) conducted experiments on the solar drying of paddy and grasses in vinyl plastic hot houses which are in wide use among farmers in Japan, indicating that the farmer should be the focus of all on-farm research efforts. Sharma et al. (1993) described the design and performance of a solar fruit and vegetable dryer comprising plastic covered flat plate collectors, drying chamber and thermally and acoustically insulated pipes joining the two. The experimental results suggest that, even under unfavorable fall weather conditions, the unit is able to produce good quality products. Moreover, due to the low investment required, the solar dryer is predestined for applications on small farms. Soponronnarit (1995) reviewed the research and development work in solar drying conducted in Thailand during the past 15 years (since 1980s). He found that, in term of techniques and economy, solar drying for some crops such as paddy, multiple crops and fruit is feasible. However, the method has not been widely accepted by farmers. Bala et al. (1995) presented a technique for optimization of natural-convection solar dryers. The optimal design was specified for the condition of Bangladesh. The optimum design was a relatively long collector, a thin grain bed and a negligible chimney height. Mumba (1995) developed a photovoltaic-powered forced-circulation grain dryer for use in the tropics. From performance testing results, it was indicated that the dryer has a capacity of 90 kg (maize) for drying from an initial moisture content 33.3% to under 20% during one day. Faborode et al. (1995) carried out experiments to investigate the effects of forced air drying and intermittent resting on the fundamental drying mechanism of fermented cocoa beans and the quality characteristics of the resulting raw cocoa. The results indicate that after an initial warm-up and surface drying period of very short duration, forced air drying of cocoa beans involves a succession of falling rate periods, the number of which appears to depend on drying temperature and post-rest moisture content if the beans are subjected to intermittent resting. Karathanos and Belessiotis (1997) reported the sun and solar air drying kinetics of some agricultural products that is, sultana grapes, currants, figs, plums and apricots. The drying rates were found for both solar and industrial drying operations. The industrial drying operation resulted in a product of superior quality compared to products dried by solar dehydration. Soponronnarit et al. (1997) states that solar natural convection drying using a block-shape dryer made of plywood coated with black colour and transparent plastic cover took 6-7days for completing one batch of fresh bananas weighing 25kg. Ekechukwu and Norton (1999) presented a comprehensive review of the various designs, details of construction and operational principles for a variety of practical solar-energy drying systems. The appropriateness of each design type for applications used by rural farmers in developing countries was discussed.

Gallali et al. (2000) compared products dried by solar dryers and natural sun drying. The study indicated that using solar dryers gives more advantages than natural sun drying, especially in terms of drying time. Aboul-Enein et al. (2000) reported a parametric study of a solar air heater with and without thermal storage for solar drying applications. An optimization process for a flat-plate solar air heater with and without thermal storage was carried out. Roosevelt et al. (2000) developed a forced type of solar dryer set-up with an electrical back-up in order to achieve faster drying rates for higher productivity and economy in drying. They conclude that by controlling the drying temperature and relative humidity in solar drying, it is possible to achieve physical characteristics closer to open drying for higher productivity and cost-effective results. Hodali and Bougard (2001) designed an adsorption unit of silica gel and integrated it in a crops solar drying installation. The integration of the adsorption unit allowed improving the quality of the dried product and permitted a cyclic operation of drying over two days by reducing the drying period from 52 to 44 h. Leon et al. (2002) presented a review of existing evaluation methods and the parameters generally considered for evaluation of solar food dryers. These parameters were classified as physical features of the dryers; thermal performance; quality of dried product; cost of dryer and payback period. Pangavhane et al. (2002) proposed a design, development and performance testing of a new convection solar dryer. The drying time of grapes was reduced by 43% compared to the open sun drying. El-Sebaii et al. (2002) reported a study of an indirect type natural convection solar which investigated experimentally and theoretically for drying grapes, figs, onions, apples, tomatoes and green peas. The drying constants for the selected crops were obtained from the experimental results and were correlated with the drying product temperature. Bena and Fuller (2002) developed a direct-type natural convection solar dryer with simple biomass burner. It was expected to be suitable for small-scale processors of dried fruits and vegetables in non-electrified areas of developing countries. Bennamoun and Azeddine (2003) studied a simple, efficient and inexpensive solar batch dryer for agricultural products through simulations. It was suggested that the study could be developed for other agricultural products and for the behaviour of solar dryer in different seasons. Bahnasawy and Shenana (2004) developed a mathematical model of direct sun and solar drying of some fermented dairy products. The model was able to predict the drying temperatures at a wide range of relative humidity values. Chen et al. (2005) developed an experimental closed-type dryer associated with a photovoltaic system. Lemon slices were dried using the closed-type solar dryer and results were compared with hot air drying at 60°C. The results indicate that the dried lemon slices using a closed-type solar dryer has better general levels of quality in terms of sensory parameters. Tunde-Akintunde et al. (2005) carried out drying of bell-pepper using sun, solar and artificial air drying methods. The drying curves and rates obtained indicated that drying of bell-pepper was done in two drying rate periods, the constant drying rate period (mainly) and the falling drying rate period. The existence of the constant drying rate period was more pronounced in the artificial air-drying method than in the other two drying methods and the drying process was also faster.

Ezekoye and Enebe (2006) successfully developed and characterized a modified passive grain solar dryer using locally available material in Nigeria. The dryer can be enlarged for large-scale drying and commercial purposes by increasing the collector size and adding more number of trays. There is no need of carrying the crops inside during the nights in order to avoid re-wetting since the dryer is sealed with glass and wood to protect the samples from dew and rain. Zhang et al. (2006) states that increasing concerns over product quality and production costs have motivated researchers to investigate and the industry to adopt combination drying technologies but current applications are limited to small categories of fruits and vegetables due to high start-up costs and relatively complicated technology as compared to conventional convection drying. Zomorodian et al. (2007) used a new approach for employing solar radiation as the main source of energy for paddy drying. The rough rice solar dryer was a cross flow and an active mixed-mode type with a new and an efficient timer assisted semi-continuous discharging system. The maximum capacity of the dryer was about 132 kg of rough rice with initially 27%db down to 13%db final moisture content in 3 h of drying period (11 - 14 with 865 W/m^2 average incident solar radiation and average ambient temperature of 25°C). Ferreiraa et al. (2007) determined the performance of a hybrid dryer in which banana slices were dried in the device and the results were compared with natural sun drying and artificial drying. Results showed that the time required by the samples to reach the desired final moisture content was lower in the hybrid dryer when compared with natural sun drying and with the artificial dryer, to similar outlet air temperatures. Tasirin et al. (2007) carried out drying kinetics of bird's eye chilli in fluidized bed dryer and to investigate the performance of the fluidized bed dryer in reducing the moisture content of dried chilli up to 16% as comparable to the dried chilli in the market. Futhermore, dried bird's eye chillies tested in order to determine the suitability of using a fluidized bed dryer compared to sun drying. Comparing the dried products, it was observed that the bird's eye chillies dried using fluidized bed dryer were better in quality compared to those dried under the sun. Tembo et al., 2008 carried out a study to evaluate the effect of different stages of ripening on the quality of *Ziziphus mauritiana* fruits during drying. The fruits were graded into green, yellowish-brown and brown categories and these formed the treatments. Some of these fruits were blanched before drying for 1, 2 and 3 weeks under the solar dryer and the open sun drying methods. Sreekumar et al. (2008) developed and tested a new type of efficient solar dryer, particularly meant for drying vegetables and fruit. The dryer was loaded with 4 kg of bitter gourd having an initial moisture content of 95% and the final desired moisture content of 5% was achieved within 6 h without losing the product colour, while it was 11 h for open sun drying. Sopian et al. (2009) has identified numerous types of solar dryers that have been designed and developed in various parts of the world, yielding varying degrees of technical performance. Basically, there are four types of solar dryers; (a) direct solar dryers, (b) indirect solar dryers, (c) mixed-mode dryers and (d) hybrid solar dryers. The paper is a review of these types of solar dryers with aspect to the product

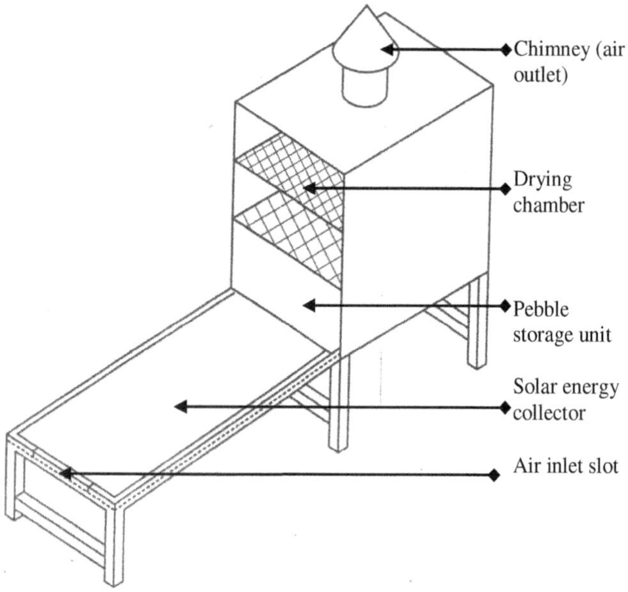

Figure 1. Laboratory-scale passive solar grain dryer.

being dried, technical and economical aspects. Giner (2009) states that food drying kinetics is usually studied by measuring the product average moisture content as a function of time, known as the drying curve, for constant air conditions. The slope of this curve is the drying rate.

In the developing countries and in rural areas the traditional open-air drying methods should be substituted by the more effective and more economic solar drying technologies (Imre and Palaniappan, 1996). This study is therefore devoted to the development, design, construction and performance of a passive solar grain dryer for small scale farmers in a tropical environment. Kisten and van Zyl (1998) defines a small scale farmer as one whose scale of operation is too small to attract the provision of the services he/she needs to be able to significantly increase his/her productivity.

MATERIALS AND METHODS

Description of passive solar grain dryer

The design concept of the solar grain dryer is to collect the solar energy through a solar collector and use it to heat up a mass of air and then pass it through a drying chamber by natural convection. Hence the heat supply to the grain is by indirect absorption of solar radiation. The solar grain dryer was developed, constructed and assembled as shown in Figure 1. The dryer consist of the following components: solar energy collector, drying chamber, storage unit, air vents and dryer stands.

The collector consists of the solar absorbing plate, which is made of corrugated 0.5 mm thick zinc roofing sheet of 1200 mm long and 600 mm wide. The collector casing is made of wood to prevent escape of heat. The top of the collector is made of one layer of 4mm thickness of colourless glass sheet 1200 mm long and 600 mm wide, which is more efficient than other materials (Sadiku et al., 2001). A gap of 0.55 mm between the glass cover and the absorber plate forms the air passage. According to Sayigh (1979), the optimum air gap between the absorber and the glass covers is over 4 cm and owing to the side shading effect caused by the collector box, less than 8 cm. The drying chamber is the place where the drying function takes place. It is fabricated from wood of 25 mm thickness to minimize heat loss. The drying chamber houses a batch tray, which would hold up to 20 kg of shelled maize or sorghum. The batch tray measuring 680 x 680 x 200 mm high is made of stainless steel wire mesh to avoid rusting. The access door for loading and off loading of the grains is pos-itioned at the side of the drying chamber to reduce shadow during handling. The storage unit is matched with the drying chamber and consist of gravel occupying a volume of 800 x 620 x 300 mm or 0.15 m^3, which is the lower limit of volume of 0.15 to 0.35 m^3 recommended by Goswani (1986). The rock pebbles for storage are painted black for better heat absorption. The function of the storage unit is to store heat during the day and release the same during off sunshine hours. There are two air-vents generally referred to as inlet air vent and outlet air vent. The air intake into the collector is through a 300 x 50 mm slot made through the col-lector wooden casing between the absorber plate and the bottom of the collector, which forms the airflow duct. The opening into the bottom of the pebble bed is 180 mm in diameter, supported by a perforated screen at the bottom to support the pebbles and allow for air circulation through the pebbles. The outlet air vent is the chimney which is situated on the top of the drying chamber. It is made of metal tube and co-rrugated roofing sheet painted black which serve the same pu-rpose as the solar collector absorber. It is circular in shape me-asuring 170 mm high and of cross sectional area of 254.4 mm^2. The chimney height is in accordance with a technique for optimization of natural-convection solar dryers presented by Bala et al. (1995). The drying chamber and solar collector stands constitute the dryer stands. The dryer stand is a set of four with 350mm height from the ground. The material selected for the stand is 40 x 40 mm hard wood because of its availability and durability. Makurdi, the Benue State Capital, lies at latitude 07°41' N, longitude 008°37' E and has an elevation of 97m, 318'. The collector slope of 7° which is an approximation of the latitude was used in the design. The solar collector was therefore mounted on a stand fabricated from a hard wood inclined at 7° to the horizontal south facing. Since the entire load is expected to come from the collector during sunshine on daytime operations and since the design is for all-year applications, this slope agrees with the recommendations given by Hsieh (1986).

Experimental procedure

The dryer was evaluated using ten kilogram (10 kg) of freshly harvested shelled maize with initial content of 32.8%wb which is slightly above the optimum moisture content range of 24 - 30%wb recommended during harvest for minimum loss (Brooker et al., 1957). The same quantity of maize at same moisture content was sun dried to allow comparison of the two drying modes. The experiment was replicated three times with the moisture content of the maize being measured daily in morning (9.00am) and in the evening (5.00pm) during the drying process to determine the number of days to attain the moisture content of 13% which is the approximate maximum for safe storage of corn for one year (Brooker et al., 1957). The ambient air, collector outlet, storage unit and drying chamber temperatures were all manually measured hourly between 9.00am to 5.00pm, which is the time sun-drying is done in Makurdi, Nigeria. The moisture content of the corn was determined using ASAE Standards, 1998. Drying rates were determined according to methods presented by Giner (2009). Temperatures were measured using mercury in bulb thermometers. Ambient air temperature was measured from a wet and dry bulb thermometer.

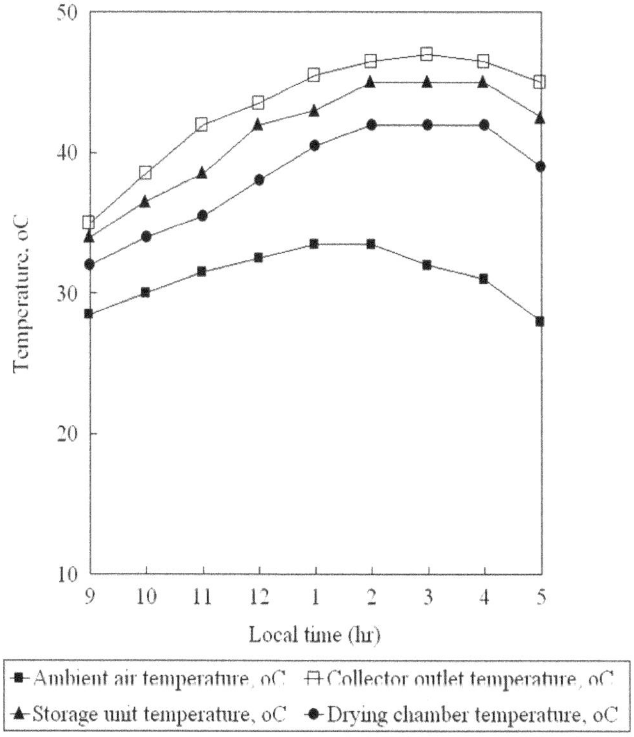

Figure 2. Typical plot of daily dryer system temperatures during the drying period.

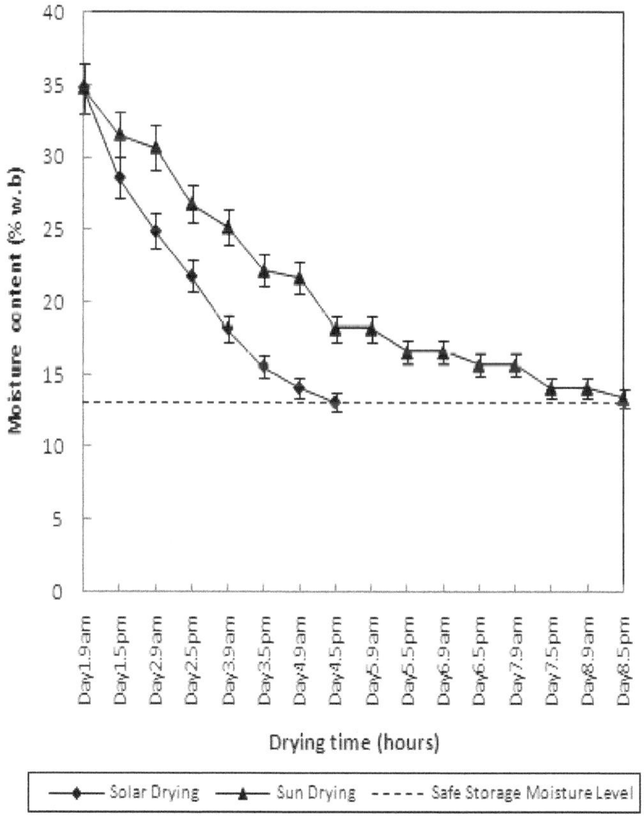

Figure 3. Variation of mean moisture content with drying time.

RESULTS AND DISCUSSION

Figure 2 is the typical plot of variation of the dryer mean system temperatures. The result shows that the average ambient temperature ranged between 28 - 33.5°C, collector outlet temperature ranged from 35 - 47°C, storage unit from 34 - 45°C and drying chamber from 32 - 42°C.

The result also shows that the air heater (collector) has the capacity to heat air by 6 - 11.5°C depending on the value of intensity of average daily solar radiation which for the period of the drying test for the month of September 2005 ranged from 37,163 - 37,837 KJ/m^2/day (Fagbenle, 1991). The storage unit has a maximum temperature of 45°C, which shows that the heat transfer coefficient between the air and the rock pebbles was very high.

The collector outlet air temperature was higher than the dryer and ambient air temperature, indicating prospects for higher drying rates over sun drying. For this research work, the purpose was to determine the number of days it will take to dry the corn to the moisture content approximating 13%. Figure 3 shows the variation of moisture content with drying time. The results show that at the end of 4-day drying period at prevailing drying chamber temperatures of 32 - 42°C, the mean moisture content of the corn was determined to be 13.1%wb giving the mean drying rate to be 0.7 kg/day per every 10 kg of corn. Comparatively, it took 8-days to sun-dry the corn to a mean moisture content of 13.4%wb under average ambient air temperatures in the range of 28 - 33.5°C, giving a mean drying rate of 0.3125 kg/day per every 10 kg of corn. This can be explained by the fact that in the case of the dryer, the heat stored in the pebble storage unit continues to effect drying in the night time while the sun-dried which is taken into the processing and storage laboratory and spread on the floor did not lose appreciable moisture to the atmosphere. This is depicted in Figure 4 where for solar drying; there is a marked difference in moisture content between the end of previous drying day (5.00pm) and start of drying on the following day (9.00am) with the difference decreasing with increase in drying days. In the case of sun drying, the difference was noticeable in the first two nights or off-sunshine hours (5.00pm - 9.00am) but subsequently no difference was observed in moisture content between end of drying on the previous day and start of drying on the following day. Ezekoye and Enebe (2006) indicate that with passive grain solar dryers, there is no need of carrying the crops inside during the nights in order to avoid re-wetting. This drying rate compares favourably to values reported in the literature. Falade et al. (1985) reported similar results on a fairly clear day with an indirect passive dryer that was used for deep-bed drying of corn.

In comparison, Mumba (1995) achieved drying of maize from 33.3%w.b to under 20% in one day using photovoltaic-powered forced-circulation dryer whereas for

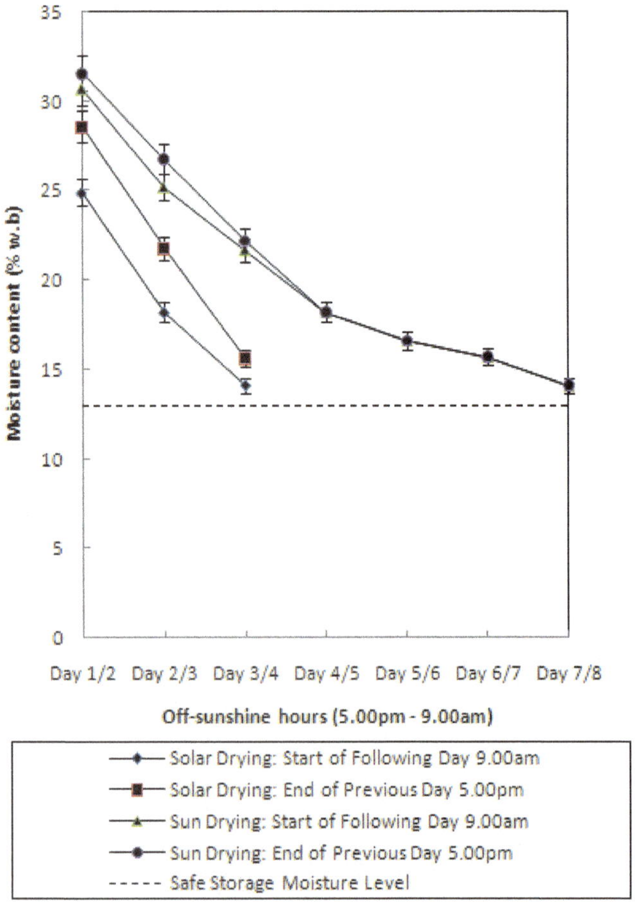

Figure 4. Effect of pebble bed storage on solar drying of corn.

this passive solar dryer, drying was achieved from 34.8 to 24.9%w.b in one day. Zomorodian et al. (2007) achieved 3 h of drying period for 132 kg of rough rice from 27 to 13%db with a cross flow and active mixed-mode type dryer equipped with a timer whereas sun drying was achieved within 11 - 14 h with 865 w/m^2 average incident solar radiation and average ambient temperature of 25°C. Soponronnarit et al. (1997) achieved 6 - 7 days drying of a batch of 25 kg of banana using solar natural convection drying whereas Sreekumar et al. (2008) dried 4 kg of bitter gourd from 95% moisture content to the desired level of 5% in 6 h while it was 11 h for open sun drying.

The potential of solar energy as a source of energy in drying agricultural produce cannot be over emphasized. The result of the study indicates that the natural convection solar grain dryer performed satisfactorily in drying freshly harvested maize from 32.8% initial moisture content, wet basis, to 13.1% in four consecutive days during harvest season. It took eight days to dry the same crop from 32.8 to 13.4%wb by the traditional sun-drying method. Thus the solar grain dryer has considerable advantages over the traditional sun-drying method in terms of faster drying rate and handling.

Savings in time was achieved by using the solar grain dryer as against traditional sun drying method. Gallali et al. (2000) reported savings in drying time for various agricultural products. Pangavhane et al. (2002) reported reduction in drying time of 43% for grapes. Similar results have been reported by Falade et al. (1985) for corn, Soponronnarit et al. (1997) and Ferreiraa et al. (2007) for banana, Zomorodian et al. (2007) for rough rice and Sreekumar et al. (2008) for bitter gourd. In terms of quality of solar dried products, Roosevelt et al. (2000) states that it is possible to achieve product physical characteristics closer to sun drying. Sharma et al. (1993) reported good quality products over sun drying for fruits and vegetables, Karathanos and Belessiotis (1997) achieved superior quality products for sultana grapes, currants, figs, plums and apricots, Chen et al. (2005) reported better general levels of quality in terms of sensory parameters, Tasirin et al. (2007) achieved same quality for bird's eye chillies with solar drying and Sreekumar et al. (2008) achieved no colour loss for bitter gourd using a new type of efficient solar dryer he developed.

Furthermore, the total estimated cost of production (materials + labour cost less profit) of the solar grain dryer was put at $79.00 (1$ = N150.00) based on Makurdi 2005 market prices, with the benefits in terms of market value of achieving higher quality products far outweighing the operational costs and initial investment with anticipated favourable discount pay-back period and benefit to cost ratio. This is desirable as Sharma et al. (1993) indicated that solar dryers are predestined for applications on small farms due to the low investment required. Soponronnarit (1995) states that in terms of economy, solar drying for some crops is feasible. With reference to combination drying technologies, Zhang et al. (2006) reported high start-up costs and relatively complicated technology as compared to conventional convection drying.

Conclusion

From the study, the following conclusions may be drawn:

1. A laboratory-scale passive solar grain dryer was developed and its performance evaluated in Makurdi metropolis, Benue State, Nigeria. With the data available, any convenient size can be produced at community level cooperative use and for any prospective investor.
2. The mean drying rate of the dryer was 0.7 kg/day per every 10 kg of corn whereas sun-drying rate was 0.3125 kg/day comparatively.
3. Savings in time was achieved by using the solar grain dryer as against traditional sun-drying method. It took 4-days to dry the corn to moisture content of 13.1%wb using the passive solar dryer while it took 8-days to dry

to 13.4%wb under sun drying.

4. For further improvement of the dryer, there is need to increase the volumetric airflow rate across the drying chamber.

REFERENCES

Aboul-Enein S, El-Sebaii AA, Ramadan MRI, El-Gohary HG (2000). Parametric study of a solar air heater with and without thermal storage for solar drying application. Renewable Energy 21: 502-522.

Arinze EA (1983). Design and performance evaluation of a commercial size natural convection solar crop dryer. Nigerian J. Solar Energy 4: 120-122.

ASAE Standards, 45th Ed. 1998. ASAE S352.2 DEC 97. Moisture measurement: unground grain and seeds, 551. St. Joseph, Mich.: ASAE. . Pp. 551.

Bahnasawy AH, Shenana ME (2004). A mathematical model of direct sun and solar drying of some fermented dairy products. J. Food Engr., 61(3):309-319.

Bala J, De-Jong K, Huang J, Vafaie H, Wechsler H (1995). Hybrid learning using genetic algorithms and decision trees for pattern classification. Proceedings of the International Joint Conference on Artificial Intelligence, Montreal, August. Pp. 719-724.

Bena B, Fuller RJ (2002). Natural convection solar dryer with biomass back-up heater. Solar Energy. 72 (1): 75-83.

Bennamoun L, Azeddine B (2003). Design and simulation of a solar dryer for agriculture products. J. Food Engr., 59 (2/3):259-266.

Brooker DB, Arkema B, Hall CW (1957). Drying and storage of grains and oil seeds. AVI Book, New York, NY 10003. . Pp. 654.

Chen H, Hernandeza CE, Huang T (2005). A study of the drying effect on lemon slices using a closed-type solar dryer. J. Food Engr., 68 (4): 439-442.

Ekechukwu OV, Norton B (1999). Review of solar-energy drying systems II: an overview of solar drying technology. Energy Conversion and Management. 40: 615-655.

El-Sebaii AA, Aboul-Enein S, Ramadan MRI, El-Gohary HG (2002). Experimental investigation of an indirect type natural convection solar dryer. Energy Conversion and Management. 43: 2251-2266.

Ezekoye BA, Enebe OM (2006). Development and performance evaluation of modified integrated passive solar grain dryer. The Pacific J. Sci. Techn., 7(2): 185 -190.

Faborode MO, Favier JF, Ajayi OA (1995). The effects of forced air drying on cocoa quality. J. Food Engr., 25(4): 455-472.

Fagbenle R (1991). On monthly average daily extraterrestrial solar radiation for Nigeria. Nig. J. Renewable Energy. 22: 1-8.

Falade A, Talabi SO, Akinsete A, Danso FK, Adejuwon AA (1985). A solar grain dryer for rural areas of Nigeria. Nig. J. Solar Energy 14:121-130.

Ferreiraa AG, Charbelb ALT, Piresb RL, Silvab JG, Malac CB (2007). Experimental analysis of a hybrid dryer. Engenharia Térmica (Thermal Engineering). 6 (2): 3-7.

Gallali YM, Abujnah YK, Bannani FK (2000). Preservation of fruits and vegetables using solar drier, a comparative study of natural and solar drying III; chemical analysis and sensory evaluation of the dried samples (grapes, figs, tomatoes and onions). Renewable Energy. 19: 203 212.

Giner SA (2009). Influence of internal and external resistances to mass transfer on the constant drying rate period in high-moisture foods. Biosystems Engineering 102 (1): 90-94.

Goswani Y (1986). Alternative energy in agriculture. CRC Press Inc., Florida. 2: Pp. 83-102.

Hodali R, Bougard J (2001). Integration of a desiccant unit in crops solar drying installation: optimization by numerical simulation. Energy Conversion and Management. 42(13): 1543-1558.

Hsieh JS (1986). Solar energy engineering. Prentice-Hall Inc., Englewood Cliffs, New Jersey. Pp. 303.

Imre EE (1995). A handbook of industrial drying. 1st Edition. New York. Pp. 550.

Irtwange SV (1991). Design and development of a passive solar air heated poultry egg incubator. Unpublished M. Eng Project. Department of Agricultural Engineering, University of Nigeria, Nsukka. Pp. 258.

Itodo IN, Adewole AM, Edemeka SK (2002). Development of active solar crop dryer: design analysis and performance evaluation. Nigeria J. Renewable Energy. 10 :1-2.

Karathanos VT, Belessiotis VG (1997). Sun and artificial air drying kinetics of some agricultural products. Journal of Food Engr.,31(1): 35-46.

Kisten JF, van J (1998). Defining small scale farmers in the South African context. Agrekon. 37(4): 560-571.

Leon MA, Kumar S, Bhattacharya SC (2002). A comprehensive procedure for performance evaluation of solar food dryers. Renewable and Sustainable Energy Reviews. 6: 367-393.

Lmre L, Palaniappan C (1996). Development of solar drying. Drying Technology. 14 (6): 1381-1388.

Mumba J (1995). Development of a photovoltaic powered forced circulation grain dryer for use in the tropics. Renewable Energy 6(7): 855-862.

Pangavhane DR, Sawhney RL, Sarsavadia PN (2002). Design, development and performance testing of a new natural convection solar dryer. Energy. (27): 579-590.

Paul VN (2001). Development of an indirect solar crop dryer for rural small scale farms. Unpublished B. Eng. Project. Department of Agricultural Engineering, University of Agriculture, Makurdi. Pp. 54.

Roosevelt DS, Sadulla S, Ramasami T (2000). Solar dryer for leather and a comparative study on the characteristics of open-, solar- and electrical-dried leathers. Renewable Energy. 19 (1-2):123-134.

Sadiku S, Oyero IO, Osunde ZD (2001). Design and evaluation of locally made solar dryers in Niger State. Technical Report. Federal University of Technology, Minna. Pp. 127.

Sato J, Irie M (1981). Solar energy utilization for grain and grass drying. Taiyo Enerugi. 7(4): 28-40.

Sayigh AAM (1979). The technology of flat-plate collector in solar energy conversion. An introductory course. Editors: AE Dixon and JD Leslie. Pergamon Press. Pp. 1324.

Sharma VK, Colangelo A, Spagna G (1993). Experimental performance of an indirect type solar fruit and vegetable dryer. Energy Conversion and Management. 34(4): 293-308.

Sopian K, Fudholi A, Ruslan MH, Alghoul MA, Sulaiman MY (2009). Review of solar dryers for agricultural and marine products. Renewable and Sustainable Energy Reviews, In Press. Uncorrected Proof, Available online 5 August. http://www.sciencedirect.com/science?_ob=ArticleURL&_udi=B6VMY-4WXPXW5-1&_user=10&_rdoc=1&_fmt=&_orig=search&_sort=d&_docanchor=&view=c&_searchStrId=1028996409&_rerunOrigin=google&_acct=C000050221&_version=1&_urlVersion=0&_userid=10&md5=4676b676657567df9174b79c55c1564c

Soponronnarit S (1995). Aeration and drying strategies for managing wet paddy. Proc. 17th ASEAN Technical Seminar on Grain Postharvest Technology, July 25-27 1995, Lumut, Perak, Malaysia. Also in Proc. Farm Machinery Conference on Technology for Agricultural Production, Pattaya (in Thai). Pp. 183-196.

Soponronnarit S, Nathakaranakule A, Limtrakool W, Rungjiwaruk W, Nathakaranakule S, Rakwichien W (1997). Banana fruit drying. RERIC Inter. Energy J., 19 (2):55-62.

Sreekumar A, Manikantan PE, Vijayakumar KP (2008). Performance of indirect solar cabinet dryer. Energy Conversion and Management. 49 (6): 1388-1395.

Tasirin SM, Kamarudin SK, Jaafar K, Lee KF (2007). The drying kinetics of bird's chillies in a fluidized bed dryer. J. Food Engr., 79 (20): 695-705.

Tembo L, Chiteka ZA, Kadzere I, Akinnifesi FK, Tagwira F (2008). Ripening stage and drying method affecting colour and quality attributes of *Ziziphus mauritiana* fruits in Zimbabwe. Afr. J. Biotech. 7 (9): 2509-2513.

Tunde-Akintunde TY, Afolabi TJ, Akintunde BO (2005). Influence of drying methods on drying of bell-pepper (*Capsicum annuum*). J. Food Engineering 68 (4): 439-442.

Zhang M, Tang J, Mujumdar AS, Wang S (2006). Trends in microwave-related drying of fruits and vegetables. Trends in Food Science

and Technology. 17 (10): 524-534.
Zomorodian A, Zare D, Ghasemkhani H (2007). Optimization and evaluation of a semi-continuous solar dryer for cereals (rice, etc). Desalination 209: 129-135.

Permissions

All chapters in this book were first published in JAERD, by Academic Journals; hereby published with permission under the Creative Commons Attribution License or equivalent. Every chapter published in this book has been scrutinized by our experts. Their significance has been extensively debated. The topics covered herein carry significant findings which will fuel the growth of the discipline. They may even be implemented as practical applications or may be referred to as a beginning point for another development.

The contributors of this book come from diverse backgrounds, making this book a truly international effort. This book will bring forth new frontiers with its revolutionizing research information and detailed analysis of the nascent developments around the world.

We would like to thank all the contributing authors for lending their expertise to make the book truly unique. They have played a crucial role in the development of this book. Without their invaluable contributions this book wouldn't have been possible. They have made vital efforts to compile up to date information on the varied aspects of this subject to make this book a valuable addition to the collection of many professionals and students.

This book was conceptualized with the vision of imparting up-to-date information and advanced data in this field. To ensure the same, a matchless editorial board was set up. Every individual on the board went through rigorous rounds of assessment to prove their worth. After which they invested a large part of their time researching and compiling the most relevant data for our readers.

The editorial board has been involved in producing this book since its inception. They have spent rigorous hours researching and exploring the diverse topics which have resulted in the successful publishing of this book. They have passed on their knowledge of decades through this book. To expedite this challenging task, the publisher supported the team at every step. A small team of assistant editors was also appointed to further simplify the editing procedure and attain best results for the readers.

Apart from the editorial board, the designing team has also invested a significant amount of their time in understanding the subject and creating the most relevant covers. They scrutinized every image to scout for the most suitable representation of the subject and create an appropriate cover for the book.

The publishing team has been an ardent support to the editorial, designing and production team. Their endless efforts to recruit the best for this project, has resulted in the accomplishment of this book. They are a veteran in the field of academics and their pool of knowledge is as vast as their experience in printing. Their expertise and guidance has proved useful at every step. Their uncompromising quality standards have made this book an exceptional effort. Their encouragement from time to time has been an inspiration for everyone.

The publisher and the editorial board hope that this book will prove to be a valuable piece of knowledge for researchers, students, practitioners and scholars across the globe.

List of Contributors

Bahaman Abu Samah
Laboratory for Rural Advancement and Agriculture Extension, Institute for Social Science Studies, Universiti of Putra, Malaysia

Jeffrey Lawrence D' Silva
Laboratory for Rural Advancement and Agriculture Extension, Institute for Social Science Studies, Universiti of Putra, Malaysia

Hayrol Azril Mohamed Shaffril
Laboratory for Rural Advancement and Agriculture Extension, Institute for Social Science Studies, Universiti of Putra, Malaysia

Kahramon Jumaboev
International Water Management Institute Tashkent, Uzbekistan

J. Mohan Reddy
International Water Management Institute Tashkent, Uzbekistan

Sh. Muhammedjanov
Scientific Information Center of the Interstate Committee for Water Coordination Tashkent, Uzbekistan

Oyture Anarbekov
International Water Management Institute Tashkent, Uzbekistan

Davron Eshmuratov
International Water Management Institute Tashkent, Uzbekistan

J. A. Adetumbi
Institute of Agricultural Research and Training, Obafemi Awolowo University, P. M. B. 5029, Moor Plantation, Ibadan, Nigeria

J. O. Saka
Institute of Agricultural Research and Training, Obafemi Awolowo University, P. M. B. 5029, Moor Plantation, Ibadan, Nigeria

B. F. Fato
Federal College of Agriculture, P. M. B. 5029, Moor Plantation, Ibadan, Nigeria

Felister Mombo
Sokoine University of Agriculture, P.O Box 3011, Morogoro, Tanzania
Ghent University, Coupure Links 653, 9000 Ghent, Belgium

Stijn Speelman
Ghent University, Coupure Links 653, 9000 Ghent, Belgium

Guido Van Huylenbroeck
Ghent University, Coupure Links 653, 9000 Ghent, Belgium

Joseph Hella
Sokoine University of Agriculture, P.O Box 3011, Morogoro, Tanzania

Munishi Pantaleo
Sokoine University of Agriculture, P.O Box 3011, Morogoro, Tanzania

Stein Moe
Norwegian University of Life Science, P.O Box 5003, Aas, Norway

Ayana Etana
Forage Agronomy Research, Adami Tulu Agricultural Research Center, P. O. Box 35, Ziway, Ethiopia

Estefanos Tadesse
Dairy Nutrition Research, Adami Tulu Agricultural Research Center, Ethiopia

Ashenafi Mengistu
Faculty of Veterinary Medicine, Addis Ababa University, Debrezeit, Ethiopia

Abubeker Hassen
Ruminant Nutrition and Pasture Science, Department of Animal and Wildlife Science, University of Pretoria, Pretoria 0002, Republic of South Africa

Thomas Kirisits
Institute of Forest Entomology, Forest Pathology and Forest Protection (IFFF), Department of Forest and Soil Sciences University of Natural Resources and Life Sciences, Vienna (BOKU), Hasenauerstraße 38, A-1190 Vienna, Austria

Peter Kritsch
Institute of Forest Entomology, Forest Pathology and Forest Protection (IFFF), Department of Forest and Soil Sciences, University of Natural Resources and Life Sciences, Vienna (BOKU), Hasenauerstraße 38, A-1190 Vienna, Austria

Katharina Kräutler
Institute of Forest Entomology, Forest Pathology and Forest Protection (IFFF), Department of Forest and Soil Sciences, University of Natural Resources and Life Sciences, Vienna (BOKU), Hasenauerstraße 38, A-1190 Vienna, Austria

Michaela Matlakova
Institute of Forest Entomology, Forest Pathology and Forest Protection (IFFF), Department of Forest and Soil Sciences, University of Natural Resources and Life Sciences, Vienna (BOKU), Hasenauerstraße 38, A-1190 Vienna, Austria

Erhard Halmschlager
Institute of Forest Entomology, Forest Pathology and Forest Protection (IFFF), Department of Forest and Soil Sciences, University of Natural Resources and Life Sciences, Vienna (BOKU), Hasenauerstraße 38, A-1190 Vienna, Austria

T. Asiamah Maxwell
Department of Agricultural Economics, Agribusiness and Extension, Kwame Nkrumah University of Science and Technology, P. M. B, Kumasi, Ashanti Region, Ghana

Patrick Dzadze
Department of Agricultural Economics, Agribusiness and Extension, Kwame Nkrumah University of Science and Technology, P. M. B, Kumasi, Ashanti Region, Ghana

Bertha Gyasi
Department of Agricultural Economics, Agribusiness and Extension, Kwame Nkrumah University of Science and Technology, P. M. B, Kumasi, Ashanti Region, Ghana

Immaculate Njuthe Maina
Kenya Agricultural Research Institute (KARI), Socioeconomics and Applied Statistics Division, Kenya

Ingrid-Ute Leonhäuser
Justus-Liebig-University Giessen, Nutrition Education and Consumer Behaviour, Germany

Siegfried Bauer
Justus-Liebig-University Giessen, Project and Regional Planning, Germany

A. K. Ango
Department of Agricultural Extension and Rural Development, Usmanu Danfodiyo University, Sokoto, Nigeria

S. A. Ibrahim
Department of Agricultural Extension and Rural Development, Usmanu Danfodiyo University, Sokoto, Nigeria

A. A. Yakubu
Department of Agricultural Economics, Usmanu Danfodiyo University, Sokoto, Nigeria

A. S. Alhaji
Department of Liberal Studies, Waziri Umaru Federal Polytechnic, B/Kebbi, Nigeria

J. Mutambara
Department of Agricultural Economics and Extension, Faculty of Agriculture, University of Zimbabwe, Box MP 167 Mt Pleasant Harare, Zimbabwe

G. B. Ayoola
Department of Agricultural Economics, College of Agricultural Economics, Extension and Management Technology, University of Agriculture, Makurdi, Benue State, Nigeria

E. P. Ejembi
Department of Agricultural Extension and Communication, College of Agricultural Economics, Extension and Management Technology, University of Agriculture, Makurdi, Benue State, Nigeria

T. Avav
Department of Crop and Environmental Protection, College of Agronomy, University of Agriculture, Makurdi, Benue State, Nigeria

J. Masvongo
Department of Agricultural Economics and Extension, Faculty of Agriculture, University of Zimbabwe, Box MP 167 Mt Pleasant Harare, Zimbabwe

S. Moyo
Department of Crop Science Faculty of Agriculture, Faculty of Agriculture, University of Zimbabwe, Box MP 167 Mt Pleasant Harare, Zimbabwe

Helyette Geman
Commodity Finance Centre, Birkbeck University of London and ESCP Europe, United Kingdom

Seth Sarfo
Commodity Finance Centre, Birkbeck University of London and ESCP Europe, United Kingdom

Chala Merera
Ambo University, P. O. Box 19, Oromia, Ethiopia

Jemal Dekebo
Bako Agricultural Research Center, P. O. Box 03, Oromia, Ethiopia

Ulfina Galmessa
Ambo University, P. O. Box 19, Oromia, Ethiopia

S. D. Y. Alfred
Department of Agricultural Economics and Extension, Federal University of Technology, Akure, Ondo State, Nigeria

A. B. Adekayode
Department of Agricultural Economics and Extension, Federal University of Technology, Akure, Ondo State, Nigeri

Yemane Asmelash
Bahir Dar University, P. O. Box 873, Ethiopia

S.O. Olatunji
Department of Agricultural Economics and Extension, University of Port-Harcourt, Nigeria

A. C. Agumagu
Department of Agricultural Economics and Extension, University of Port-Harcourt, Nigeria

O. M. Adesope
Department of Agricultural Economics and Extension, University of Port-Harcourt, Nigeria

Bedada Begna
Ethiopian Institute of Agricultural Research (EIAR), Kulumsa Agricultural Research Center (KARC), P. O. Box 489, Assela, Ethiopia

Messay Yami
Ethiopian Institute of Agricultural Research (EIAR), Kulumsa Agricultural Research Center (KARC), P. O. Box 489, Assela, Ethiopia

Eshetu Lemma
Ethiopian Institute of Agricultural Research (EIAR), Kulumsa Agricultural Research Center (KARC), P. O. Box 489, Assela, Ethiopia

Tesfaye Solomom
Ethiopian Institute of Agricultural Research (EIAR), Kulumsa Agricultural Research Center (KARC), P. O. Box 489, Assela, Ethiopia

Tarekegn Etana
Ethiopian Institute of Agricultural Research (EIAR), Kulumsa Agricultural Research Center (KARC), P. O. Box 489, Assela, Ethiopia

Bernice Mireille Tekombo Kingha
Department of Biological Sciences, Faculty of Sciences, University of Ngaoundéré, P.O. Box 454 Ngaoundere, Cameroon

Fernand-Nestor Tchuenguem Fohouo
Department of Biological Sciences, Faculty of Sciences, University of Ngaoundéré, P.O. Box 454 Ngaoundere, Cameroon

Albert Ngakou
Department of Biological Sciences, Faculty of Sciences, University of Ngaoundéré, P.O. Box 454 Ngaoundere, Cameroon

Dorothea Brückner
Forschungsstelle für Bienenkunde, Universität Bremen, FB2, Postfach 33 04 40, 28334 Bremen, Germany

Hélyette Geman
Birkbeck, University of London, London, England
Johns Hopkins University, United States

Pedro Vergel Eleuterio
Birkbeck, University of London, London, England

ODUNAYA Adewale Adeleke
Ogun State Agricultural Development Programme, P. M. B. 2122, Sapon, Abeokuta, Ogun State, Nigeria

Beliyu Limenih
Ethiopia Institute of Agricultural Research, HARC, P. O. Box, 2003, Addis Ababa, Ethiopia

Tederose Tefera
College of Agriculture, Hawassa University, P. O. Box, 05, Hawassa, Ethiopia

Hosea Mpogole
Department of Agricultural Economics, Sokoine University of Agriculture, Tanzania

Melchior E. Mlambiti
St. Augustine University of Tanzania, Mwanza, Tanzania

Reuben M. J. Kadigi
Department of Agricultural Economics, Sokoine University of Agriculture, Tanzania

Wasiu Olalekan Akinbode
Department of Agricultural Economics, Faculty of Agriculture, Obafemi Awolowo University, Ile-Ife, Osun State, Nigeria

Adebayo Simeon Bamire
Department of Agricultural Economics, Faculty of Agriculture, Obafemi Awolowo University, Ile-Ife, Osun State, Nigeria

D. O. Torimiro
Department of Agricultural Economics, Education and Extension, Botswana College of Agriculture, Gaborone

O. J. Yusuf
Department of Agricultural Economics and Extension Services, Kwara State University, Malete, Nigeria

S. K. Subair
Department of Agricultural Economics and Extension Services, Kwara State University, Malete, Nigeria

B. J. Amujoyegbe
Department of Crop Production and Protection, Obafemi Awolowo University, Ile-Ife, Nigeria

N. Tselaesele
Department of Agricultural Economics, Education and Extension, Botswana College of Agriculture, Gaborone

J. O. Ayinde
Department of Crop Production and Protection, Obafemi Awolowo University, Ile-Ife, Nigeria

B. J. Amujoyegbe
Department of Crop Production and Protection, Faculty of Agriculture, Obafemi Awolowo University, Ile-Ife, Osun State, Nigeria

O. S. Alabi
Department of Agricultural Extension and Rural Development, Faculty of Agriculture, Obafemi Awolowo University, Ile-Ife, Osun State, Nigeria

Daniel Y. Opare-Atakora
CSIR-Savanna Agricultural Research Institute, Tamale, Ghana

Samuel A. Donkoh
Department of Agricultural and Resource Economics, University for Development Studies, Tamale, Ghana

Amin Alhassan
Department of Communication Sciences, University for Development Studies, Tamale, Ghana

Swapnal Sisodia
Department of Botany, Aligarh Muslim University, Aligarh-20 2002, India

M. Badruzzaman Siddiqui
Department of Botany, Aligarh Muslim University, Aligarh-20 2002, India

Isaac Mwaura Njuguna
Department of Applied Community Development Studies, Faculty of Education and Community Studies, Egerton University, Njoro, Kenya

Catherine Ngéndo Munyua
Department of Applied Community Development Studies, Faculty of Education and Community Studies, Egerton University, Njoro, Kenya

Susan Kamuru Makal
Department of Applied Community Development Studies, Faculty of Education and Community Studies, Egerton University, Njoro, Kenya

S. V. Irtwange
Department of Agricultural and Environmental Engineering, University of Agriculture, Makurdi, Nigeria

S. Adebayo
Department of Agricultural and Environmental Engineering, University of Agriculture, Makurdi, Nigeria